Automated Design of Control Systems

Information and Systems Theory

A series edited by J. Richalet, *C. E. R. A.*, P. Vidal, *Faculté des Sciences de Lille*, T. E. Stern, *Department of Electrical Engineering, Columbia University* and P. C. Parks, *School of Engineering Science, University of Warwick.*

P. VIDAL *Non-linear Sampled-data Systems*

P. VIDAL *Non-linear Sampled-data Systems – Exercises and Problems*

S. THELLIEZ *Introduction to the Study of Ternary Switching Structures*

C. W. MERRIAM III *Automated Design of Control Systems*

Other volumes in preparation

Automated Design of Control Systems

C. W. MERRIAM III

Department of Electrical Engineering
The University of Rochester

GORDON AND BREACH SCIENCE PUBLISHERS
NEW YORK LONDON PARIS

Gordon and Breach, Science Publishers, Inc.
1 Park Avenue
New York, N.Y. 10016

Editorial office for the United Kingdom

Gordon and Breach, Science Publishers Ltd.
42 William IV Street
London W.C.2

Editorial office for France

Gordon & Breach
7–9 rue Emile Dubois
Paris 75014

Library of Congress catalog card number 73-86001. ISBN 0 677 04440 2 (*hardbound*); 0 677 04445 3 (*paperbound*). Printed in Great Britain

To my Parents

Information and systems theory

Series introduction

The unceasing development of science and technology in recent years has caused industry to be in closer contact with the universities. Engineers, in effect, are obliged to keep their knowledge up-to-date, and a "recirculation" appears to be essential for those who wish to remain in the forefront of scientific progress.

The present series, comprising a number of monographs on mathematics and applied physics, has been created with this in mind. The range of material dealt with in the works of this series is therefore wide, but we shall concern ourselves in particular with systems theory, which embraces the two main disciplines of control theory and information theory.

These volumes are intended for senior students working for a diploma of advanced studies or a doctorate, for students at schools of engineering and for already qualified engineers working in research and development departments.

By careful production and a wide distribution of the works, some of which are available simultaneously in English and French editions, we hope to achieve our purpose, which is to inform theoreticians and practitioners alike of the latest progress in their special subjects, as well as to provide students with useful texts.

T. E. STERN P. C. PARKS
J. RICHALET P. VIDAL

Preface

This book is the outgrowth of courses taught at Cornell University in the general area of control theory. These courses follow typical undergraduate courses in mathematics and introductory systems theory. The author assumes that the reader is familiar with solutions to linear constant-coefficient differential equations in vector form, properties of matrices, introductory aspects of linear vector spaces, transforms, flow graphs, and stability theory of linear time-invariant systems. Moreover, some familiarity with the goals, problems, and design specifications of feedback control is assumed.

The goal of this book is to present the reader with material that forms an introduction to the automated design of continuous control systems. Automated design requires systematic selection of design specifications in an *a priori* fashion, recourse to efficient and reliable computational methods, and a practical means for system implementation. Success in achieving fully automated design is, of course, only partial. However, important advances in this area have been achieved recently and involve such diverse topics as transfer-function matrix synthesis, incomplete state-variable feedback, sensitivity reduction, human operator simulation in a design context, and stochastic stability.

The author believes that linearization and optimization techniques constitute the most fundamental and powerful approach to the automated design of control systems. Furthermore, the author believes that these techniques, as a means of achieving automated design, will in fact rapidly replace many classical design techniques in engineering practice. This book also fulfils a long neglected need for a systematic and yet compact transition between classical control theory and advanced optimal control theory. This transition is based on differential calculus and hence can be taught in depth in the context of automated design at either the fourth or fifth year level.

The theorem-proof format is used almost exclusively in an effort to present the material in the sharpest possible focus. However, theoretical results are motivated and evaluated throughout by practical considerations which are associated with automated design requirements. Motivation and evaluation are in large part based on the presentation of a number of reasonably elaborate design problems. The author also believes that digital computer studies should be undertaken throughout a serious study of this book because the full impact of automated design techniques will only be realized from the use of computers in realistic design studies. A large number of suggested problems is included in order to facilitate an understanding and appreciation of the implications of this material.

Chapter 1 is intended to be an introduction to notation and conventions used throughout the book. In addition, this chapter serves as an introduction to the formulation of both deterministic and stochastic design problems which arise in control. Analytical design methods for parameter optimization are also discussed here by way of introduction.

Chapter 2 serves as a self-contained introduction to the theory of minimizing functions of many variables. Functions which are twice continuously differentiable are emphasized because they apply to a wide variety of control problems and lend themselves to gradient methods of successive approximation. The theory of equality and inequality constraints is also included, and the Lagrange multiplier formulation is emphasized for eventual computational purposes. Also discussed are geometrical and vector space interpretations of Lagrange multipliers.

Chapter 3 presents a class of search methods that have particular utility in automated design. These methods are based on gradient and, in some cases, hessian calculations. The corresponding computational algorithms are, for the most part, just computer implementations of the theory presented in the previous chapter.

Chapter 4 emphasizes the formulation of large-scale deterministic design problems which are characterized by linear constant-coefficient state equations and an infinite interval of operation. In addition, computational methods required to perform gradient searches are emphasized. Excess pole and pole suppression specifications are of particular importance. Moreover, incomplete state-variable feedback and relationships to linear optimal control problems are worthy of note.

Chapter 5 emphasizes the formulation of large-scale stochastic design problems which are again characterized by linear constant-coefficient state equations and an infinite interval of operation. Additional computational methods required to solve equations arising in stochastic stability theory are also emphasized. Stochastic models of the human operator are introduced for the purpose of simulating manual control in the context of automated design. Topics in filtering and prediction theory which arise in control are also discussed.

Chapter 6 is based on design problems with finite intervals of operation and state equations that may be nonlinear. Relationships between solutions to infinite and finite time-interval problems are emphasized. Computational methods for solving two-point boundary-value problems are also included as well as system implementation techniques which are based on linearization. This chapter is also intended to serve as an introduction to advanced theories of approximation and optimization that have an impact on automated design of control systems.

Appendix A includes a library of APL/360 computer programs which are primarily intended for student use. These programs share the inherent instructional advantages of an interactive programming language. However, use of these programs requires only an introductory knowledge of APL/360. Typical methods of successive approximation and substitution are included as well as methods for solving matrix Riccati equations.

I am indebted to many students and professional associates for ideas and encouragement. To mention only a few, Professor D. Jordan of the University of Connecticut was a co-worker on much of the material appearing in Chapter 4, and Mr. J. McDaniel of Cornell University participated in evaluating computational methods discussed in Chapter 3. I am especially indebted to Mr. D. F. Makers and Mr. M. Marx of the General Electric Company in Binghamton, N.Y., for their long-standing encouragement to develop and apply automated design techniques to meaningful flight control problems. Finally, I wish to express my appreciation to Mrs. Lucy Mozeleski for her careful and diligent typing of the manuscript.

C. W. M.

Contents

CHAPTER 1

Introduction

Development of optimization techniques for control system design has experienced a truly remarkable growth since the end of World War II. The first such development [1] concerned adjustment of parameters in a linear system having a fixed configuration. These parameter optimization techniques are primarily analytical and are briefly summarized later in this chapter. The second such development [2, 3] concerned impulse-response optimization of a system having a free configuration. These impulse-response optimization techniques are also primarily analytical and in essence constituted a new application of the celebrated filter optimization techniques of Wiener [4] and Lee [5]. The third such development concerned optimization of feedback control equations [6] and was based on either dynamic programming techniques of Bellman [7] or extensions to the calculus of variations contributed by Pontryagin [8]. The prodigious development in feedback-control optimization techniques and potential applications were made possible primarily by the availability of digital computers.

At the present time, however, optimization techniques generally are not being put to use in the large area of control system design that has been traditionally associated with classical trial-and-error techniques. This phenomenon seems to have arisen because many optimization techniques have not been reformulated so as to be amenable to modern computational methods. Moreover, optimization problems have not been reformulated so as to incorporate design criteria such as the synthesis of transfer-function matrices, incomplete state-variable feedback, and sensitivity to parametric disturbances.

This book attempts to introduce the reader to the optimization techniques that are particularly germane to the automated design of control systems. Computational methods employed in the solution of these problems are emphasized throughout, and the serious reader is encouraged to solve representative design problems that require use of the digital computer. In addition, considerable background is supplied for the formulation of criteria for automated design problems arising in control as well as for the simulation of human operators. This restricted approach to the general problem of automated design has a number of pedagogical advantages. First of all, this material involves only differential calculus and the background obtained in traditional and often required undergraduate courses in systems theory.‡ Second, the class of control system design problems that can be solved by methods presented here may well be the most important class for which optimization techniques have actually resulted in engineering implementation of the system. The third advantage is derived from the surprisingly general interpretations of more inclusive optimization problems that can be given to the automated design-problem formulation presented here.

1.1 VECTOR-MATRIX NOTATION

Vector-matrix notation is used throughout this book for algebraic and differential equations. This method of representation is closely allied to pertinent computational procedures and also facilitates vector space interpretations of many significant results. A very brief summary of the notational conventions employed in this book follows here.

Vertical one-dimensional arrays, referred to as *column vectors*, are denoted by lower case letters $\mathbf{f}(t)$, $\mathbf{g}(t)$, etc. and are written either as

$$\mathbf{f} = \begin{bmatrix} f_1 \\ f_2 \\ \vdots \\ f_m \end{bmatrix} \tag{1-1}$$

or as

$$\mathbf{f} = [f_i] \text{ with } \dim \mathbf{f} = m. \tag{1-2}$$

In addition, two-dimensional arrays, referred to as *matrices*, are denoted by capital letters $\mathbf{F}$, $\mathbf{G}$, etc. and are written either as

$$\mathbf{F} = \begin{bmatrix} f_{11} & f_{12} & \cdots & f_{1n} \\ f_{21} & f_{22} & \cdots & f_{2n} \\ \cdots & \cdots & \cdots & \cdots \\ f_{m1} & f_{m2} & \cdots & f_{mn} \end{bmatrix} \tag{1-3}$$

‡For instance, see References [9,10].

or as

$$\mathbf{F} = [f_{ij}] \text{ with } \dim \mathbf{F} = m \times n. \tag{1-4}$$

A similar and completely compatible notation is used for the partitioned form of vectors and matrices. In particular, vector **f** is subdivided into groups of rows called *vector partitions* $\mathbf{f}_i$ so that

$$\mathbf{f} = \begin{bmatrix} \mathbf{f}_1 \\ \mathbf{f}_2 \\ \vdots \\ \mathbf{f}_k \end{bmatrix} \text{ with } \dim \mathbf{f}_i = m_i \text{ and } \dim \mathbf{f} = \sum_{i=1}^{k} m_i. \tag{1-5}$$

Moreover, matrix **F** is subdivided into groups of rows and columns so that *matrix partition* $\mathbf{F}_{ij}$ is defined by

$$\mathbf{F} = \begin{bmatrix} \mathbf{F}_{11} & \mathbf{F}_{12} & \dots & \mathbf{F}_{1l} \\ \mathbf{F}_{21} & \mathbf{F}_{22} & \dots & \mathbf{F}_{2l} \\ \dots & \dots & \dots & \dots \\ \mathbf{F}_{k1} & \mathbf{F}_{k2} & \dots & \mathbf{F}_{kl} \end{bmatrix} \tag{1-6}$$

with

$$\dim \mathbf{F}_{ij} = m_i \times n_j \text{ and } \dim \mathbf{F} = \left(\sum_{i=1}^{k} m_i \right) \times \left(\sum_{j=1}^{l} n_j \right).$$

This choice of notation is adopted so that the special case of $\dim \mathbf{f}_i = 1$ and $\dim \mathbf{F}_{ij} = 1 \times 1$ for each i, j reverts to the normal representation of vectors and matrices.

The matrix formed by interchanging rows and columns of **F** is called the *transpose* of **F** and is denoted by $\mathbf{F}'$ where

$$\mathbf{F}' = \begin{bmatrix} f_{11} & f_{21} & \dots & f_{m1} \\ f_{12} & f_{22} & \dots & f_{m2} \\ \dots & \dots & \dots & \dots \\ f_{1n} & f_{2n} & \dots & f_{mn} \end{bmatrix} \tag{1-7}$$

corresponds to (1-3). In a similar fashion, a *row vector* denoted by $\mathbf{f}'$ is defined as

$$\mathbf{f}' = [f_1 \quad f_2 \quad \dots \quad f_m] \tag{1-8}$$

in correspondence with (1-1). A frequently occurring form of partitioned matrices is by either columns or rows. Matrix **F** is often written as

$$\mathbf{F} = [\mathbf{f}_1 \quad \mathbf{f}_2 \quad \ldots \mathbf{f}_n] \tag{1-9}$$

when partitioned by columns and is written either as

$$\mathbf{F} = \begin{bmatrix} \mathbf{f}_1' \\ \mathbf{f}_2' \\ \vdots \\ \mathbf{f}_m' \end{bmatrix} \quad \text{or as } \mathbf{F}' = [\mathbf{f}_1 \quad \mathbf{f}_2 \quad \ldots \quad \mathbf{f}_m] \tag{1-10}$$

when partitioned by rows. Matrices are called *square* when $m = n$, and square matrices are called *symmetric* when $\mathbf{F} = \mathbf{F}'$.

The determinant of a square matrix **F** arises frequently in this book and is denoted as det **F**. An important property of the determinant is the equivalence

$$\det \mathbf{F} = 0 \Leftrightarrow \exists\, \mathbf{m} \neq \mathbf{0} \ni \mathbf{F}\mathbf{m} = \mathbf{0}, \tag{1-11}$$

where **m** may be complex when **F** is complex. Symbols ⇔ for "implies and is implied by", ∃ for "there exists", and ∋ for "such that" appear in this book but are used as sparingly as possible. An alternate expression for $\mathbf{Fm} = \mathbf{0}$ is of course

$$\sum_{j=1}^{n} \mathbf{f}_j m_j = \mathbf{0}$$

so that $\det \mathbf{F} = 0$ is equivalent to a linear dependence among the columns of **F**. Because $\det \mathbf{F} = \det \mathbf{F}'$ even when **F** is not symmetric, a linear dependence between the columns of **F** is equivalent to a linear dependence between the rows of **F**.

Another important property of the determinant is derived from the relationship

$$(\text{cof } \mathbf{F})'\mathbf{F} = \mathbf{F}(\text{cof } \mathbf{F})' = (\det \mathbf{F})\mathbf{I}, \tag{1-12}$$

where cof **F** denotes the matrix of cofactors of **F**. Specifically, the *ij*-element of cof **F** is the *ij*-cofactor of **F** and is equal to $(-1)^{i+j}$ times the determinant of the matrix formed by removing the *i*th row and *j*th column of **F**. Matrix **I** is referred to as the *identity matrix* and is

$$\mathbf{I} = \begin{bmatrix} 1 & 0 & \ldots & 0 \\ 0 & 1 & \ldots & 0 \\ \ldots & \ldots & \ldots & \ldots \\ 0 & 0 & \ldots & 1 \end{bmatrix} \tag{1-13}$$

with unit major-diagonal elements and zero off-diagonal elements. The $n \times n$ identity matrix sometimes is partitioned by columns as

$$\mathbf{I} = [\mathbf{i}_1 \quad \mathbf{i}_2 \quad \ldots \quad \mathbf{i}_n],$$

and column vector $\mathbf{i}_j$ is referred to as the jth *standard basis vector.* A matrix $\mathbf{F}$ is called *singular* if and only if $\det \mathbf{F} = 0$. A nonsingular matrix $\mathbf{F}$ possesses the *inverse* $\mathbf{F}^{-1}$ that can be identified from (1-12) as

$$\mathbf{F}^{-1} = \frac{1}{\det \mathbf{F}} \operatorname{cof} \mathbf{F}'. \tag{1-14}$$

Eigenvalues and eigenvectors of a square matrix $\mathbf{F}$ often are denoted as $\lambda\{\mathbf{F}\}$ and $\mathbf{m}\{\mathbf{F}\}$ respectively, and they are defined as any complex scalar λ and vector $\mathbf{m} \neq \mathbf{0}$ for which

$$\mathbf{Fm} = \lambda \mathbf{m}. \tag{1-15}$$

By virtue of (1-11), eigenvalues of $\mathbf{F}$ are roots of the *characteristic polynomial* $\det(\lambda \mathbf{I} - \mathbf{F})$ of $\mathbf{F}$. In addition, eigenvalues of an $n \times n$ matrix $\mathbf{F}$ satisfy

$$\det \mathbf{F} = \prod_{i=1}^{n} \lambda_i\{\mathbf{F}\} \tag{1-16}$$

and

$$\operatorname{tr} \mathbf{F} = \sum_{i=1}^{n} \lambda_i\{\mathbf{F}\} \tag{1-17}$$

where $\operatorname{tr} \mathbf{F}$ denotes the trace of $\mathbf{F}$.

Vector products arise frequently in this book in a number of contexts. In particular, *inner product* $\mathbf{f}'\mathbf{g}$ and *outer product* $\mathbf{fg}'$ for real vectors $\mathbf{f}$ and $\mathbf{g}$ are defined compatibly with standard matrix multiplications. One context in which inner products arise concerns properties of matrices. For instance, a real symmetric matrix $\mathbf{F}$ is called *positive semi-definite* if and only if

$$\mathbf{f}'\mathbf{Ff} \geqslant 0 \;\forall \text{ real } \mathbf{f}, \tag{1-18}$$

where the symbol $\forall$ denotes "for all". All eigenvalues of real symmetric matrices are real, and all eigenvalues of positive semi-definite matrices are nonnegative. In addition, a real symmetric matrix is called *positive definite* if and only if

$$\mathbf{f}'\mathbf{Ff} > 0 \;\forall \text{ real } \mathbf{f} \neq \mathbf{0}. \tag{1-19}$$

Positive semi-definite matrices are positive definite if and only if they are nonsingular.

Norms of vectors and matrices also arise frequently in this book. A scalar function $\|\cdot\|$ defined on an appropriate vector space is called a *norm* if and only if

$$\begin{aligned}&\text{(i)}\ \ \mathbf{f} = \mathbf{0} \Rightarrow \|\mathbf{f}\| = 0 \text{ and } \mathbf{f} \neq \mathbf{0} \Rightarrow \|\mathbf{f}\| > 0,\\ &\text{(ii)}\ \ \|c\mathbf{f}\| = |c|\ \|\mathbf{f}\|,\end{aligned} \tag{1-20}$$

and

$$\text{(iii)}\ \ \|\mathbf{f} + \mathbf{g}\| \leqslant \|\mathbf{f}\| + \|\mathbf{g}\|$$

hold for all $\mathbf{f}$ and $\mathbf{g}$ belonging to the vector space. Symbol $\Rightarrow$ denotes "implies". Typical examples of norms, defined on the space $\mathscr{C}^n$ of all complex vectors with $\dim \mathbf{f} = n$, are

$$\|\mathbf{f}\|_1 = \sum_{i=1}^{n} |f_i|,\ \|\mathbf{f}\|_2 = (\mathbf{f}^\dagger \mathbf{f})^{1/2}, \text{ and } \|\mathbf{f}\|_\infty = \max_i |f_i|. \tag{1-21}$$

Operation $(\cdot)^\dagger$ denotes the complex conjugate transpose which is defined in terms of the complex conjugate $(\cdot)^*$ as

$$\mathbf{f}^\dagger = (\mathbf{f}^*)' = (\mathbf{f}')^*. \tag{1-22}$$

Norm $\|\mathbf{f}\|_2$ for real vectors $\mathbf{f}$ is sometimes called the Euclidean length of $\mathbf{f}$ and is denoted as $|\mathbf{f}|$. Algebraic properties of a norm are extended to matrices by the definition

$$\|\mathbf{F}\| = \inf \{c\colon \|\mathbf{Ff}\| \leqslant c\,\|\mathbf{f}\|\ \forall\ \mathbf{f} \in \mathscr{C}^n\} \tag{1-23}$$

where $\{c\colon \ldots\}$ denotes the set of all c subject to restrictions "..." and $\in$ denotes "belongs to". The greatest-lower-bound, denoted by inf, appearing in (1-23) can be replaced by the equivalent definition

$$\|\mathbf{F}\| = \sup \{\|\mathbf{Ff}\| : \|\mathbf{f}\| = 1 \text{ and } \mathbf{f} \in \mathscr{C}^n\} \tag{1-24}$$

employing the least-upper-bound denoted by sup. Norms of matrices corresponding to the examples given in (1-21) for vectors are

$$\|\mathbf{F}\|_1 = \max_j \left(\sum_{i=1}^{m} |f_{ij}| \right),\ \|\mathbf{F}\|_2 = \lambda_{\max}^{1/2} \{\mathbf{F}^\dagger \mathbf{F}\},$$

and

$$\|\mathbf{F}\|_\infty = \max_i \left(\sum_{j=1}^{n} |f_{ij}| \right) \tag{1-25}$$

respectively when $\dim \mathbf{F} = m \times n$. Symbol $\lambda_{\max}$ is used to denote the maximum eigenvalue of a positive semi-definite matrix. The three mathematical properties given in (1-20) for norms of vectors apply to norms of matrices as well, and the property

$$\|\mathbf{FG}\| \leqslant \|\mathbf{F}\|\ \|\mathbf{G}\| \tag{1-26}$$

can also be derived [11].

As is customary with square matrices, functions of matrices are defined analogously to the corresponding functions of a scalar. For instance, polynomial

$$p(x) = \sum_{k=0}^{l} p_k x^k \tag{1-27}$$

is used to define the matrix polynomial

$$p(\mathbf{F}) = \sum_{k=0}^{l} p_k \mathbf{F}^k. \tag{1-28}$$

In a similar fashion, an infinite series such as

$$e^x = \sum_{k=0}^{\infty} \frac{1}{k!} x^k \tag{1-29}$$

is used to define the matrix exponential

$$e^{\mathbf{F}t} = \sum_{k=0}^{\infty} \frac{1}{k!} \mathbf{F}^k t^k. \tag{1-30}$$

The sum appearing in (1-30) converges uniformly on every finite closed interval of t and also converges absolutely [11] at every finite t. Moreover, functions of matrices often share many properties with their scalar counterparts such as

$$\frac{d}{dt}(e^{\mathbf{F}t}) = \mathbf{F}(e^{\mathbf{F}t}); \quad e^{\mathbf{0}} = \mathbf{I} \tag{1-31}$$

and

$$(e^{\mathbf{F}t})^{-1} = e^{-\mathbf{F}t}. \tag{1-32}$$

Constant matrices $\mathbf{F}$ are called *asymptotically stable* if and only if there exist constants $\Gamma, \gamma > 0$ for which

$$\| e^{\mathbf{F}t} \| \leqslant \Gamma e^{-\gamma t} \ \forall\ t \geqslant 0. \tag{1-33}$$

This definition is also motivated by similarities between (1-31) and the scalar differential equation satisfied by e^{ft}. However, restrictions such as

$$e^{(\mathbf{F}+\mathbf{G})t} = e^{\mathbf{F}t} e^{\mathbf{G}t} \ \forall\ t \Leftrightarrow \mathbf{F} \text{ and } \mathbf{G} \text{ commute} \tag{1-34}$$

do not arise with the corresponding scalar function.

Notational conventions concerning functions in vector-matrix form are summarized here entirely in terms of vectors. In particular, vectors which are parameterized in terms of t such as $\mathbf{f}(t)$ appear in a vector of derivatives as

$$\dot{\mathbf{f}} = \frac{\mathrm{d}}{\mathrm{d}t}\mathbf{f} = \begin{bmatrix} \frac{\mathrm{d}f_1}{\mathrm{d}t} \\ \frac{\mathrm{d}f_2}{\mathrm{d}t} \\ \vdots \end{bmatrix} \tag{1-35}$$

and in a vector of integrals as

$$\int_0^t \mathbf{f}(\xi)\,\mathrm{d}\xi = \begin{bmatrix} \int_0^t f_1(\xi)\,\mathrm{d}\xi \\ \int_0^t f_2(\xi)\,\mathrm{d}\xi \\ \vdots \end{bmatrix} \tag{1-36}$$

Moreover, vectors of one-sided Laplace transforms appear frequently in this book and are denoted as

$$\tilde{\mathbf{f}}(s) = \int_0^\infty \mathbf{f}(t)\,\mathrm{e}^{-st}\,\mathrm{d}t, \tag{1-37}$$

where

$$\tilde{\mathbf{f}}(s) = \begin{bmatrix} F_1(s) \\ F_2(s) \\ \vdots \end{bmatrix}. \tag{1-38}$$

In a similar fashion, a vector of statistical expectations is written as

$$E\{\mathbf{f}(t)\} = \begin{bmatrix} E\{f_1(t)\} \\ E\{f_2(t)\} \\ \vdots \end{bmatrix} \tag{1-39}$$

where $E\{\cdot\}$ denotes the expectation operation [12]. Notational conventions for functions arranged in matrix form such as $\mathbf{F}(t)$ are completely analogous to those presented here for vector functions.

Partial derivatives of scalar and vector functions of many variables also appear frequently in this book, and partial derivatives generally are denoted by subscripts. For instance, let f be the differentiable scalar function $f(\mathbf{c})$ with dim $\mathbf{c} = m$. Then first partials are written as

$$f_{c_i} = \frac{\partial f}{\partial c_i} \quad \text{or} \quad f_{\mathbf{c}} = \left[\frac{\partial f}{\partial c_i}\right] \quad \text{with dim } f_{\mathbf{c}} = m, \tag{1-40}$$

and vector $f_{\mathbf{c}}$ is called the *gradient* of f with respect to $\mathbf{c}$. This definition is extended to scalar functions f of a matrix $\mathbf{C}$ with dim $\mathbf{C} = m \times n$ as

$$f_{\mathbf{C}} = \left[\frac{\partial f}{\partial c_{ij}}\right] \text{ with dim } f_{\mathbf{C}} = m \times n. \tag{1-41}$$

In a similar fashion, let f be the twice-differentiable scalar function $f(\mathbf{c}, \mathbf{d})$ with dim $\mathbf{c} = m$ and dim $\mathbf{d} = n$. Then second partials are written as

$$f_{c_i d_j} = \frac{\partial^2 f}{\partial c_i \partial d_j} \quad \text{or} \quad f_{\mathbf{cd}} = f'_{\mathbf{dc}} = \left[\frac{\partial^2 f}{\partial c_i \partial d_j}\right] \text{ with dim } f_{\mathbf{cd}} = m \times n, \tag{1-42}$$

and matrix $f_{\mathbf{cd}}$ is called the *hessian* of f with respect to $\mathbf{c}$ and $\mathbf{d}$. This notation is also specialized to hessians with respect to a single vector variable as

$$f_{\mathbf{cc}} = \left[\frac{\partial^2 f}{\partial c_i \partial c_j}\right]. \tag{1-43}$$

so that $f_{\mathbf{cc}}$ may be symmetric. Partial derivatives of differentiable vector functions $\mathbf{f}(\mathbf{c})$ with dim $\mathbf{f} = m$ and dim $\mathbf{c} = n$ occur frequently as well. The *Jacobian* of $\mathbf{f}$ with respect to $\mathbf{c}$ is the matrix $\mathbf{f_c}$ defined by

$$\mathbf{f_c} = \left[\frac{\partial f_i}{\partial c_j}\right] \text{ with dim } \mathbf{f_c} = m \times n. \tag{1-44}$$

Additional notational conventions are used in the contexts of various individual sections of this book. These conventions, however, are somewhat less standardized and hence are defined in the appropriate sections.

1.2 LINEAR CONSTANT-COEFFICIENT STATE EQUATIONS

A major portion of this book concerns design problems for which open-loop dynamics can be modeled by linear constant-coefficient equations. Forcing functions of these equations, however, are deterministic in some problems and stochastic in other problems. Problem formulations differ for these two types of forcing functions, and hence a brief introduction to these formulations is given here.

In the case of deterministic forcing functions, state equations for open-loop dynamics of linear time-variant systems are written as

$$\dot{\mathbf{x}} = \mathbf{F}\mathbf{x} + \mathbf{G}\mathbf{u}; \quad \mathbf{x}(0) = \mathbf{x}_0, \tag{1-45}$$

where $\mathbf{x}(t)$ is called the state vector and $\mathbf{u}(t)$ is called either the control vector or the forcing-function vector depending on the context. Response equations are also introduced for many problems and generally are written as

$$\mathbf{y} = \mathbf{H}\mathbf{x}, \tag{1-46}$$

where $\mathbf{y}(t)$ is called the response vector. Coefficient matrices $\mathbf{F}$, $\mathbf{G}$ and $\mathbf{H}$ are independent of time t unless stated otherwise. However, these matrices frequently are taken to be functions $\mathbf{F}(\mathbf{c})$, $\mathbf{G}(\mathbf{c})$, and $\mathbf{H}(\mathbf{c})$ of a vector of adjustable parameters $\mathbf{c}$. Adjustment of $\mathbf{c}$ is employed in order to achieve the goals of parameter optimization.

Solutions to linear constant-coefficient state equations are well known and can be written in terms of the matrix exponential $e^{\mathbf{F}t}$. In particular, state vector

$$\mathbf{x}(t) = e^{\mathbf{F}t}\mathbf{x}_0 + \int_0^t e^{(t-\xi)\mathbf{F}}\mathbf{G}\mathbf{u}(\xi)\, d\xi \tag{1-47}$$

is obtained directly from the theory of ordinary differential equations.

In the case of stochastic forcing functions, state equations for open-loop dynamics of linear time-invariant systems are given in the somewhat less well-known form

$$d\mathbf{x} = (\mathbf{F}\mathbf{x})\, dt + \mathbf{G}\, d\mathbf{u}; \quad \mathbf{x}(0) = \mathbf{x}_0 \tag{1-48}$$

used in stochastic calculus [13]. Vector $\mathbf{x}(t)$ denotes the stochastic state vector, and vector $\mathbf{u}(t)$ denotes either the stochastic disturbance or forcing-function vector depending on the context. Stochastic state equations are always written with the understanding that

$$E\{\mathbf{x}_0\} = \mathbf{0} \quad \text{and} \quad E\{\mathbf{u}(t)\} = \mathbf{0}\ \forall\ t \tag{1-49}$$

so that $\mathbf{x}(t)$ also has zero mean for all time. Moreover, forcing-function vector $\mathbf{u}(t)$ is always taken to be Brownian motion [13] with covariance matrix

$$E\{\mathbf{u}(t)\mathbf{u}'(\tau)\} = \mathbf{V} \min\{t, \tau\}. \tag{1-50}$$

Response equations for stochastic problems are again taken to be those given in (1-46). In addition, coefficient matrices $\mathbf{F}$, $\mathbf{G}$, and $\mathbf{H}$ are subject to the same assumptions as involved in the deterministic case. Similarly, covariance coefficient matrix $\mathbf{V}$ is taken to have no dependence on time although it may be a function of adjustable parameters $\mathbf{c}$ and other covariance variables.

Solutions to these stochastic state equations can also be written in terms of $e^{\mathbf{F}t}$. In particular, stochastic state vector

$$\mathbf{x}(t) = e^{\mathbf{F}t}\mathbf{x}_0 + \int_0^t e^{(t-\xi)\mathbf{F}}\mathbf{G}\,d\mathbf{u}(\xi) \tag{1-51}$$

is obtained in terms of a stochastic integral [13] which is in Stieltjes form instead of Riemann form.

Purely for the purposes of comparison, stochastic state equations are rewritten here in the form commonly presented in engineering text books. Brownian motion $\mathbf{u}$ is related to so-called *white noise* $\mathbf{v}$ by the stochastic integral

$$\mathbf{u}(t) = \int_0^t d\mathbf{v}(\xi). \tag{1-52}$$

Introduction of white noise variables leads to rewriting (1-48) in the symbolic form

$$\dot{\mathbf{x}} = \mathbf{F}\mathbf{x} + \mathbf{G}\mathbf{v}; \quad \mathbf{x}(0) = \mathbf{x}_0 \tag{1-53}$$

and replacing (1-50) by the symbolism

$$E\{\mathbf{v}(t)\mathbf{v}'(\tau)\} = \mathbf{V}\delta_0(t-\tau), \tag{1-54}$$

where $\delta_0(x)$ denotes the so-called impulse functional [14]. In addition, introduction of white-noise variables leads to rewriting (1-51) in the symbolic form

$$\mathbf{x}(t) = e^{\mathbf{F}t}\mathbf{x}_0 + \int_0^t e^{(t-\xi)\mathbf{F}}\mathbf{G}\mathbf{v}(\xi)\,d\xi. \tag{1-55}$$

Equations (1-53), (1-54), and (1-55) are not used in any mathematical manipulations in order to avoid questions concerning mathematical rigor. However, the state equation form appearing in (1-53) is used frequently for the purpose of merely identifying coefficient matrices $\mathbf{F}$ and $\mathbf{G}$ in the formulation of specific design problems.

1.3 TRANSFER FUNCTIONS

One-sided Laplace transforms are of course very useful in the analysis of linear constant-coefficient systems, and the algebra of transfer functions is used throughout this book in the formulation of specific design problems. Therefore, a brief introduction to frequently used frequency-domain relationships is given here.

For the purposes of this book, only transforms of piecewise continuous functions [15] need be considered. A real vector-function $\mathbf{f}(t)$ is *piecewise continuous* on $[0, \infty)$ if and only if, given any finite interval $[0, T]$,

i) $\mathbf{f}(t)$ is continuous except at a finite number of points of $(0, T)$

and

ii) $\mathbf{f}^+(t) = \lim_{\epsilon \to 0^+} \mathbf{f}(t + \epsilon)$ exists at every point of $[0, T)$, and

$\mathbf{f}^-(t) = \lim_{\epsilon \to 0^+} \mathbf{f}(t - \epsilon)$ exists at every point of $(0, T]$.

Piecewise continuous functions $\mathbf{f}(t)$ are said to be of *exponential order* if and only if there exist constants Σ and σ for which

$$\| f(t) \| \leqslant \Sigma \, e^{\sigma t} \; \forall \; t \geqslant 0.$$

The Laplace transform, namely the integral appearing in (1-37), converges absolutely and uniformly for all s with $\text{Re}\{s\} > \sigma$. Henceforth $\tilde{\mathbf{f}}(s)$ is used to denote only one-sided Laplace transforms of time functions of exponential order as defined here. Notation $\text{Re}\{\cdot\}$ and $\text{Im}\{\cdot\}$ denotes real and imaginary parts respectively.

Two basic properties of one-sided Laplace transforms are used throughout this book. The first property concerns time shifting of time functions [15] and is stated as follows:‡

Proposition 1-1 *Suppose* $\mathbf{f}(t)$ *is of exponential order and let*

$$\mathbf{g}(t) = \begin{Bmatrix} \mathbf{0} & \text{for } 0 \leqslant t < T \\ \mathbf{f}(t - T) & \text{for } t \geqslant T \end{Bmatrix}.$$

Then

$$\tilde{\mathbf{g}}(s) = e^{-sT} \tilde{\mathbf{f}}(s). \tag{1-56}$$

The second property concerns differentiation of time functions and is stated as follows:

Proposition 1-2 *Suppose* $\mathbf{f}(t)$ *is continuous and* $\dot{\mathbf{f}}(t)$ *is of exponential order. Then*

$$\int_0^\infty \dot{\mathbf{f}}(t) \, e^{-st} \, dt = s\tilde{\mathbf{f}}(s) - \lim_{t \to 0^+} \mathbf{f}(t). \tag{1-57}$$

---

‡ Fundamental results which are not proved in this book are stated as propositions.

Although there are many other important properties of one-sided Laplace transforms, only these involve the basic assumptions concerning the admissibility of time functions that are needed in this book.

One-sided Laplace transforms, of course, can be applied directly to deterministic state equations with forcing-functions of exponential order. The Laplace transforms of both sides of (1-45) yields

$$s\tilde{\mathbf{x}} = \mathbf{x}_0 + \mathbf{F}\tilde{\mathbf{x}} + \mathbf{G}\tilde{\mathbf{u}}$$

which can be further rewritten as

$$\tilde{\mathbf{x}} = (s\mathbf{I} - \mathbf{F})^{-1}(\mathbf{x}_0 + \mathbf{G}\tilde{\mathbf{u}}). \tag{1-58}$$

A comparison of (1-58) and (1-47) reveals that $(s\mathbf{I} - \mathbf{F})^{-1}$ is the Laplace transform of $e^{\mathbf{F}t}$ and $(s\mathbf{I} - \mathbf{F})^{-1}\mathbf{G}\tilde{\mathbf{u}}$ is the Laplace transform of the integral appearing in (1-47). In a similar fashion, the Laplace transform of both sides of (1-46) and the elimination of $\tilde{\mathbf{x}}$ using (1-58) yields

$$\tilde{\mathbf{y}} = \mathbf{H}(s\mathbf{I} - \mathbf{F})^{-1}(\mathbf{x}_0 + \mathbf{G}\tilde{\mathbf{u}}) \tag{1-59}$$

for response variables. Elements of the matrix

$$\mathbf{T}(s) = \mathbf{H}(s\mathbf{I} - \mathbf{F})^{-1}\mathbf{G} \tag{1-60}$$

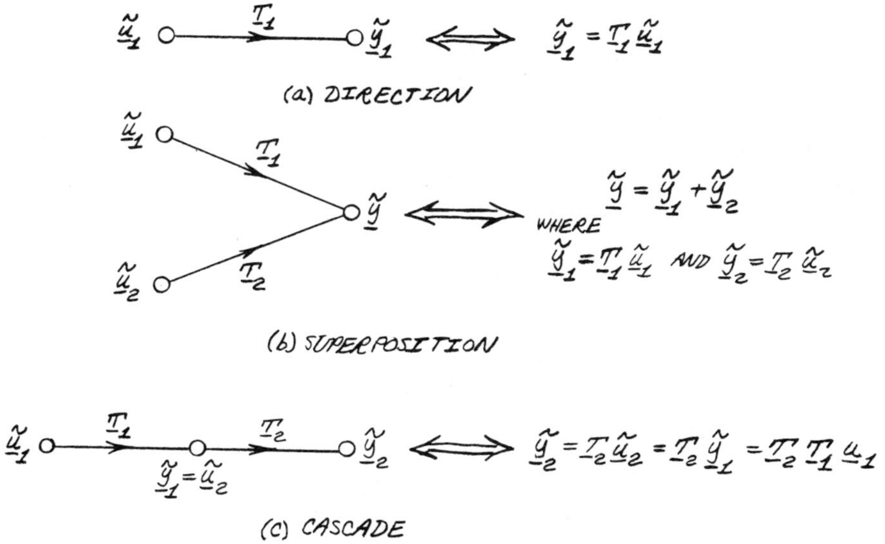

Figure 1.1 Vector flow graph conventions.

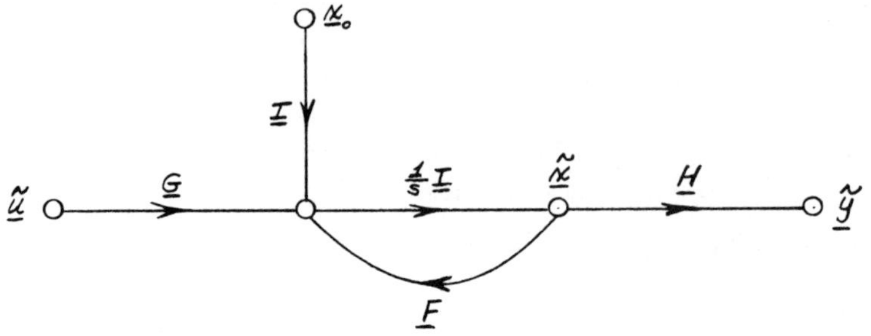

Figure 1-2 Vector flow graph of open-loop state and response equations.

are called *transfer functions* and completely characterize *zero-state response*, that is, the response obtained with $\mathbf{x}_0 = \mathbf{0}$, in terms of transforms of controls $\tilde{\mathbf{u}}$.

In order to clarify example problems, system transfer-functions are often depicted graphically in terms of flow graphs [10]. Although not explicitly discussed here, topological reduction methods can also be employed to compute composite transfer functions from scalar flow graphs.

Conventions stated in Figure 1-1 serve to define a vector flow-graph in terms of Laplace transforms of excitation variables $\tilde{\mathbf{u}}_1$ and $\tilde{\mathbf{u}}_2$, transforms of response variables $\tilde{\mathbf{y}}_1$ and $\tilde{\mathbf{y}}_2$, and transfer-function matrices $\mathbf{T}_1$ and $\mathbf{T}_2$. Vectors of Laplace transforms are represented by graph nodes, and transfer-function matrices are represented by graph branches. Excitation and response variables must conform to legitimate applications of superposition and homogeneity, and hence branch directions corresponding to the roles of these variables cannot be interchanged in general. By way of example, the vector flow-graph shown in Figure 1-2 directly depicts (1-45) and (1-46). In addition, removal of the self-loops appearing in Figure 1-2 yields the partially reduced vector flow graph appearing in Figure 1-3 which corresponds to (1-59).

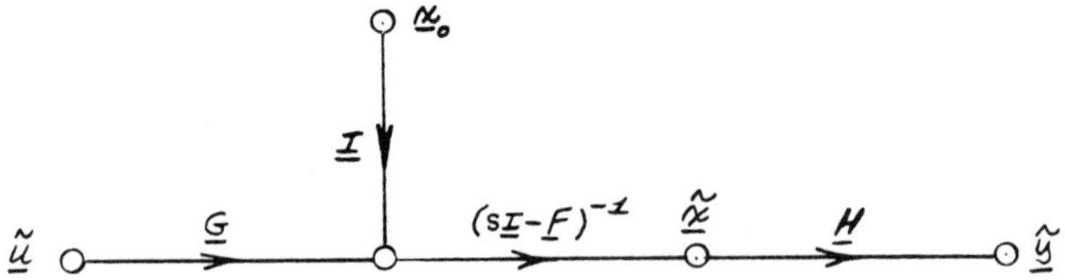

Figure 1-3 Partial reduction of flow graph appearing in Figure 1-2.

An example of scalar flow graphs is shown in Figure 1-4. This flow graph corresponds to the *companion matrix*

$$\mathbf{F} = \begin{bmatrix} 0 & 1 & 0 & \dots & 0 \\ 0 & 0 & 1 & \dots & 0 \\ \dots & \dots & \dots & \dots & \dots \\ 0 & 0 & 0 & \dots & 1 \\ -p_0 & -p_1 & -p_2 & \dots & -p_{n-1} \end{bmatrix} \tag{1-61}$$

and matrices

$$\mathbf{G}' = [0 \quad 0 \quad \dots \quad 1], \quad \mathbf{H} = [q_0 \quad q_1 \quad \dots \quad q_{n-2} \quad q_{n-1}]. \tag{1-62}$$

Elimination of self-loops such as illustrated by Figures 1-2 and 1-3 or computation by Mason's gain formula [10] results in

$$\frac{Y_1}{U_1} = \frac{\sum\limits_{k=0}^{n-1} q_k s^k}{\sum\limits_{k=0}^{n} p_k s^k} \text{ with } p_n = 1. \tag{1-63}$$

The flow graph shown in Figure 1-4 thus serves as a canonical realization of a single transfer function which has the property of vanishing at $|s| = \infty$.

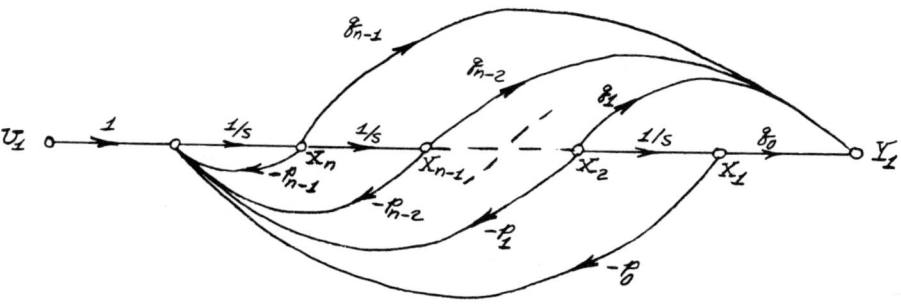

Figure 1-4 Flow graph realization of a transfer function.

1.4 PERFORMANCE INTEGRALS

Formulation of all design problems of course requires selection of an appropriate model of system dynamics. Formulation of almost all problems to be solved by automated design methods requires, in addition, selection of a performance functional. The purpose of this section is to introduce the notion of a performance functional in the form of a performance integral. Moreover, performance integrals of the type discussed here are evaluated in the tabulated form of Phillips [1].

A simplified class of deterministic problems is introduced first for the purposes of discussion. Specifically, state equations given in (1-45) are assumed to represent closed-loop dynamics with forcing functions taken to be $\mathbf{u} = \mathbf{0}$ for all time. The system state vector then becomes

$$\mathbf{x}(t) = e^{\mathbf{F}t}\mathbf{x}_0 \tag{1-64}$$

in accordance with (1-47). In addition, a response vector is introduced as an error vector $\mathbf{e}$ and is written as

$$\mathbf{e} = \mathbf{H}\mathbf{x}, \tag{1-65}$$

where $\mathbf{H} = \mathbf{H}(\mathbf{c})$. This error vector can also be written as

$$\mathbf{e}(t) = \mathbf{H}\, e^{\mathbf{F}t}\mathbf{x}_0 \tag{1-66}$$

for any initial state $\mathbf{x}_0$. Parameters $\mathbf{c}$ presumably are selected in an attempt to achieve the condition $\mathbf{e}(t) \simeq \mathbf{0}$. Because this goal is seldom achieved exactly, a measure of error must be introduced so that a relative evaluation of various selections of $\mathbf{c}$ can be made. Thus, an integral of error is defined as

$$E(T) = \int_0^T \{\mathbf{e}'(t)\mathbf{e}(t)\}\, dt \tag{1-67}$$

for convenience. Moreover, condition $\mathbf{e}(\infty) = \mathbf{0}$ is desired for most regulator problems so that performance integral

$$I = \lim_{T \to \infty} E(T) \tag{1-68}$$

is adopted. In order to assure that the limit appearing in (1-68) exists as a finite number, the condition that $\mathbf{F}$ be asymptotically stable often must be imposed.

A simplified class of stochastic problems is introduced here for the purposes of comparison. Specifically, state equations given in (1-48) are assumed to represent closed-loop dynamics, and assumptions $\mathbf{x}_0 = \mathbf{0}$ and rank $\mathbf{G} = 1$ are also imposed. In addition, response equations given in (1-46) and expectation $E\{\mathbf{y}'(T)\mathbf{y}(T)\}$ are introduced for this class of stochastic problems. This expectation can then be reduced to the form of a performance integral as follows. First, properties of the trace yield

$$E\{\mathbf{y}'(T)\mathbf{y}(T)\} = \mathrm{tr}[E\{\mathbf{y}(T)\mathbf{y}'(T)\}]. \tag{1-69}$$

Second, covariance matrix $E\{\mathbf{x}(T)\mathbf{x}'(T)\}$ is given by the well-known [13] expression

$$E\{\mathbf{x}(T)\mathbf{x}'(T)\} = \int_0^T \{e^{\mathbf{F}t}\mathbf{GVG}'\,e^{\mathbf{F}'t}\}\,dt, \tag{1-70}$$

where $\mathbf{V}$ is the covariance coefficient matrix appearing in (1-50). Third, matrix $(\mathbf{GVG}')$ is positive semi-definite and has rank one. Hence, this matrix can always be written as

$$\mathbf{GVG}' = \mathbf{x}_0\mathbf{x}_0', \tag{1-71}$$

and this decomposition of $(\mathbf{GVG}')$ defines an equivalent deterministic initial-state $\mathbf{x}_0$. Equations (1-46), (1-70), and (1-71) then combine to yield

$$E\{\mathbf{y}(T)\mathbf{y}'(T)\} = \int_0^T \{\mathbf{e}(t)\mathbf{e}'(t)\}\,dt \tag{1-72}$$

using (1-66) to define an equivalent deterministic error vector $\mathbf{e}(t)$. In addition, properties of the trace are used again to obtain

$$E\{\mathbf{y}'(T)\mathbf{y}(T)\} \int_0^T \{\mathbf{e}'(t)\mathbf{e}(t)\}\,dt, \tag{1-73}$$

and the performance integral

$$I = \lim_{T\to\infty} E\{\mathbf{y}'(T)\mathbf{y}(T)\} \tag{1-74}$$

is adopted for this class of stochastic problems assuming that $\mathbf{F}$ is asymptotically stable when necessary.

For both deterministic and stochastic problems discussed here, *performance integrals* become the functional

$$I = \int_0^\infty \{\mathbf{e}'(t)\mathbf{e}(t)\}\,dt. \tag{1-75}$$

This form of performance integral is particularly useful because analytical evaluation of these integrals is possible. In particular, the one-sided Laplace transform can be specialized to $s = j\omega$ for systems with $\mathbf{e}(\infty) = \mathbf{0}$ so that

$$\tilde{\mathbf{e}}(j\omega) = \int_0^\infty \mathbf{e}(t)\,e^{-j\omega t}\,dt \tag{1-76}$$

and

$$\mathbf{e}(t) = \frac{1}{2\pi j}\int_{-j\infty}^{j\infty} \tilde{\mathbf{e}}(j\omega)\,e^{j\omega t}\,d(j\omega) \tag{1-77}$$

is the corresponding transform pair. Moreover, Parseval's theorem [14] yields an equivalence for performance integral I appearing in (1-75); namely,

$$I = \frac{1}{2\pi j} \int_{-j\infty}^{j\infty} \{\tilde{\mathbf{e}}'(j\omega)\tilde{\mathbf{e}}(-j\omega)\}\, \mathrm{d}(j\omega). \tag{1-78}$$

This form of the performance integral does not involve time domain analysis. Instead, only the one-sided Laplace transform of error vector $\mathbf{e}(t)$ is needed and is found from (1-66) to be

$$\tilde{\mathbf{e}}(s) = \mathbf{H}(s\mathbf{I} - \mathbf{F})^{-1}\mathbf{x}_0. \tag{1-79}$$

Thus, transform $\tilde{\mathbf{e}}(s)$ can be computed directly using transfer-function methods such as flow graph reduction.

The performance integral appearing in (1-78) can be evaluated term-by-term as follows. An integral I_n is written in terms of one-sided Laplace transform $E_n(s)$ of error signal $e_n(t)$ as

$$I_n = \frac{1}{2\pi j} \int_{-j\infty}^{j\infty} E_n(j\omega)E_n(-j\omega)\, \mathrm{d}(j\omega). \tag{1-80}$$

Error signal $e_n(t)$ is defined by the one-sided Laplace transform

$$E_n(s) = \frac{Q_n(s)}{P_n(s)}, \tag{1-81}$$

where

$$Q_n(s) = \sum_{k=0}^{n-1} q_k s^k \tag{1-82}$$

and

$$P_n(s) = \sum_{k=0}^{n} p_k s^k. \tag{1-83}$$

Phillips [1] tabulated I_n up to degree $n = 10$ under the conditions that $P_n(s)$ is of minimal degree and is *strictly Hurwitz* having zeros located in the left-half plane only. The first four of these integrals are useful for the purposes of this book and are listed as

$$I_1 = \frac{q_0^2}{2p_0p_1}, \tag{1-84}$$

$$I_2 = \frac{q_1^2p_0 + q_0^2p_2}{2p_0p_1p_2}, \tag{1-85}$$

$$I_3 = \frac{q_2^2p_0p_1 + (q_1^2 - 2q_0q_2)p_0p_3 + q_0^2p_2p_3}{2p_0p_3(-p_0p_3 + p_1p_2)}, \tag{1-86}$$

and

$$I_4 = \frac{q_3^2(-p_0^2 p_3 + p_0 p_1 p_2) + (q_2^2 - 2q_1 q_3)p_0 p_1 p_4 + (q_1^2 - 2q_0 q_2)p_0 p_3 p_4 + q_0^2(-p_1 p_4^2 + p_2 p_3 p_4)}{2p_0 p_4(-p_0 p_3^2 - p_1^2 p_4 + p_1 p_2 p_3)}. \tag{1-87}$$

Values of the performance integral given in (1-75) thus can be determined using only frequency domain methods. Moreover, the analytical form of (1-84) through (1-87) permits direct analytical evaluation of gradients and hessians for specific design problems.

1.5 EXAMPLE WITH DETERMINISTIC SIGNALS

An example is introduced here in order to illustrate some of the important aspects of analytical methods presented earlier for problems with deterministic signals. This example is intended to illustrate use of transform techniques and the tabulated integrals given previously. In addition, this example is intended to serve as an introduction to parameter optimization methods.

The flow graph shown in Figure 1-5 depicts dynamics of a typical inner-loop design problem. Specifically, performance integral

$$I = \int_0^\infty \{e_1^2(t) + e_2^2(t)\}\,dt$$

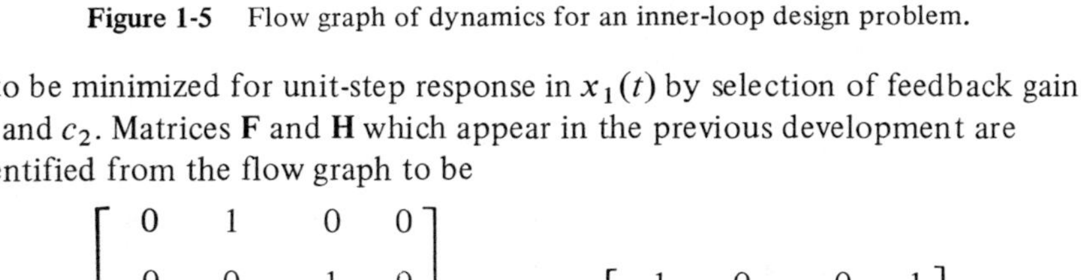

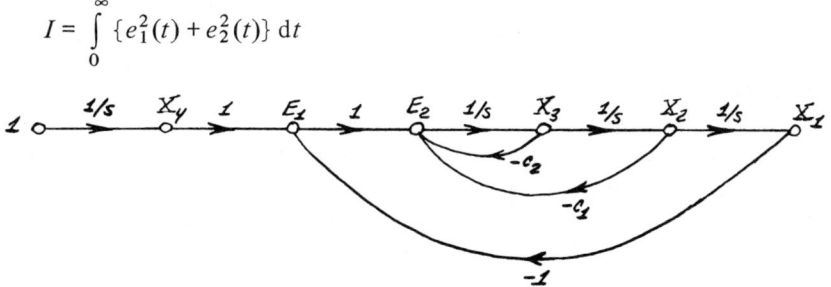

Figure 1-5 Flow graph of dynamics for an inner-loop design problem.

is to be minimized for unit-step response in $x_1(t)$ by selection of feedback gains c_1 and c_2. Matrices **F** and **H** which appear in the previous development are identified from the flow graph to be

$$\mathbf{F} = \begin{bmatrix} 0 & 1 & 0 & 0 \\ 0 & 0 & 1 & 0 \\ -1 & -c_1 & -c_2 & 1 \\ 0 & 0 & 0 & 0 \end{bmatrix} \quad \text{and } \mathbf{H} = \begin{bmatrix} -1 & 0 & 0 & 1 \\ -1 & -c_1 & -c_2 & 1 \end{bmatrix}.$$

In addition, the initial state depicted in Figure 1-5 is

$$\mathbf{x}_0' = [0 \quad 0 \quad 0 \quad 1].$$

Error signal $e_1(t)$ thus is noted to be the difference between a unit-step command $x_4(t)$ and position response $x_1(t)$. Error signal $e_2(t)$ nominally would also be control $u_1(t)$.

Transforms $\tilde{\mathbf{e}}(s)$ of course can be computed using matrix methods indicated in (1-79). However, these transforms can also be computed directly from the flow graph appearing in Figure 1-5 by inspection methods. Such computations yield

$$\tilde{\mathbf{e}}(s) = \frac{1}{s^3 + c_2 s^2 + c_1 s + 1} \begin{bmatrix} (s^2 + c_2 s + c_1) \\ s^2 \end{bmatrix}.$$

Moreover, use of (1-86) for both terms of the performance integral for this problem yields

$$I = \frac{c_2(c_2 + c_1^2)}{2(c_1 c_2 - 1)} \tag{1-88}$$

assuming

$$1 < c_1 c_2 < \infty \quad \text{and} \quad 0 < c_2 < \infty$$

so that $(s^3 + c_2 s^2 + c_1 s + 1)$ is strictly Hurwitz.

Inner-loop design is accomplished by minimizing this performance integral with respect to feedback gains c_1 and c_2. Minimization for this simple problem can be performed graphically; namely, a sketch of the locus of points in the $c_1 - c_2$ plane for which I is equal to some constant can be made for a number of such constants until the minimum possible value of these constants has been determined. The locus of points in the $c_1 - c_2$ plane for which I is equal to a constant is called a *contour* of I. This procedure is equivalent to solving (1-88) for c_1, thereby obtaining

$$c_1 = I \left[1 \pm \sqrt{1 - \frac{c_2^2 + 2I}{I^2 c_2}} \right],$$

where

$$\frac{I^2}{2} \left[1 - \sqrt{1 - \frac{8}{I^3}} \right] \leqslant c_2 \leqslant \frac{I^2}{2} \left[1 + \sqrt{1 - \frac{8}{I^3}} \right]$$

because c_1 must be real and where

$$I \geqslant 2$$

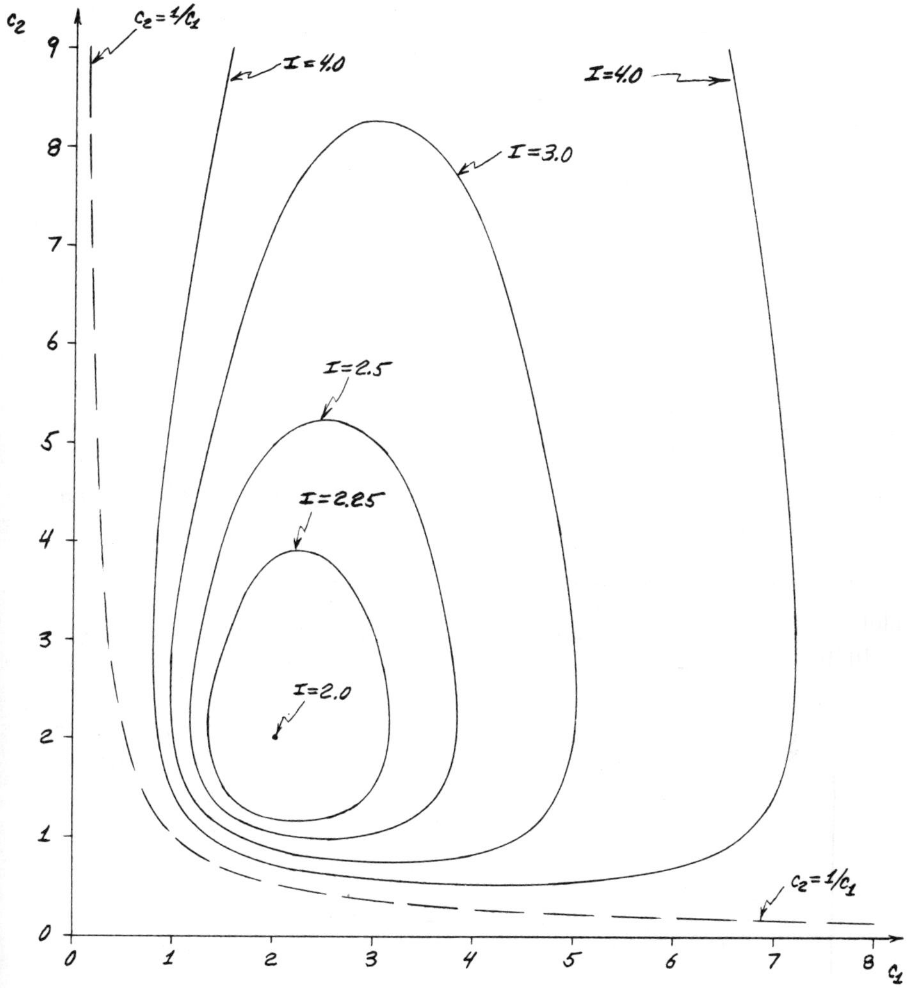

Figure 1-6 Contours for I given in (1-88) and **c** admissible.

because c_2 must be real and I must be nonnegative for asymptotically stable systems. Contours corresponding to these relationships are plotted in Figure 1-6, and feedback gains $c_1 = c_2 = 2$ are seen to yield the relative minimum $I = 2$.

The performance functional appearing in (1-88) can be regarded as a two-dimensional surface $I(c_1, c_2)$ which is depicted in Figure 1-7. A plane which is tangent to $I(c_1, c_2)$ at $c_1 = c_2 = 2$ is also depicted in Figure 1-7, and this plane has zero slopes because I has a relative minimum at the point of tangency. That is, slopes in the c_1- and c_2- directions become

$$\frac{\partial I}{\partial c_1} = 0 \quad \text{and} \quad \frac{\partial I}{\partial c_2} = 0$$

respectively at the relative minimum. For the performance integral given in (1-88), these partial derivatives, which are easily found to be

$$I_{c_1} = c_2 \frac{c_1(c_1c_2 - 2) - c_2^2}{2(c_1c_2 - 1)^2}$$

and

$$I_{c_2} = \frac{c_2(c_1c_2 - 2) - c_1^2}{2(c_1c_2 - 1)^2},$$

of course vanish at $c_1 = c_2 = 2$. Moreover, condition $I_\mathbf{c} = \mathbf{0}$ gives two equations which can be solved for c_1 and c_2; namely,

$$c_1(c_1c_2 - 2) = c_2^2 \quad \text{and} \quad c_2(c_1c_2 - 2) = c_1^2.$$

The procedure of solving equations such as these can often be used as an indirect method for locating a relative minimum of I as opposed to the tedious job of plotting contours of I.

Indirect methods of solving minimization problems such as merely satisfying the condition $I_\mathbf{c} = \mathbf{0}$ may lead to erroneous results and hence must be approached

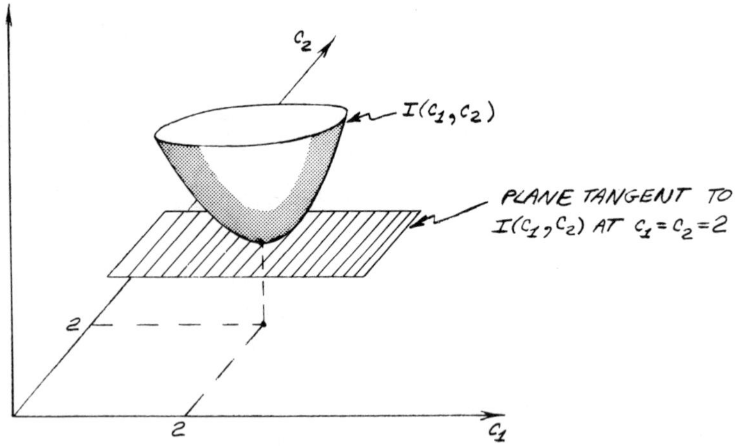

Figure 1-7 Sketch of $I(c_1, c_2)$ for (1-88).

with great care. For instance, both $c_1 = c_2 = -1$ and $c_1 = c_2 = 0$ are also solutions to the algebraic equations given previously. These values of c_1 and c_2 of course are inadmissible because $(s^3 + c_2 s^2 + c_1 s + 1)$ is no longer strictly Hurwitz. Moreover, these values of c_1 and c_2 do not correspond to a relative minimum of the function I given in (1-88). Contours sketched in Figure 1-8 indicate that I has an inflection point at $c_1 = c_2 = 0$ and that I is not even well-defined at $c_1 = c_2 = -1$

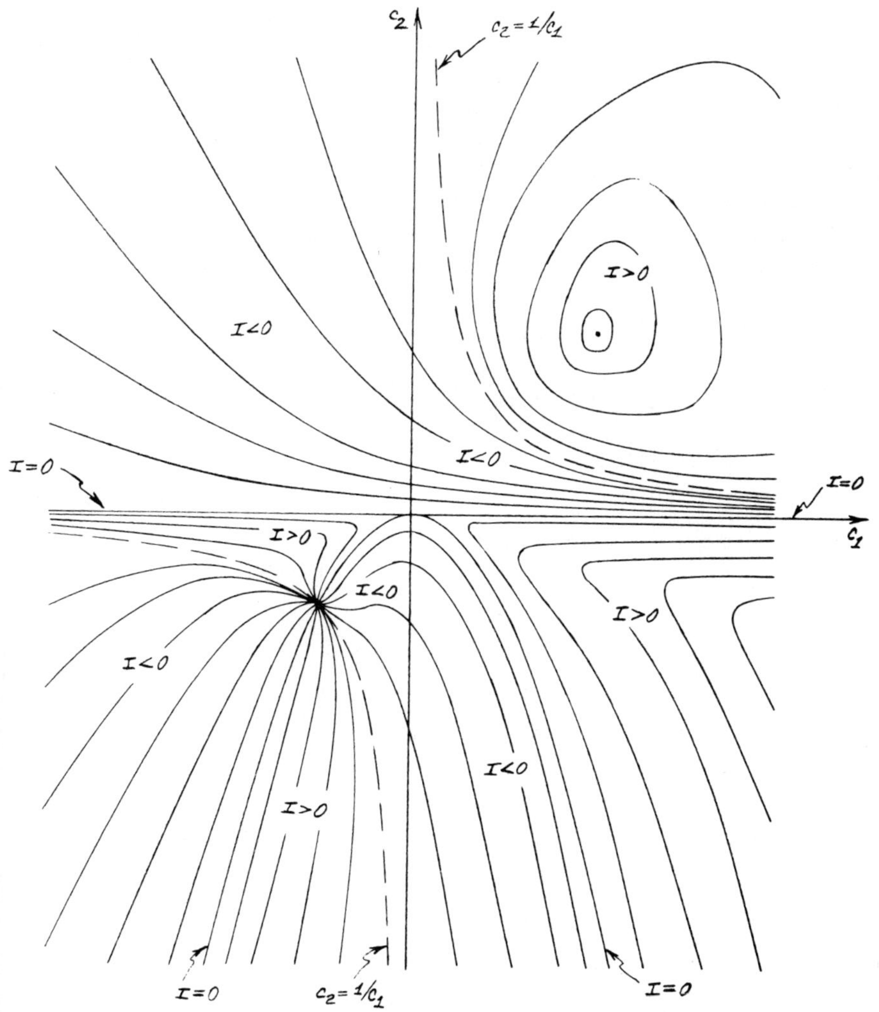

Figure 1-8 Sketch of contours for I given in (1-88).

as well as all other points for which $c_1 c_2 = 1$. This figure also illustrates the complicated geometrical situation that may arise even for very simple problems. In fact, this figure illustrates contours that are not closed, contours that terminate at infinity, as well as contours that terminate at finite points.

1.6 EXAMPLE WITH STOCHASTIC SIGNALS

Another example is introduced here in order to illustrate some of the important aspects of analytical methods presented earlier for problems with stochastic signals. This example is intended to give a simple illustration of the initial-state equivalent of a stochastic problem as well as to point out difficulties that may arise with parameter optimization methods.

The flow graph shown in Figure 1-9 depicts a position controller with a stochastic input signal. The input signal is represented by white noise with a covariance coefficient equal to 2. The design problem is to minimize

$$I = \lim_{t \to \infty} E\{x_1^2(t)\}$$

subject to some restrictions. Namely, constraint

$$\lim_{t \to \infty} E\{x_2^2(t)\} \leqslant 1$$

is imposed by electronic considerations, and constraint

$$c_2 \leqslant \omega_m$$

is imposed by hydraulic considerations.

In order to use transform methods for this stochastic problem, the initial-state equivalent of the system depicted in Figure 1-9 is introduced first. This

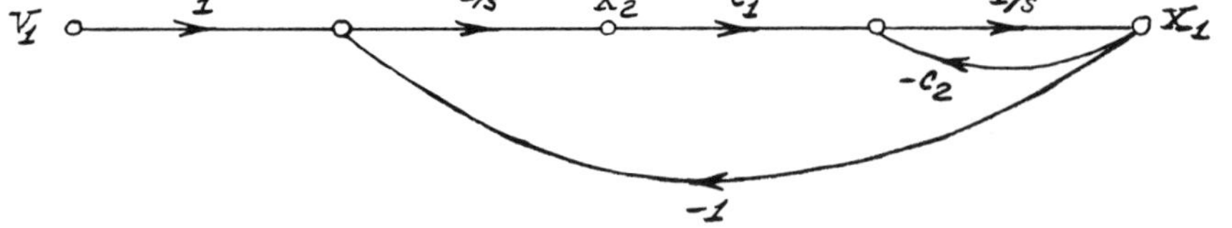

Figure 1-9 Flow graph of a compensated controller with a white-noise input signal.

deterministic equivalent corresponds to the initial state defined by (1-71) and is depicted in Figure 1-10. Transform methods now result in

$$E_1 = \frac{\sqrt{2}c_1}{s^2 + c_2 s + c_1} \quad \text{and} \quad B = \frac{\sqrt{2}\,(s + c_2)}{s^2 + c_2 s + c_1}.$$

Use of the tabulated integral given in (1-85) then yields

$$I = \frac{c_1}{c_2}.$$

This performance functional is to be minimized subject to

$$\frac{c_2^2 + c_1}{c_1 c_2} \leqslant 1 \quad \text{and} \quad c_2 \leqslant \omega_m$$

and under the assumption

$$0 < c_1, \quad c_2 < \infty$$

so that $(s^2 + c_2 s + c_1)$ is strictly Hurwitz.

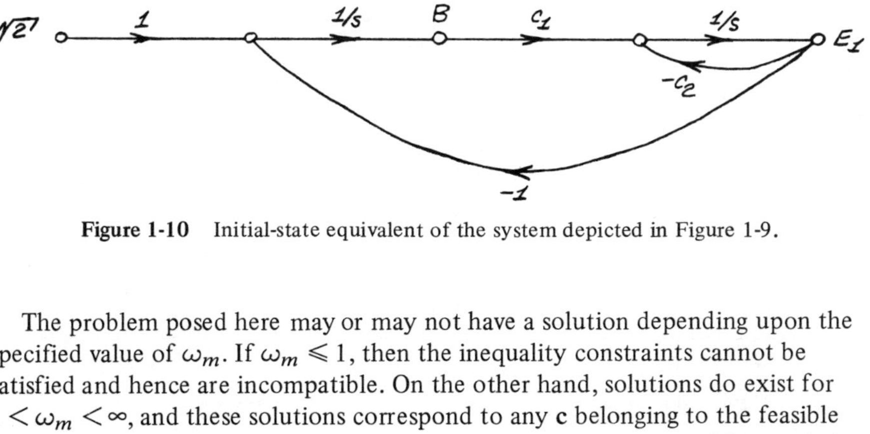

Figure 1-10 Initial-state equivalent of the system depicted in Figure 1-9.

The problem posed here may or may not have a solution depending upon the specified value of ω_m. If $\omega_m \leqslant 1$, then the inequality constraints cannot be satisfied and hence are incompatible. On the other hand, solutions do exist for $1 < \omega_m < \infty$, and these solutions correspond to any **c** belonging to the feasible region depicted in Figure 1-11 with $|\mathbf{c}| < \infty$. Moreover, contours of I are radial with slope inversely proportional to I so that

$$c_1 = \frac{\omega_m^2}{\omega_m - 1} \text{ and } c_2 = \omega_m$$

yield

$$I = \frac{\omega_m}{\omega_m - 1}$$

which is minimal for $1 < \omega_m < \infty$. This result can be determined graphically from Figure 1-11.

Equality and inequality constraints such as illustrated by the problem discussed here arise frequently in control system design. However, graphical methods such as used here become intractable even for relatively simple problems. Indirect methods such as the use of Lagrange multipliers, thus, are vitally important to the practicability of parameter optimization techniques for automated control system design.

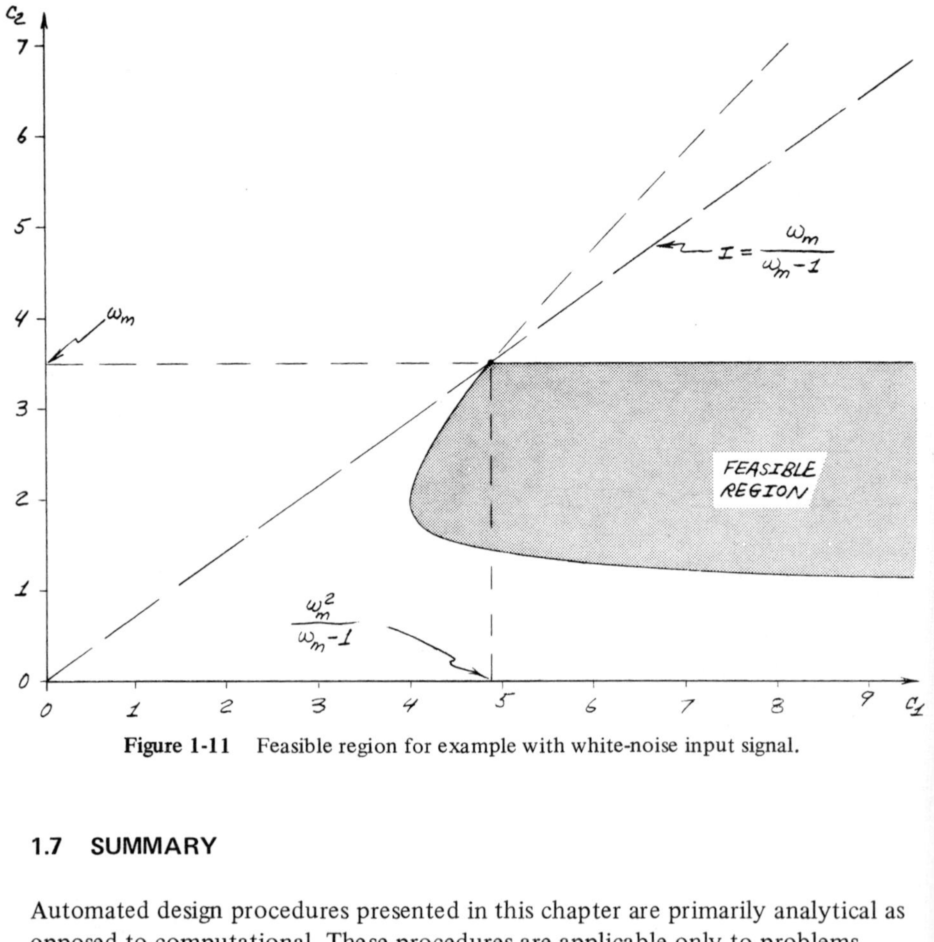

Figure 1-11 Feasible region for example with white-noise input signal.

1.7 SUMMARY

Automated design procedures presented in this chapter are primarily analytical as opposed to computational. These procedures are applicable only to problems that have linear constant-coefficient dynamics. Also, these procedures are applicable only to problems that have performance functionals in the form of suitably defined performance integrals. Techniques presented in this chapter for the minimization of performance integrals primarily are graphical and hence are severely restricted to relatively simple problems.

Formulation of deterministic design problems presented in this chapter do not explicitly include input and other signals. All problems in this category are based on the performance functional given in (1-75) where error vector $\mathbf{e}(t)$ is the initial-state response given in (1-66). However, this formulation does permit the inclusion of any signal that can be generated as the initial-state solution of linear constant-coefficient ordinary differential equations. These signals can be incorporated into the problem by suitable choice of coefficient matrices $\mathbf{H}$ and $\mathbf{F}$ and by suitable choice of initial state $\mathbf{x}_0$.

A number of somewhat more general deterministic design problems arise as follows. Performance considerations depend upon more than one initial-state vector. Specifically, system response to each initial state $\mathbf{x}_0 = \mathbf{b}_j$ for $j = 1, 2, \ldots, m$ must be weighted. The performance functional used for these problems are of the form

$$I = \sum_{j=1}^{m} \int_0^\infty \{\mathbf{e}_j'(t)\mathbf{e}_j(t)\}\, \mathrm{d}t, \tag{1-89}$$

where

$$\mathbf{e}_j(t) = \mathbf{H}\, \mathrm{e}^{\mathbf{F}t}\mathbf{b}_j \quad \text{for} \quad j = 1, 2, \ldots, m. \tag{1-90}$$

Tabulated integrals of course can be used to evaluate each term appearing in this expression for I.

Formulation of stochastic design problems presented in this chapter do not explicitly include input and other signals that are not white-noise signals corresponding to Brownian-motion processes. However, more general Gaussian signals with rational spectral densities can easily be incorporated into these problems by associating a portion of the system state equations with stochastic signal generation. Flow graphs for these problems which include dynamics corresponding to both signal generation and the control system can of course be put in state-equation form. In particular, this stochastic system is depicted in Figure 1-2 when stochastic initial-state $\mathbf{x}_0 = \mathbf{0}$ is used and when forcing functions $\mathbf{u}$ correspond to white-noise signals which are normally denoted by $\mathbf{v}$.

Performance functionals given in (1-74) for these stochastic design problems can always be evaluated using transforms and tabulated integrals when covariance coefficient matrix $\mathbf{V}$ is given numerically. This evaluation is accomplished by introducing m deterministic initial-state equivalents of the stochastic problem. Matrices $\mathbf{F}$, $\mathbf{G}$, and $\mathbf{H}$ of the equivalent deterministic problems are identical to those of the stochastic problem which has been reduced to state-equation form.

Equivalent initial states $\mathbf{x}_0 = \mathbf{b}_j$ for $j = 1, 2, \ldots, m$ are determined from $(\mathbf{GVG}')$. Specifically, every positive semi-definite matrix $(\mathbf{GVG}')$ of rank less than or equal to m can be written in the form

$$\mathbf{GVG}' = \sum_{j=1}^{m} \mathbf{b}_j \mathbf{b}_j'. \tag{1-91}$$

The performance integral given in (1-74) then is reduced to the form appearing in (1-89) where equivalent deterministic error vectors $\mathbf{e}_j(t)$ are defined by (1-90). A number of methods, such as finding an orthogonal modal matrix [10], can be employed for this decomposition of $(\mathbf{GVG}')$. Moreover, vectors $\mathbf{b}_j$ are not unique for $m > 1$. Unfortunately, this decomposition cannot actually be performed in general unless $\mathbf{V}$ is given numerically.

BIBLIOGRAPHY

1. James, H. M., *et al.: Theory of Servomechanisms*, McGraw-Hill Book Company, Inc., New York, 1947.
2. Newton, G. C., Jr., *et al.: Analytical Design of Linear Feedback Controls*, John Wiley and Sons, Inc., New York, 1957.
3. Chang, S. S. L.: *Synthesis of Optimum Control Systems*, McGraw-Hill Book Company, Inc., New York, 1961.
4. Wiener, N.: *Extrapolation, Interpolation, and Smoothing of Stationary Time Series*, The Technology Press of the Massachusetts Institute of Technology, Cambridge, Mass., 1949.
5. Lee, Y. W.: *Statistical Theory of Communication*, John Wiley and Sons, Inc., New York, 1960.
6. Merriam, C. W., III: *Optimization Theory and the Design of Feedback Control Systems*, McGraw-Hill Book Company, Inc., New York, 1964.
7. Bellman, R.: *Dynamic Programming*, Princeton University Press, Princeton, N.J., 1957.
8. Pontryagin, L. S., *et al.: The Mathematical Theory of Optimal Processes*, Interscience Publishers, New York, 1962.
9. DeRusso, P. M., *et al.: State Variables for Engineers*, John Wiley and Sons, Inc., New York, 1965.
10. Merriam, C. W., III: *Analysis of Lumped Electrical Systems*, John Wiley and Sons, Inc., New York, 1969.
11. Zadeh, L. A. and C. A. Desoer: *Linear System Theory*, McGraw-Hill Book Company, New York, 1963.
12. Papoulis, A.: *Probability, Random Variables, and Stochastic Processes*, McGraw-Hill Book Company, Inc., New York, 1965.
13. Bucy, R. S. and P. D. Joseph: *Filtering for Stochastic Processes with Applications to Guidance*, Interscience Publishers, New York, 1968.
14. Papoulis, A.: *The Fourier Integral and Its Applications*, McGraw-Hill Book Company, Inc., New York, 1962.
15. Kreider, D. L., *et al.: An Introduction to Linear Analysis*, Addison-Wesley Publishing Company, Inc., Reading, Mass., 1966.

PROBLEMS

1-1 The state transition matrix of a linear constant-coefficient system is often written in closed form as

$$e^{\mathbf{F}t} = \sum_{i=1}^{n} \mathbf{C}_i f_i(t).$$

a) Explain how functions $f_i(t)$ are identified from the matrix **F**.
b) Explain how coefficient matrices $\mathbf{C}_i$ can be calculated in terms of powers of **F**.

1-2 For

$$\mathbf{F} = \begin{bmatrix} -2 & -1 & -1 \\ -1 & -1 & -1 \\ 1 & 1 & 0 \end{bmatrix},$$

a) Find all eigenvalues $\lambda\{\mathbf{F}\}$.
b) Find a maximal set of linearly independent eigenvectors $\mathbf{m}\{\mathbf{F}\}$.
c) Find $e^{\mathbf{F}t}$ in the form of Problem 1-1.

1-3 Repeat Problem 1-2 for

$$\mathbf{F} = \begin{bmatrix} -1 & 1 & \frac{3}{4} & -\frac{1}{4} \\ 1 & -1 & \frac{1}{4} & -\frac{3}{4} \\ -\frac{3}{2} & \frac{1}{2} & -1 & 1 \\ -\frac{1}{2} & \frac{3}{2} & 1 & -1 \end{bmatrix}.$$

1-4 For the matrix **F** given in Problem 1-2,

a) Compute $\dfrac{d}{dt}(e^{\mathbf{F}t})$.
b) Compute $(e^{\mathbf{F}t})^{-1}$.
c) Compute $e^{(\lambda\mathbf{I} - \mathbf{F})t}$.

1.5 Repeat Problem 1-4 for the matrix **F** given in Problem 1-3.

1-6
a) Draw the flow graph corresponding to initial state $\mathbf{x}_0 = \mathbf{0}$ and coefficient matrices

$$\mathbf{F} = \begin{bmatrix} 0 & 0 & \cdots & 0 & -p_0 \\ 1 & 0 & \cdots & 0 & -p_1 \\ 0 & 1 & \cdots & 0 & -p_2 \\ \cdots & \cdots & \cdots & \cdots & \cdots \\ 0 & 0 & \cdots & 1 & -p_{n-1} \end{bmatrix}, \quad \mathbf{G} = \begin{bmatrix} q_0 \\ q_1 \\ \vdots \\ q_{n-1} \end{bmatrix}, \quad \text{and} \quad \mathbf{H} = [0 \;\; 0 \;\; \cdots \;\; 0 \;\; 1].$$

b) Find transfer function Y_1/U_1.

1-7

a) Draw the flow graph corresponding to an arbitrary initial state $\mathbf{x}_0$ and coefficient matrices

$$\mathbf{F} = \begin{bmatrix} 0 & 1 & 0 \\ -1 & -2 & 0 \\ 0 & 0 & -1 \end{bmatrix}, \quad \mathbf{G} = \begin{bmatrix} 0 \\ 0 \\ 1 \end{bmatrix}, \quad \text{and} \quad \mathbf{H} = \begin{bmatrix} 2 & 1 & 0 \\ 1 & 0 & 1 \\ 1 & 0 & -1 \end{bmatrix}.$$

b) Using only flow-graph reduction techniques, compute $(s\mathbf{I} - \mathbf{F})^{-1}$ and $\mathbf{H}(s\mathbf{I} - \mathbf{F})^{-1}\mathbf{G}$.

c) Compute $e^{\mathbf{F}t}$ in the form of Problem 1-1 using the results of part b).

1-8

a) Find coefficient matrices **F, G,** and **H** for the flow graph shown in Figure 1-12.

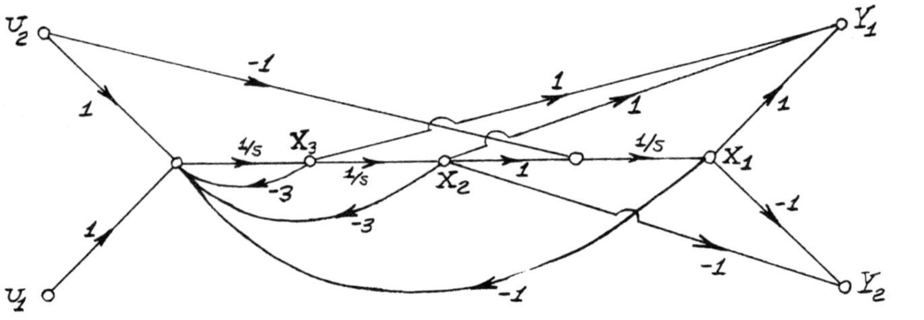

Figure 1-12

b) Find the matrix of transfer functions $\mathbf{H}(s\mathbf{I} - \mathbf{F})^{-1}\mathbf{G}$.

1-9 Repeat Problem 1-8 for the flow graph shown in Figure 1-13.

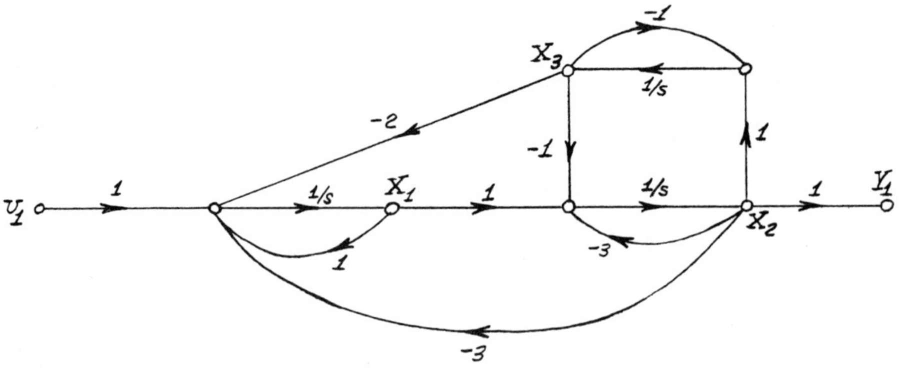

Figure 1-13

1-10 Let $p(s)$ be any polynomial

$$p(s) = \sum_{k=0}^{n} p_k s^k$$

with real coefficients and $p_n > 0$. Also let $\mathbf{P}$ be the $n \times n$ matrix

$$\mathbf{P} = \begin{bmatrix} p_{n-1} & p_n & 0 & 0 & \cdots & 0 & 0 & 0 \\ p_{n-3} & p_{n-2} & p_{n-1} & p_n & \cdots & 0 & 0 & 0 \\ \cdots & \cdots & \cdots & \cdots & \cdots & \cdots & \cdots & \cdots \\ 0 & 0 & 0 & 0 & \cdots & p_0 & p_1 & p_2 \\ 0 & 0 & 0 & 0 & \cdots & 0 & 0 & p_0 \end{bmatrix}.$$

Then $p(s)$ is strictly Hurwitz if and only if the principle minors of $\mathbf{P}$, namely

$$\det [p_{n-1}], \det \begin{bmatrix} p_{n-1} & p_n \\ p_{n-3} & p_{n-2} \end{bmatrix}, \text{etc.}$$

are positive. Determine whether the following polynomials are strictly Hurwitz.
a) $s^4 + 3s^3 + 4s^2 + 3s + 1$
b) $s^5 + s^4 + 2s^3 + 2s^2 + s + 1$.

1-11 Find the values of c for which the following polynomials are strictly Hurwitz.
a) $cs^3 + ca_1 s^2 + (1 + c)s + (a_1 + a_2)$ with $a_1, a_2 > 0$
b) $3s^3 + \dfrac{2c}{1+c}s^2 + \dfrac{2+c}{1+c}s + \dfrac{c}{1+c}$.

1-12 Evaluate $f_{\mathbf{c}}$ for each of the following examples of $f(\mathbf{c})$.
a) $\mathbf{a}'\mathbf{c}$
b) $\frac{1}{2}\mathbf{c}'\mathbf{A}\mathbf{c}$
c) $h[g(\mathbf{c})]$
d) $\boldsymbol{\lambda}'\mathbf{g}(\mathbf{c})$.

1-13 Evaluate $f_{\mathbf{cc}}$ for each of the examples of $f(\mathbf{c})$ given in Problem 1-12.

1-14 For square matrices $\mathbf{A}$ and $\mathbf{B}$, verify
a) $\operatorname{tr}\{\mathbf{A}'\} = \operatorname{tr}\{\mathbf{A}\}$
b) $\dfrac{\partial}{\partial c}\operatorname{tr}\{\mathbf{A}\} = \operatorname{tr}\left\{\dfrac{\partial}{\partial c}\mathbf{A}\right\}$
c) $\operatorname{tr}\{\mathbf{A}+\mathbf{B}\} = \operatorname{tr}\{\mathbf{A}\} + \operatorname{tr}\{\mathbf{B}\}$
d) $\operatorname{tr}\{\mathbf{a}'\mathbf{A}\mathbf{a}\} = \operatorname{tr}\{(\mathbf{a}\mathbf{a}')\mathbf{A}\}$
e) $\operatorname{tr}\{\mathbf{AB}\} = \operatorname{tr}\{\mathbf{BA}\}$.

1-15 Evaluate $f_{\mathbf{C}}$ for each of the following examples of $f(\mathbf{C})$.
a) $\operatorname{tr}\{\mathbf{C}\}$
b) $\operatorname{tr}\{\mathbf{ACB}\}$
c) $\operatorname{tr}\{\mathbf{AC}^l\mathbf{B}\}$ for some integer $l > 1$.

1-16 Evaluate $\dfrac{d}{dc_j}f$ for $f = f[\mathbf{X}(\mathbf{c})]$.

1-17 Evaluate $\dfrac{d}{dc_j}f_{\mathbf{X}}$ for $f = \operatorname{tr}\{\mathbf{X}^l(\mathbf{c})\}$ and integer $l > 1$.

1-18 For a deterministic system without forcing functions,
a) Show that (1-67) can be rewritten in the form

$$E(T) = \mathbf{x}_0'\mathbf{X}(T)\mathbf{x}_0$$

and give an expression for $\mathbf{X}(T)$.
b) Find the first order matrix differential equation and boundary condition that can be used to compute $\mathbf{X}(T)$.
c) Bound $\|\mathbf{X}(T)\|$ from above by an explicit function of T assuming $\mathbf{F}$ is asymptotically stable.
d) Give the matrix algebraic equation satisfied by $\mathbf{X}(\infty)$ assuming $\mathbf{F}$ is asymptotically stable.

1-19 For a stochastic system subjected to Brownian-motion disturbances,
a) Derive the covariance matrix

$$\mathbf{X}(T) = E\{\mathbf{x}(T)\mathbf{x}'(T)\}$$

assuming that stochastic initial state $\mathbf{x}_0$ is not always equal to zero but is statistically independent of white-noise vector $\mathbf{v}$.
b) Repeat part b) of Problem 1-18.
c) Repeat part c) of Problem 1-18.
d) Repeat part d) of Problem 1-18.

1-20
a) Compute the performance integral I defined by Figure 1-14.

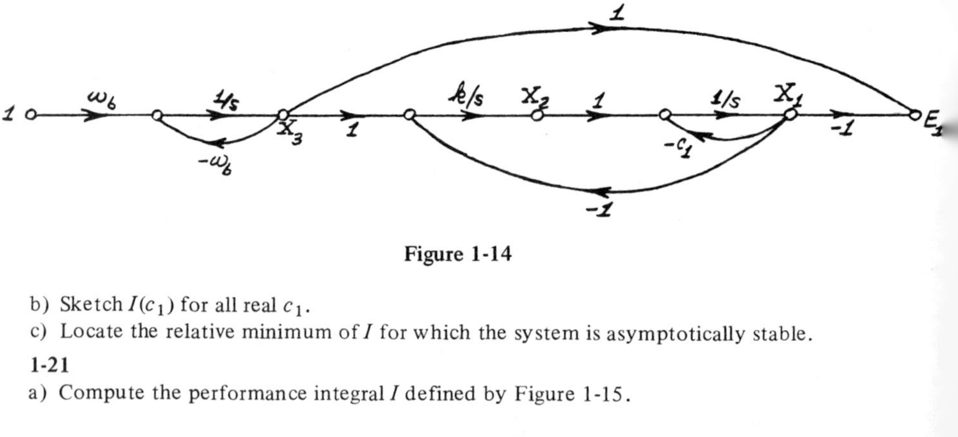

Figure 1-14

b) Sketch $I(c_1)$ for all real c_1.
c) Locate the relative minimum of I for which the system is asymptotically stable.

1-21
a) Compute the performance integral I defined by Figure 1-15.

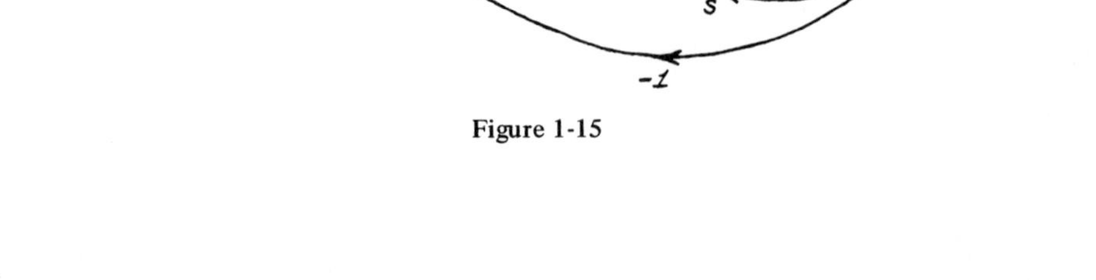

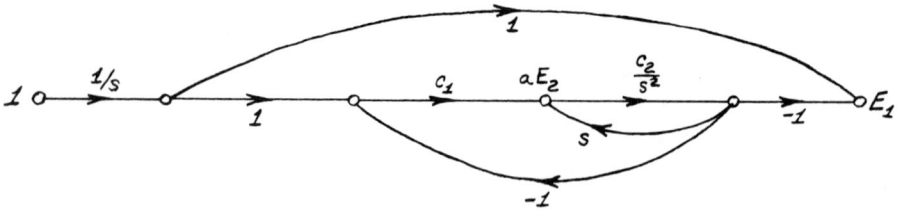

Figure 1-15

b) Sketch contours of $I(c_1, c_2)$.
c) Locate the relative minimum of I for which the system is asymptotically stable.

1-22
a) Compute the performance integral I defined by the flow graph shown in Figure 1-16 and also compute

$$I_b = \int_0^{\infty} b^2(t)\,dt.$$

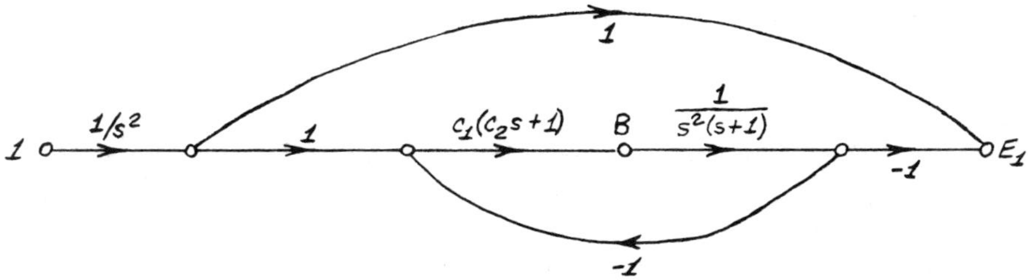

Figure 1-16

b) Plot the contours of I, and plot the feasible region corresponding to $I_b \leqslant 7$.
c) Determine the value of **c** that minimizes I subject to $I_b \leqslant 7$.

1-23 A system subjected to a white-noise signal with unit covariance coefficient is depicted in Figure 1-17.

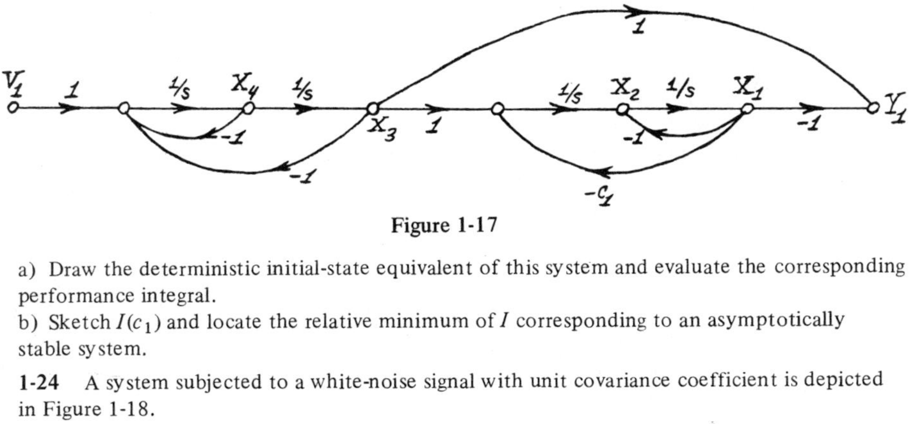

Figure 1-17

a) Draw the deterministic initial-state equivalent of this system and evaluate the corresponding performance integral.
b) Sketch $I(c_1)$ and locate the relative minimum of I corresponding to an asymptotically stable system.

1-24 A system subjected to a white-noise signal with unit covariance coefficient is depicted in Figure 1-18.

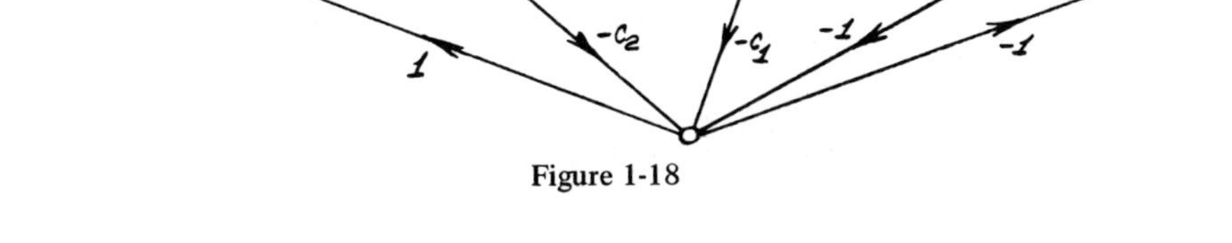

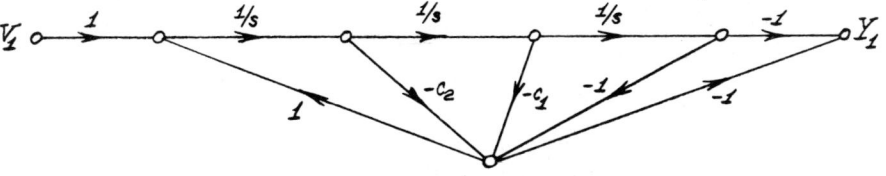

Figure 1-18

a) Draw the deterministic initial-state equivalent of this system and evaluate the corresponding performance integral.
b) Sketch contours of $I(c_1, c_2)$ and locate the relative minimum of I corresponding to an asymptotically stable system.

1-25 Let $\mathbf{E} = [\mathbf{e}_1 \quad \mathbf{e}_2 \quad \ldots \quad \mathbf{e}_m]$ and $\mathbf{B} = [\mathbf{b}_1 \quad \mathbf{b}_2 \quad \ldots \quad \mathbf{b}_m]$.
a) Show that the performance integral defined by (1-89) and (1-90) can be rewritten to obtain

$$I = \int_0^\infty \operatorname{tr}\{\mathbf{E}'(t)\mathbf{E}(t)\}\, dt.$$

b) Repeat part a) to obtain

$$I = \operatorname{tr}\{\mathbf{B}'\mathbf{X}\mathbf{B}\}$$

for asymptotically stable systems and give an expression for **X**.

1-26 Verify that performance functionals given in (1-74) for stochastic problems can be written in the form of (1-89).

1-27 Real symmetric matrices **F** can always be written in terms of real eigenvectors $\mathbf{m}_j\{\mathbf{F}\}$ and eigenvalues $\lambda_j\{\mathbf{F}\}$ as

$$\mathbf{F} = \mathbf{M}\Lambda\mathbf{M}',$$

where

$$\mathbf{M} = [\mathbf{m}_1 \quad \mathbf{m}_2 \quad \ldots] \quad \text{and} \quad \Lambda = \begin{bmatrix} \lambda_1 & 0 & \ldots \\ 0 & \lambda_2 & \ldots \\ \ldots & \ldots & \ldots \end{bmatrix}.$$

Matrix **M** is an orthogonal ($\mathbf{M}^{-1} = \mathbf{M}'$) modal matrix of **F** and Λ is the corresponding diagonal eigenvalue matrix of **F**.
a) Determine restrictions on eigenvalues of **F** when **F** is positive semi-definite.
b) Repeat part a) when **F** is positive definite.

1-28 Find **M** and Λ in the form of Problem 1-27 for

$$\mathbf{F} = \begin{bmatrix} \frac{1}{2} & -\frac{1}{2} & 0 \\ -\frac{1}{2} & \frac{1}{2} & 0 \\ 0 & 0 & 2 \end{bmatrix}.$$

1-29 Repeat Problem 1-28 for

$$\mathbf{F} = \begin{bmatrix} 1 & 1 & 1 \\ 1 & 1 & 1 \\ 1 & 1 & 1 \end{bmatrix}.$$

1-30 Give an explicit expression for $\mathbf{B} = [\mathbf{b}_1 \quad \mathbf{b}_2 \quad \ldots \quad \mathbf{b}_m]$, which is defined by (1-91), in terms of an orthogonal modal matrix and corresponding diagonal eigenvalue matrix (see Problem 1-27) of $(\mathbf{G}\mathbf{V}\mathbf{G}')$.

1-31 Suppose $\mathbf{G} = \mathbf{G}(\mathbf{c})$ is an $n \times m$ matrix. Show that $(\mathbf{GVG}')$ can be written explicitly in the form of (1-91) when $\mathbf{V}$ is given numerically as a positive semi-definite matrix even though rank $(\mathbf{GVG}') < m$.

1-32 Give an interpretation of the performance integral

$$I = \mathrm{tr}\left\{ \int_0^\infty e^{\mathbf{F}'t}\mathbf{H}'\mathbf{H}\, e^{\mathbf{F}t}\, dt \right\}$$

for a system with deterministic initial states.

1-33 Repeat Problem 1-32 for a system with white-noise disturbances.

1-34

a) Compute performance integral (1-89) for the deterministic system depicted in Figure 1-19 when $\mathbf{b}_j$ is equal to the jth standard basis vector for $j = 1, 2,$ and 3.

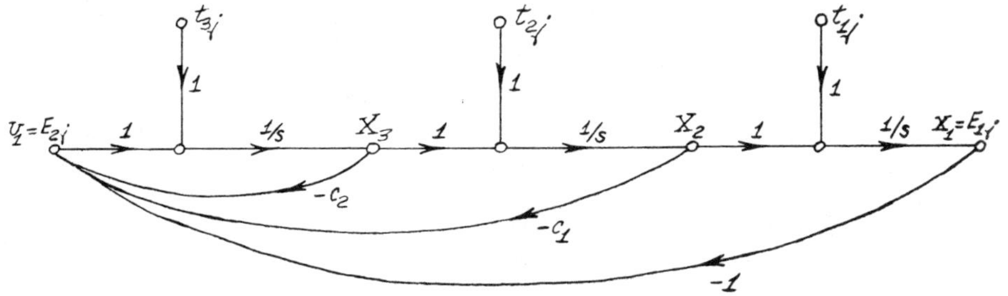

Figure 1-19

b) Locate the relative minimum of I for which the system is asymptotically stable.

1-35 A filter system is depicted in Figure 1-20 where white-noise signals have covariance coefficient matrix

$$\mathbf{V} = \begin{bmatrix} 1 & 1 \\ 1 & 2 \end{bmatrix}.$$

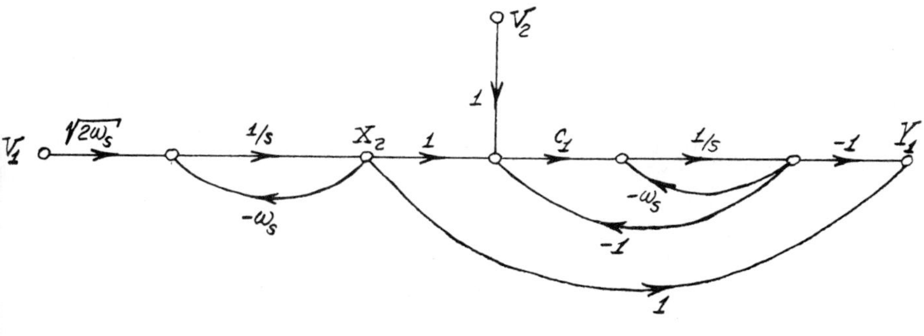

Figure 1-20

a) Draw and label the deterministic initial-state equivalents of this system.
b) Evaluate performance integral I, and locate the relative minimum of I for which the system is asymptotically stable.

1-36 A control system is depicted in Figure 1-21 where white-noise signals have covariance coefficient matrix $\mathbf{V} = \mathbf{I}$.

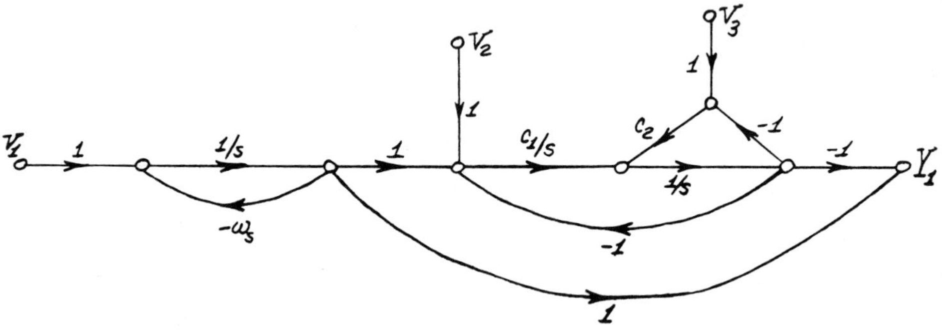

Figure 1-21

a) Draw and label the deterministic initial-state equivalents of this system.
b) Evaluate performance integral I, and locate the relative minimum of I for which $\omega_s = 0$ and the system is stable.

CHAPTER 2

Function minimization and constraints

Practical optimization techniques are constructed on the basis of asssumed mathematical properties of the function to be minimized. More specifically, indirect methods for locating minima are constructed on the basis of one or more mathematical conditions that must be satisfied at a minimum of these functions. A rudimentary knowledge of relationships between assumed properties of functions and conditions for a minimum thus is essential to any systematic evaluation of optimization techniques. These relationships are rather obvious for functions of one variable but are somewhat more complicated and less intuitive for functions of more than one variable.

Many practical problems involving the minimization of a function also involve one or more constraints. Both equality and inequality constraints occur. An introduction to both the geometric and algebraic character of constraints is presented here.

This chapter is intended to be a concise and yet precise summary of well-known and useful results concerning minima of functions of a vector variable with and without constraints. Goals of this chapter seem best served by both stating a result and proving the result when the proof is directly related to the formulation of automated design techniques.

2.1 MINIMA OF FUNCTIONS

Confusion concerning minima of functions often arises because there are in fact more than one kind of minimum. Various kinds of minima are defined in terms of various subsets of $\mathscr{R}^n$ which is the space of all real $\mathbf{c}$ with dim $\mathbf{c} = n$. Throughout this and subsequent sections of this chapter, an arbitrary subset of $\mathscr{R}^n$ is denoted by S. Also, the *neighborhood* of $\mathbf{c}_0$ corresponding to some $\delta > 0$ is denoted by N and is defined as

$$N = \{\mathbf{c}: |\mathbf{c} - \mathbf{c}_0| < \delta \quad \text{and} \quad \mathbf{c}, \mathbf{c}_0 \in \mathscr{R}^n\}. \tag{2-1}$$

Definitions of various types of minima can now be defined in terms of S, N, and their intersection which is denoted as $S \cap N$. Specifically, these definitions are stated as

Definition 2-1 Suppose there exists a point $\mathbf{c}_0 \in S$ for which

$$\mathbf{c} \in S \Rightarrow f(\mathbf{c}_0) \leqslant f(\mathbf{c}).$$

Then $f(\mathbf{c}_0)$ is called the *absolute minimum* of f on S. If in addition the equality holds only when $\mathbf{c} = \mathbf{c}_0$, then the absolute minimum at $\mathbf{c}_0$ is called *strict.*

and as

Definition 2-2 Suppose there exists a neighborhood N of $\mathbf{c}_0$ for which

$$\mathbf{c} \in S \cap N \Rightarrow f(\mathbf{c}_0) \leqslant f(\mathbf{c}).$$

Then $f(\mathbf{c}_0)$ is called a *relative minimum* of f in S. If in addition the equality holds only when $\mathbf{c} = \mathbf{c}_0$, then the relative minimum at $\mathbf{c}_0$ is called *strict.*

An absolute minimum at $\mathbf{c}_0$ is a global property of f because all other points in S are involved. On the other hand, a relative minimum at $\mathbf{c}_0$ is a local property of f because only all other points in $S \cap N$ are involved and N can be an arbitrarily small neighbourhood of $\mathbf{c}_0$. Of course, every (strict) absolute minimum is also a (strict) relative minimum but the converse is not true.

The main problems concerning minima that arise are whether a given function has a minimum on a given set and then locating one or more minima of the function in this set. The first of these problems is largely mathematical and hence is not pursued here. For example, the result

Proposition 2-1 *Every continuous function which is defined on a closed and bounded subset of* $\mathscr{R}^n$ *has an absolute minimum on that subset.*

can be proved by techniques beyond the scope of this book [1]. However, this knowledge generally is of little or no help in determining the location of points at which the minimum occurs or even how many such points occur in the set. The problem of locating just one minimum of a given function on a given set is the first demand imposed by the desire to use optimization techniques for automated design. In fact, practical considerations generally dictate that only local investigations of the function can be made and hence only relative minima can be sought.

2.2 CONDITIONS FOR RELATIVE MINIMA

Conditions for relative minima are of course local and hence differentiable functions are of particular interest. Of course, not all functions of interest are differentiable. For instance, the three norms given in (1-21) have an absolute minimum on $\mathscr{R}^n$ at $\mathbf{0}$ but are not differentiable at that point. Most functions that arise in parameter optimization problems of control, however, are differentiable on the sets of interest.

Definitions concerning differentiation of functions of a vector variable are summarized here in terms of standard definitions for partial derivatives [2]. Specifically,

Definition 2-3 Function f is *differentiable* at $\mathbf{c}$ if and only if $f_{\mathbf{c}}$ exists and

$$\lim_{\alpha \to 0} \left[\frac{f(\mathbf{c} + \alpha \mathbf{d}) - f(\mathbf{c})}{\alpha} - \mathbf{d}' f_{\mathbf{c}} \right] = 0 \ \forall \ \mathbf{d} \in \mathscr{R}^n. \tag{2-2}$$

The inner product $\mathbf{d}'f_{\mathbf{c}}$ appearing in this definition occurs frequently in subsequent developments and is referred to as the derivative of f at $\mathbf{c}$ in the direction of $\mathbf{d}$. This definition for differentiability leads directly to the useful property [1].

Proposition 2-2 *If f is differentiable at* $\mathbf{c}$. *then f is also continuous at* $\mathbf{c}$.

The first necessary condition for a relative minimum, namely *stationarity*, can now be stated as

Theorem 2-1 *Suppose f is differentiable at* $\mathbf{c}$, *Then* $f(\mathbf{c})$ *is a relative minimum of f in an open set containing* $\mathbf{c}$ *only if* $f_{\mathbf{c}} = \mathbf{0}$.

Proof Assume $f(\mathbf{c})$ is a relative minimum of f in an open set S. Let N be any neighborhood of $\mathbf{c}$ for which $N \subset S$, and let $g(\alpha)$ be defined as

$$g(\alpha) = \begin{cases} \dfrac{f(\mathbf{c} + \alpha \mathbf{d}) - f(\mathbf{c})}{\alpha} & \text{for} \quad \alpha \neq 0 \\ \mathbf{d}' f_{\mathbf{c}} & \text{for} \quad \alpha = 0 \end{cases}$$

given any α, $\mathbf{d}$ for which $(\mathbf{c} + \alpha \mathbf{d}) \in N$. By assumption of differentiability at $\mathbf{c}$ and the previous proposition, function g is continuous at 0. Now assume $g(0) \neq 0$. Then, by the sign preserving property of continuous functions, there is an open interval containing 0 on which $g(\alpha)$ is of the same sign as $g(0)$. Parameter α thus can always be chosen so that $\alpha g(\alpha) < 0$ and hence $f(\mathbf{c})$ could not have been a relative minimum in S. Because this construction holds with $\mathbf{d} = \mathbf{i}_j$ for each j, every element of $f_{\mathbf{c}}$ must vanish.

Stationarity is illustrated geometrically in Figure 1-7. Of course, stationarity can also occur at relative maxima and *saddle points* at which $f_{\mathbf{c}} = \mathbf{0}$ but are neither relative minima nor maxima.

Additional conditions which are often applicable to relative minima are based on second partial derivatives referred to in the definition

Definition 2-4 Function f is *twice differentiable* at $\mathbf{c}_0$ if and only if f_{ci} is differentiable at $\mathbf{c}_0$ for each i.

In other words, a function is twice differentiable at $\mathbf{c}$ if and only if $f_{\mathbf{cc}}$ exists and

$$\lim_{\alpha \to 0} \left[\frac{f_{(\mathbf{c}+\alpha\mathbf{d})} - f_{\mathbf{c}}}{\alpha} - f_{\mathbf{cc}}\mathbf{d} \right] = \mathbf{0} \ \forall \ \mathbf{d} \in \mathscr{R}^n. \tag{2-3}$$

By Proposition 2-2, gradient $f_{\mathbf{c}}$ is continuous at $\mathbf{c}_0$ whenever f is twice differentiable at $\mathbf{c}_0$. Additional continuity conditions are also frequently applicable. Specifically, functions referred to in the definition

Definition 2-5 Function f is *continuously differentiable* (or *twice continuously differentiable*) at $\mathbf{c}_0$ when f is differentiable (or twice differentiable) and $f_{\mathbf{c}}$ (or $f_{\mathbf{cc}}$) is continuous on a neighborhood of $\mathbf{c}_0$.

arise frequently. Twice continuously differentiable functions have the convenient property [1]

Proposition 2-3 *If f is twice continuously differentiable at* $\mathbf{c}$, *then $f_{\mathbf{cc}}$ is symmetric.*

Theorems concerning minima of twice continuously differentiable functions are based on a form of Taylor's theorem for functions of a scalar variable. This theorem is stated as

Proposition 2-4 *Suppose $g(\alpha)$ is twice differentiable on (a, b) and g, g_α are continuous at a, b. Then*

$$g(b) = g(a) + (b - a)g_a + \tfrac{1}{2}(b - a)^2 g_{\alpha\alpha} \tag{2-4}$$

holds for some $\alpha \in (a, b)$.

and is extended to functions of a vector variable by introducing the set

$$(\mathbf{a}, \mathbf{b}) = \{\alpha\mathbf{b} + (1 - \alpha)\mathbf{a}: \alpha \in (0, 1)\} \tag{2-5}$$

called the *line segment* between $\mathbf{a}$ and $\mathbf{b}$. This set permits a convenient statement of the result.‡

Lemma 2-1 *Suppose f is twice differentiable on* $(\mathbf{a}, \mathbf{b})$ *and f, $f_{\mathbf{c}}$ are continuous at* $\mathbf{a}, \mathbf{b}$. *Then*

$$f(\mathbf{b}) = f(\mathbf{a}) + (\mathbf{b} - \mathbf{a})'f_{\mathbf{a}} + \tfrac{1}{2}(\mathbf{b} - \mathbf{a})'f_{\mathbf{cc}}(b - a) \tag{2-6}$$

‡ Preliminary results, not of direct interest but worthy of proof by virtue of the techniques employed, are called lemmas in this book.

holds for some $\mathbf{c} \in (\mathbf{a}, \mathbf{b})$.

Proof Let

$$\mathbf{c} = \alpha\mathbf{b} + (1 - \alpha)\mathbf{a} \quad \text{and} \quad g(\alpha) = f(\mathbf{c})$$

so that $g(\alpha)$ satisfies all the preconditions of Proposition 2-4 with $a = 0$ and $b = 1$. In addition, differentiation yields

$$g_\alpha = (\mathbf{b} - \mathbf{a})' f_\mathbf{c} \quad \text{and} \quad g_{\alpha\alpha} = (\mathbf{b} - \mathbf{a})' f_{\mathbf{cc}}(\mathbf{b} - \mathbf{a})$$

so that (2-6) follows directly from substitutions into (2-4).

The second necessary condition for a relative minimum, namely *convexity*, can now be stated as

Theorem 2-2 *Suppose f is twice continuously differentiable at* $\mathbf{c}$ *and* $f_\mathbf{c} = \mathbf{0}$. *Then* $f(\mathbf{c})$ *is a relative minimum of f in an open set containing* $\mathbf{c}$ *only if* $f_{\mathbf{cc}}$ *is positive semi-definite.*

Proof Assume $f(\mathbf{c})$ is a relative minimum of f in an open set S, and assume $f_{\mathbf{cc}}$ has one or more negative eigenvalues. Then [3], by continuity of $f_{\mathbf{cc}}$, there exist a neighborhood $N \subset S$ and some vector $\mathbf{d}$ such that $\mathbf{c} + \mathbf{d} \in N$ and $\mathbf{d}' f_{\boldsymbol{\gamma\gamma}} \mathbf{d} < 0$ holds for all $\boldsymbol{\gamma} = \mathbf{c} + \alpha\mathbf{d}$ corresponding to $\alpha \in (0, 1)$. Use of (2-6) with $f_\mathbf{c} = \mathbf{0}$ then yields

$$f(\mathbf{c} + \alpha\mathbf{d}) < f(\mathbf{c}) \ \forall \ \alpha \in (0, 1),$$

and hence there is no neighborhood of $\mathbf{c}$ on which the inequality required for a relative minimum holds.

Stationarity and convexity, of course, do not combine to form a sufficient condition for a relative minimum. For example, the function $f(c) = -c^4$ illustrates this point.

Stationarity, convexity, and $f_{\mathbf{cc}}$ nonsingular do combine, however, to form the sufficient condition

Theorem 2-3 *Suppose f is twice continuously differentiable at* $\mathbf{c}$, $f_\mathbf{c} = \mathbf{0}$, *and* $f_{\mathbf{cc}}$ *is positive definite. Then* $f(\mathbf{c})$ *is a strict relative minimum of f in an open set containing* $\mathbf{c}$.

Proof Let N be a neighborhood of $\mathbf{c}$ on which f is twice continuously differentiable. Then, by continuity of $f_{\mathbf{cc}}$, there exists an open set $S \subset N$ for which $\gamma \in S \Rightarrow f_{\gamma\gamma}$ positive definite. Use of (2-6) with $f_\mathbf{c} = \mathbf{0}$ then yields

$$f(\gamma) > f(\mathbf{c})$$

given any $\boldsymbol{\gamma} \neq \mathbf{c}$ which belongs to S.

This sufficient condition is not necessary for strict relative minima or even for relative minima without the strictness property as examples

$$f(c) = c^4 \quad \text{and} \quad f(c) = \begin{cases} (c-1)^3 & \text{for } c > 1 \\ 0 & \text{for } -1 \leqslant c \leqslant 1 \\ -(c+1)^3 & \text{for } c < -1 \end{cases}$$

clearly illustrate.

Theorems 2-1, 2-2, and 2-3 are fundamental to almost all practical methods of parameter optimization used in control system design. Of particular importance are relative minima for which f_{cc} is nonsingular. Relative minima in an open set for which f_{cc} is singular are called *degenerate* and are almost always caused by incorrect problem formulation.

2.3 CONVEX SETS AND FUNCTIONS

Conditions for absolute minima are expressed in terms of global properties of functions. Convex functions arise most frequently in this context and have the general structural properties depicted in Figure 2-1 for one dimension. In particular, function $g(c)$ is the linear interpolation of $f(c)$ given by

$$g(c) = \frac{\beta - c}{\beta - \alpha} f(\alpha) + \frac{c - \alpha}{\beta - \alpha} f(\beta)$$

for points α and β. Functions of one variable which are convex on (a, b) have the property that, given any $\alpha, \beta \in (a, b)$,

$$f(c) \leqslant g(c) \;\forall\; c \in (\alpha, \beta).$$

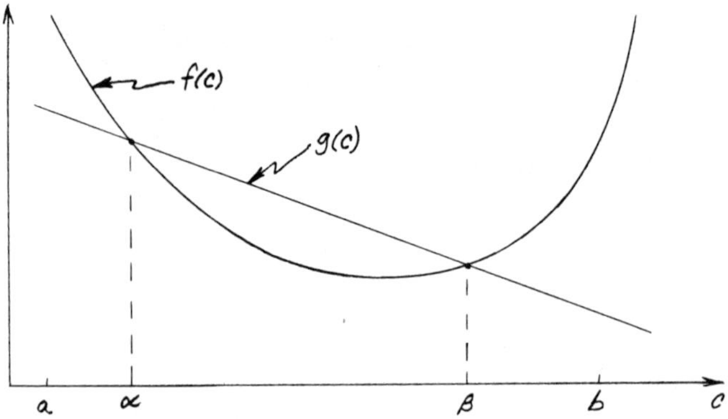

Figure 2-1 Construction of a convex function of one variable.

Functions of one variable which are convex on (a, b) and have a point $c \in (a, b)$ at which $f_c = 0$ possess an absolute minimum at c.

Convex functions of a vector variable must be defined in a somewhat more careful fashion using the notion of a convex set; namely,

Definition 2-6 A set S is *convex* if and only if

$$\mathbf{a}, \mathbf{b} \in S \Rightarrow \alpha\mathbf{b} + (1 - \alpha)\mathbf{a} \in S \;\forall \alpha \in (0, 1).$$

A typical example of a convex set is the neighborhood defined in (2-1). Convex functions of a vector variable then are defined as

Definition 2-7 Function f is *convex* on a convex set S if and only if

$$\mathbf{a}, \mathbf{b} \in S \Rightarrow f(\alpha\mathbf{b} + (1 - \alpha)\mathbf{a}) \leqslant \alpha f(\mathbf{b}) + (1 - \alpha)f(\mathbf{a}) \;\forall\; \alpha \in (0, 1).$$

If in addition the equality holds only for $\mathbf{a} = \mathbf{b}$, the convex function is called *strict.*

A typical function which is strictly convex on $\mathscr{R}^n$ is the quadratic form

$$f(\mathbf{c}) = \tfrac{1}{2}\mathbf{c}'\mathbf{F}\mathbf{c}$$

where $\mathbf{F}$ is positive definite.

Convex sets and functions have many interesting and useful properties [4] which are not explored in detail here. For instance, the result

Proposition 2-5 *If f is convex on some open convex set S, then f is continuous on S.*

follows directly from the definition of a convex function [1]. In addition, properties summarized in

Proposition 2-6 *Suppose f is twice continuously differentiable on some open convex set. Then*

i) f is convex on S if and only if $f_{\mathbf{cc}}$ is positive semi-definite on S

and

ii) f is strictly convex on S if $f_{\mathbf{cc}}$ is positive definite on S.

follow directly from the definition of convex functions and (2-6). These properties are useful in testing for the convexity of functions and also yield the relationships between relative minima and convexity for functions which are twice continuously differentiable.

2.4 CONDITIONS FOR ABSOLUTE MINIMA

Conditions for absolute minima of differentiable functions follow in an elementary fashion from the fundamental properties of convex functions stated in

Lemma 2-2 *Suppose f is differentiable on a convex set S. Then f is convex on S if and only if*

$$f(\mathbf{c}+\mathbf{d}) \geqslant f(\mathbf{c}) + \mathbf{d}'f_{\mathbf{c}} \tag{2-7}$$

holds for every $\mathbf{c}, \mathbf{c}+\mathbf{d} \in S$. *In addition, f is strictly convex if and only if the equality appearing in* (2-7) *holds only for* $\mathbf{d} = \mathbf{0}$ *given any* $\mathbf{c} \in S$.

Proof Suppose f is convex on S so that

$$f(\mathbf{c}+\alpha\mathbf{d}) \leqslant \alpha f(\mathbf{c}+\mathbf{d}) + (1-\alpha)f(\mathbf{c})$$

and hence

$$\frac{f(\mathbf{c}+\alpha\mathbf{d}) - f(\mathbf{c})}{\alpha} - \mathbf{d}'f_{\mathbf{c}} \leqslant f(\mathbf{c}+\mathbf{d}) - f(\mathbf{c}) - \mathbf{d}'f_{\mathbf{c}}$$

holds for all $\alpha \in (0, 1)$ and $\mathbf{c}, \mathbf{c}+\mathbf{d} \in S$. The limit of both sides of this equation as $\alpha \to 0$ then yields the desired result for differentiable functions in accordance with (2-2). Now suppose f is strictly convex on S so that

$$(\alpha\mathbf{d})'f_{\mathbf{c}} < f(\mathbf{c}+\alpha\mathbf{d}) - f(\mathbf{c}) \leqslant \alpha\,[f(\mathbf{c}+\mathbf{d}) - f(\mathbf{c})]$$

holds by virtue of (2-7) for strictly convex functions and what has already been shown for convex functions. The desired result is finally obtained when $\alpha > 0$ is eliminated from both sides of this equation by division. Conversely, suppose (2-7) holds so that

$$f(\mathbf{c}) \geqslant f(\mathbf{c}+\alpha\mathbf{d}) - \alpha\mathbf{d}'f_{(\mathbf{c}+\alpha\mathbf{d})}$$

and

$$f(\mathbf{c}+\mathbf{d}) \geqslant f(\mathbf{c}+\alpha\mathbf{d}) + (1-\alpha)\mathbf{d}'f_{(\mathbf{c}+\alpha\mathbf{d})}$$

also hold for all $\alpha \in (0, 1)$ and $\mathbf{c}, \mathbf{c}+\mathbf{d} \in S$. Then

$$(1-\alpha)f(\mathbf{c}) + \alpha f(\mathbf{c}+\mathbf{d}) \geqslant f(\mathbf{c}+\alpha\mathbf{d})$$

is finally obtained and is the property required of convex functions. The strict inequality corresponding to strict convexity follows from a similar construction without equalities.

The expression given in (2-7) has a number of interpretations. In geometrical terms, a differentiable convex function f cannot have values lying below any *tangent hyperplane* of f which is expressed as the function

$$h(\mathbf{c}+\mathbf{d}) = f(\mathbf{c}) + \mathbf{d}'f_{\mathbf{c}} \tag{2-8}$$

for point of tangency $\mathbf{c}$. In a similar fashion, a differentiable strictly convex function must lie above every tangent hyperplane except at the point of tangency.

The main result concerning convex functions and minima is stated as

Theorem 2-4 *Suppose f is differentiable and (strictly) convex on an open convex set S. Then f has a (strict) absolute minimum on S at* **c** *if and only if* $f_{\mathbf{c}} = \mathbf{0}$.

Proof Necessity of $f_{\mathbf{c}} = \mathbf{0}$ follows from the assumption of differentiability and Theorem 2-1 because a (strict) absolute minimum on S must also be a (strict) relative minimum in S. Sufficiency of $f_{\mathbf{c}} = \mathbf{0}$ for a (strict) absolute minimum on S follows from (2-7).

An additional result of considerable significance is stated as

Theorem 2-5 *A function which is differentiable and strictly convex on an open convex set S has at most one point* $\mathbf{c} \in S$ *at which* $f_{\mathbf{c}} = \mathbf{0}$.

Proof Suppose f is strictly convex on S and has an absolute minimum at distinct points $\mathbf{a}, \mathbf{b} \in S$ so that

$$f(\mathbf{c}) \geqslant f(\mathbf{a}) = f(\mathbf{b}) \ \forall \ \mathbf{c} \in S$$

holds by definition of an absolute minimum. Let $\mathbf{c} = \alpha\mathbf{b} + (1 - \alpha)\mathbf{a}$ with $\alpha \in (0, 1)$ so that $\mathbf{c} \in S$ and

$$f(\mathbf{c}) < f(\mathbf{a}) = f(\mathbf{b})$$

holds by the property of strict convexity. This inequality contradicts the assumption of an absolute minimum at both **a** and **b**, and hence both $f_{\mathbf{a}} = \mathbf{0}$ and $f_{\mathbf{b}} = \mathbf{0}$ cannot occur by Theorem 2-4.

Although these results concerning absolute minima are seldom used explicitly in automated design problems, they provide insight and illustrate properties of an important class of functions which occur frequently.

2.5 EQUALITY CONSTRAINTS

Numerous automated design problems arising in control involve one or more equality constraint functions that must be satisfied in order to achieve an admissible solution. This section is intended to provide both an intuitive introduction to relative minima constrained by equality constraints and also a mathematical basis for the first condition necessary to have a constrained relative minimum.

By way of an intuitive introduction, suppose the function $f(\mathbf{c})$ is to be minimized subject to equality constraint $g(\mathbf{c}) = 0$. Functions f and g are assumed to be continuously differentiable at **c**, and parameterized curves $\boldsymbol{\gamma}(\alpha)$ are introduced for convenience. These curves are entirely arbitrary except that $g[\boldsymbol{\gamma}(\alpha)] = 0$ for all

$\alpha \in (-\epsilon, \epsilon)$, $\boldsymbol{\gamma}$ is differentiable at 0, and $\boldsymbol{\gamma}(0) = \mathbf{c}$. Now let $\mathbf{t}$ denote $\dot{\boldsymbol{\gamma}}$ at $\alpha = 0$. Then the total derivative of $g[\boldsymbol{\gamma}(\alpha)]$ with respect to α is zero at $\alpha = 0$ and hence

$$\mathbf{t}' g_{\mathbf{c}} = 0. \tag{2-9}$$

In other words, admissible derivative vectors $\mathbf{t}$ must be *orthogonal* to gradient $g_{\mathbf{c}}$ for each point at which g is continuously differentiable and $g(\mathbf{c}) = 0$.

Now suppose $f(\mathbf{c})$ is a relative minimum of f in a set S on which g vanishes and suppose $\boldsymbol{\gamma}(\alpha) \in S$ for all $\alpha \in (-\epsilon, \epsilon)$. Then function

$$h(\alpha) = f[\boldsymbol{\gamma}(\alpha)]$$

has a relative minimum at $\alpha = 0$ and hence $\dot{h}$ vanishes at $\alpha = 0$. Evaluation of $\dot{h}$ at $\alpha = 0$ then must yield

$$\mathbf{t}' f_{\mathbf{c}} = 0 \tag{2-10}$$

at a constrained relative minimum. In other words, admissible derivative vectors $\mathbf{t}$ must also be orthogonal to gradient $f_{\mathbf{c}}$ for any point at which f is continuously differentiable and $f(\mathbf{c})$ is a relative minimum in S.

If the problem of finding a constrained relative minimum is not trivial in the sense that the admissible derivative vector at $\mathbf{c}$ can be chosen so that $\mathbf{t} \neq \mathbf{0}$, then gradient vectors $f_{\mathbf{c}}$ and $g_{\mathbf{c}}$ must be colinear. That is, a scalar λ, which is called a Lagrange multiplier, must exist so that

$$f_{\mathbf{c}} + \lambda g_{\mathbf{c}} = \mathbf{0}. \tag{2-11}$$

This condition will be found to have a very convenient form for representing the effective reduction in the number of arbitrary elements of $\mathbf{c}$ that results from constraint $g(\mathbf{c}) = 0$.

The condition for a constrained relative minimum given in (2-11) is depicted geometrically in Figure 2-2 for dim $\mathbf{c} = 2$. Admissible derivative vectors at $\mathbf{c}$ lie in the tangent plane of contour $g = 0$ and gradient $g_{\mathbf{c}}$ is normal to the tangent plane. If $f(\mathbf{c})$ is a constrained relative minimum, then gradient $f_{\mathbf{c}}$ is also normal to the tangent plane as shown. For the geometry depicted in Figure 2-2, the Lagrange multiplier is positive.

This rather intuitive view of constrained relative minima extends readily to more than one equality constraint. Each function in the vector constraint equation $\mathbf{g}(\mathbf{c}) = \mathbf{0}$ yields an admissibility condition of the form appearing in (2-9). These admissibility conditions also can be arranged in vector form as

$$\mathbf{g}_{\mathbf{c}} \mathbf{t} = \mathbf{0} \tag{2-12}$$

when the Jacobian matrix defined in (1-44) is employed. If dim $\mathbf{g} = m$ and $\mathbf{g}_{\mathbf{c}}$ has rank m, then in fact there exists a Lagrange multiplier vector $\boldsymbol{\lambda}$ such that $f_{\mathbf{c}}$

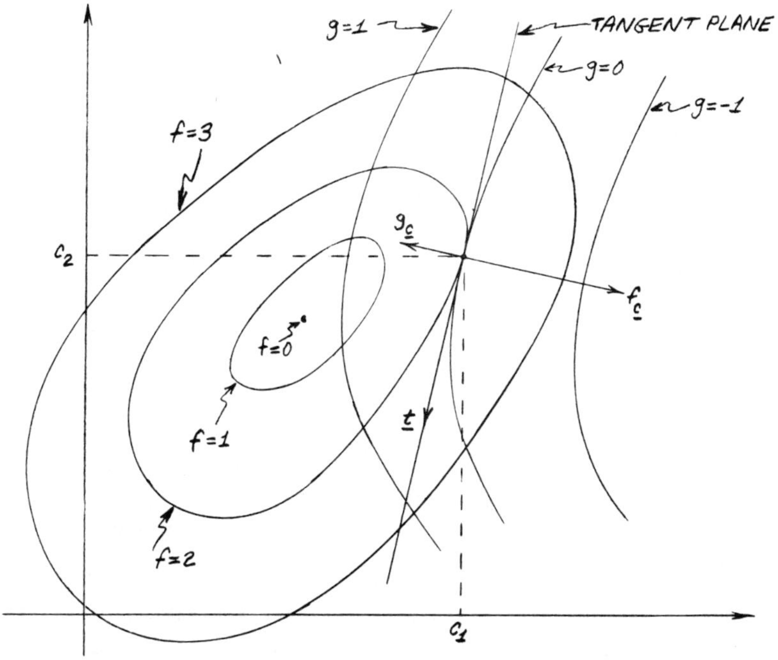

Figure 2-2 Geometry of a constrained relative minimum in two dimensions.

satisfying (2-10) for all admissible $\mathbf{t}$ can be expressed as

$$f_\mathbf{c} + \mathbf{g}_\mathbf{c}'\boldsymbol{\lambda} = \mathbf{0}. \tag{2-13}$$

The necessary condition given in (2-13) for a relative minimum constrained by $\mathbf{g}(\mathbf{c}) = \mathbf{0}$ is used frequently and is given a mathematical basis in the sequel.

In order to properly formulate notions concerning equality constraints, appropriate restrictions must be placed on subsets of contour $\{\mathbf{c}: \mathbf{g}(\mathbf{c}) = \mathbf{0}\}$. Specifically, only manifolds are considered. These sets are defined by

Definition 2-8 Let functions $\mathbf{g}(\mathbf{c})$ with dim $\mathbf{g} = m$ and dim $\mathbf{c} = n$ be continuously differentiable on some open set S, and let

$$M = \{\mathbf{c}: \mathbf{c} \in S, \mathbf{g}(\mathbf{c}) = \mathbf{0}, \quad \text{and} \quad \mathbf{g}_\mathbf{c} \text{ has rank } m\}. \tag{2-14}$$

If M is not empty, then set M is called an $(n - m)$-dimensional *manifold.*

Vectors tangent to these subsets of contours now can be defined as

Definition 2-9 Let M be a manifold and $\mathbf{c} \in M$. Then $\mathbf{t}$ is called a *tangent vector*

to M at $\mathbf{c}$ if and only if there exist an $\epsilon > 0$ and a vector function $\boldsymbol{\gamma}(\alpha) \in M$ for all $\alpha \in (-\epsilon, \epsilon)$ such that $\boldsymbol{\gamma}(0) = \mathbf{c}$ and $\dot{\boldsymbol{\gamma}}(0) = \mathbf{t}$.

Also, the set of all tangent vectors to M at $\mathbf{c}$ is called the *tangent space* at $\mathbf{c}$ and henceforth is denoted by $T(\mathbf{c})$. The main result concerning relative minima constrained by equality constraints is stated as

Theorem 2-6 *Suppose M is the manifold of* $\mathbf{g}$, $T(\mathbf{c})$ *is the tangent space at* $\mathbf{c} \in M$, *and f is differentiable at* $\mathbf{c}$. *Then $f(\mathbf{c})$ is a relative minimum of f in M only if* $\mathbf{t}'f_{\mathbf{c}} = 0$ *for all* $\mathbf{t} \in T(\mathbf{c})$.

Proof Suppose $f(\mathbf{c})$ is a relative m ıimum in M but $\mathbf{t}'f_{\mathbf{c}} \neq 0$ for some $\mathbf{t} \in T(\mathbf{c})$. Let curve $\boldsymbol{\gamma}(\alpha)$ correspond to this derivative vector in accordance with Definition 2-9. Then, by construction, function

$$h(\alpha) = f[\boldsymbol{\gamma}(\alpha)]$$

also has a relative minimum at $\alpha = 0$ because $\boldsymbol{\gamma}(\alpha) \in M$ on $(-\epsilon, \epsilon)$, and hence $\dot{h}(0) = 0$ must occur. However, differentiation yields $\dot{h}(0) = \mathbf{t}'f_{\mathbf{c}} \neq 0$ for some choice of $\mathbf{t} \in T(\mathbf{c})$, and hence the assumption of a relative minimum in M at $\mathbf{c}$ is contradicted.

This necessary condition for a constrained relative minimum requires only that $f_{\mathbf{c}}$ be orthogonal to every tangent vector $\mathbf{t}$. If tangent space $T(\mathbf{c})$ is not $\mathscr{R}^n$ but instead only a subspace of $\mathscr{R}^n$, then $f_{\mathbf{c}}$ of course does not have to vanish at the relative minimum.

Elements of tangent space $T(\mathbf{c})$ can be given the following interpretation. Relationship

$$\mathbf{g}[\boldsymbol{\gamma}(\alpha)] = \mathbf{0} \;\forall \alpha \in (-\epsilon, \epsilon)$$

follows from Definition 2-8 by construction so that differentiation then yields

$$\frac{d}{d\alpha}\mathbf{g}[\boldsymbol{\gamma}(\alpha)] = \mathbf{g}_{\boldsymbol{\gamma}}\dot{\boldsymbol{\gamma}}(\alpha) = 0$$

at every point of $(-\epsilon, \epsilon)$ at which $\boldsymbol{\gamma}$ is differentiable. Moreover, derivative $\dot{\boldsymbol{\gamma}}$ exists at $\alpha = 0$ by construction, and hence the admissibility condition given in (2-12) is obtained at $\alpha = 0$. In other words, $T(\mathbf{c}) \subset N(\mathbf{g_c})$ where $N(\mathbf{g_c})$ is defined as

$$N(\mathbf{g_c}) = \{\mathbf{t}\colon \mathbf{t} \in \mathscr{R}^n \quad \text{and} \quad \mathbf{g_c t} = \mathbf{0}\} \tag{2-15}$$

and is called the real *null space* of $\mathbf{g_c}$. If dim $\mathbf{g} = m$ so that dim $\mathbf{g_c} = m \times n$, the null space of $\mathbf{g_c}$ is a subset of $\mathscr{R}^n$ which is also a linear vector space [5] having dimension $(n - m)$ when $\mathbf{g_c}$ is of rank m. The full significance of a tangent space is derived from the fact that the implicit function theorem [6] can be used to show that $T(\mathbf{c}) \supset N(\mathbf{g_c})$ also holds. In other words, the result [1]

Proposition 2-7 *Tangent space $T(\mathbf{c})$ and null space $N(\mathbf{g_c})$ are identical.*

establishes that tangent space $T(\mathbf{c})$ as defined is a linear vector space of dimension $(n - m)$.

2.6 LAGRANGE MULTIPLIER RULE FOR EQUALITY CONSTRAINTS

Linear vector space concepts can now be used to prove the Lagrange multiplier rule in an elementary fashion. The set

$$R(\mathbf{g_c'}) = \{\mathbf{g_c'}\boldsymbol{\lambda} : \boldsymbol{\lambda} \in \mathscr{R}^m\} \tag{2-16}$$

is called the real *range space* of $\mathbf{g_c'}$, and the columns of $\mathbf{g_c'}$ span this linear vector space. Range space $R(\mathbf{g_c'})$ hence has dimension m when $\mathbf{g_c}$ is of rank m. In addition, the set defined as

$$N_\perp(\mathbf{g_c}) = \{\mathbf{n} : \mathbf{n} \in \mathscr{R}^n \quad \text{and} \quad \mathbf{t} \in N(\mathbf{g_c}) \Rightarrow \mathbf{n}'\mathbf{t} = 0\} \tag{2-17}$$

is called the *orthogonal complement* of $N(\mathbf{g_c})$. Set $N_\perp(\mathbf{g_c})$ is a linear vector space having a number of useful properties which are summarized in

Lemma 2-3 *Given some real matrix* $\mathbf{G}$ *of rank m with dim* $\mathbf{G} = m \times n$, *let N denote the real null space of* $\mathbf{G}$, $N_\perp$ *denote the orthogonal complement of N, and R denote the real range space of* $\mathbf{G}'$. *Then*

i) $N_\perp \cap N = \{\mathbf{0}\}$

ii) $N_\perp + N = \{\mathbf{n} + \mathbf{t} : \mathbf{n} \in N_\perp \quad \textit{and} \quad \mathbf{t} \in N\} = \mathscr{R}^n$

and

iii) $R = N_\perp$.

Proof Because $N_\perp$ and N are linear vector spaces, set $N_\perp \cap N$ contains $\mathbf{0}$. Let $\{\mathbf{t}_j : j = 1, 2, \ldots, (n - m)\}$ be an orthonormal basis set of N. Then every $\mathbf{t} \in N$ can be expressed as

$$\mathbf{t} = \sum_{j=1}^{n-m} (\mathbf{t}'\mathbf{t}_j)\mathbf{t}_j.$$

However, if $\mathbf{n} \in N_\perp$, then $\mathbf{n}'\mathbf{t}_j = 0$ for each j by construction of $N_\perp$ so that $\mathbf{n}$ cannot be an element of N and i) is proved. Each row of $\mathbf{G}$ is an element of $N_\perp$ so that the dimension of $N_\perp$ is at least m. Then ii) is proved because N has dimension $(n - m)$ and i) holds. Finally iii) is proved because $N_\perp$ has dimension m, by virtue of i) and ii), and hence the rows of $\mathbf{G}$ are a basis set of $N_\perp$ as well as R.

The so-called Lagrange multiplier rule for equality constraints can now be stated and proved in the form

Theorem 2-7 Let M be the manifold of $\mathbf{g}$, *f be differentiable at* $\mathbf{c} \in M$, *and*

$$L = f(\mathbf{c}) + \boldsymbol{\lambda}'\mathbf{g}(\mathbf{c}). \tag{2-18}$$

Then $f(\mathbf{c})$ is a relative minimum of f in M only if $L_{\mathbf{c}} = \mathbf{0}$ for some real $\boldsymbol{\lambda}$.

Proof By Theorem 2-6, $f(\mathbf{c})$ is a relative minimum in M only if $\mathbf{t}'f_{\mathbf{c}} = 0$ for all $\mathbf{t} \in N(\mathbf{g_c})$. Thus $f_{\mathbf{c}} \in N_{\perp}(\mathbf{g_c})$ and hence $f_{\mathbf{c}} \in R(\mathbf{g}'_{\mathbf{c}})$ by Lemma 2-2. In other words, gradient $f_{\mathbf{c}}$ can always be expressed as a linear combination of the columns of $\mathbf{g}'_{\mathbf{c}}$, such as given in (2-13), at a relative minimum of f in M. Direct computation of $L_{\mathbf{c}}$ for $\boldsymbol{\lambda}$ satisfying (2-13) then yields $L_{\mathbf{c}} = \mathbf{0}$ at every relative minimum in M.

Scalar L defined in (2-18) is referred to as a *Lagrangian.*

2.7 LAGRANGIAN FOR EQUALITY CONSTRAINTS

The Lagrangian defined in (2-18) leads to a number of useful concepts related to relative minima which are constrained by equality constraints. If this scalar is taken to be the function $L = L(\mathbf{c}, \boldsymbol{\lambda})$, then necessary condition $L_{\mathbf{c}} = \mathbf{0}$ appearing in Theorem 2-7 and admissibility condition $L_{\boldsymbol{\lambda}} = \mathbf{g}(\mathbf{c}) = \mathbf{0}$ are equivalent to stationarity of the Lagrangian. In other words, the first necessary condition for a constrained relative minimum is stationarity of the Lagrangian with respect to the partitioned vector $[\mathbf{c}' \;\; \boldsymbol{\lambda}']'$. At a point where the Lagrangian is stationary, Lagrange multipliers $\boldsymbol{\lambda}$ are seen from (2-13) to be the expansion coefficients of vector $(-f_{\mathbf{c}})$ in range space $R(\mathbf{g}'_{\mathbf{c}})$ using the basis set formed by the columns of $\mathbf{g}'_{\mathbf{c}}$. In fact, Lagrange multipliers can be determined uniquely by first premultiplying both sides of (2-13) by $\mathbf{g_c}$ to obtain

$$(\mathbf{g_c g'_c})\,\boldsymbol{\lambda} = \mathbf{g_c}(-\mathbf{f_c}).$$

The so-called constraint qualification [7] $\det(\mathbf{g_c g'_c}) \neq 0$ has already been imposed by virtue of the assumption that $\mathbf{g_c}$ is of rank m with $\dim \mathbf{g_c} = m \times n$. Therefore, Lagrange multipliers are given by

$$\boldsymbol{\lambda} = (\mathbf{g_c g'_c})^{-1}\mathbf{g_c}(-\mathbf{f_c}). \tag{2-19}$$

Elements of $R(\mathbf{g}'_{\mathbf{c}})$ are orthogonal to tangent space $T(\mathbf{c})$ and hence are referred to as *normal vectors* to the manifold M at $\mathbf{c}$. At a point where the Lagrangian is stationary, therefore, gradient $f_{\mathbf{c}}$ must be normal to the manifold at $\mathbf{c}$, and the geometry depicted in Figure 2-2 generalizes conceptually to any case where $1 \leqslant \dim \mathbf{g} < \dim \mathbf{c}$.

Linear vector space concepts inherent in the Lagrange multiplier formulation can be used to give a further interpretation of the Lagrange multiplier rule. Suppose direction vector $\mathbf{d}$ is an arbitrary element of $\mathscr{R}^n$. Then there exist unique $\mathbf{t} \in N(\mathbf{g_c})$ and $\mathbf{n} \in N_\perp(\mathbf{g_c})$ such that

$$\mathbf{d} = \mathbf{t} + \mathbf{n}. \tag{2-20}$$

Uniqueness of $\mathbf{t}$ and $\mathbf{n}$ follows from i) and ii) of Lemma 2-2. There also exists a unique vector $\mathbf{m} \in \mathscr{R}^m$ such that

$$\mathbf{n} = \mathbf{g_c'm}$$

because the columns of $\mathbf{g_c'}$ are a basis set of $N_\perp(\mathbf{g_c})$. By construction, the rows of $\mathbf{g_c}$ are orthogonal to $\mathbf{t}$ so that premultiplication of (2-20) by $\mathbf{g_c}$ yields

$$\mathbf{g_c d} = (\mathbf{g_c g_c'})\mathbf{m}$$

and hence

$$\mathbf{m} = (\mathbf{g_c g_c'})^{-1}\mathbf{g_c d}.$$

Thus, for any $\mathbf{d} \in \mathscr{R}^n$, normal vector $\mathbf{n}$ can be written as

$$\mathbf{n} = (\mathbf{I} - \mathbf{P})\mathbf{d} \tag{2-21}$$

and tangent vector $\mathbf{t}$ can be written as

$$\mathbf{t} = \mathbf{Pd} \tag{2-22}$$

where

$$\mathbf{P} = \mathbf{I} - \mathbf{g_c'}(\mathbf{g_c g_c'})^{-1}\mathbf{g_c}. \tag{2-23}$$

Moreover, the necessary condition appearing in Theorem 2-6 can be stated as

$$\mathbf{d}'(\mathbf{P}f_\mathbf{c}) = 0 \ \forall \ \mathbf{d} \in \mathscr{R}^n,$$

and hence

$$\mathbf{P}f_\mathbf{c} = \mathbf{0} \tag{2-24}$$

is an alternate form of the first necessary condition for a relative minimum constrained by equality constraints.

Matrix $\mathbf{P}$ is called the *projection operator* normal to $\mathbf{g_c}$ because

$$\mathbf{P} = \mathbf{P}', \ \mathbf{P}^2 = \mathbf{P}, \quad \text{and} \quad \mathbf{g_c P} = \mathbf{0}. \tag{2-25}$$

In addition, vectors $\mathbf{t}$ and $\mathbf{n}$ are called *projections* of direction vector $\mathbf{d}$ into $N(\mathbf{g_c})$ and $N_\perp(\mathbf{g_c})$ respectively. At a constrained relative minimum, therefore, the projection of gradient $f_\mathbf{c}$ into tangent space $T(\mathbf{c})$ must vanish.

Perhaps the most useful aspects of the Lagrangian concerns additional conditions for constrained relative minima which are based on [8]

Proposition 2-8 *Suppose* $\mathbf{g}$ *is twice continuously differentiable at* $\mathbf{c}$. *Then, given any tangent vector* $\mathbf{t}$ *to manifold* M *at* $\mathbf{c}$, *there exists a vector function* $\boldsymbol{\gamma}(\alpha)$ *that has the properties specified in Definition* 2-9 *and is twice continuously differentiable on* $(-\delta, \delta)$ *for some* $\delta > 0$.

For example, convexity of the Lagrangian is the second necessary condition for a constrained relative minimum as stated in

Theorem 2-8 *Suppose*

i) M *is the manifold of* $\mathbf{g}$ *and* $T(\mathbf{c})$ *is the tangent space at* $\mathbf{c} \in M$

ii) f *and* $\mathbf{g}$ *are twice continuously differentiable at* $\mathbf{c}$

and

iii) $L_{\mathbf{c}} = \mathbf{0}$ *for some real* $\boldsymbol{\lambda}$.

Then $f(\mathbf{c})$ *is a relative minimum of* f *in* M *only if* $\mathbf{t}'L_{\mathbf{cc}}\mathbf{t} \geqslant 0$ *for all* $\mathbf{t} \in T(\mathbf{c})$.

Proof Function $l(\alpha) = f[\boldsymbol{\gamma}(\alpha)] = L[\boldsymbol{\gamma}(\alpha), \boldsymbol{\lambda}]$ is defined on $(-\epsilon, \epsilon)$, and differentiation yields

$$\dot{l}(\alpha) = L_{\boldsymbol{\gamma}}\dot{\boldsymbol{\gamma}}(\alpha) \quad \text{and} \quad \ddot{l}(\alpha) = L_{\boldsymbol{\gamma}}\ddot{\boldsymbol{\gamma}} + \dot{\boldsymbol{\gamma}}'L_{\boldsymbol{\gamma\gamma}}\dot{\boldsymbol{\gamma}}$$

at points where $\boldsymbol{\gamma}$ is twice differentiable including $\alpha = 0$. Moreover, function $\ddot{l}(\alpha)$ is continuous at $\alpha = 0$ so that there exists an open neighborhood of 0 on which $\ddot{l}(\alpha)$ has the same sign as $\ddot{l}(0)$ when $\ddot{l}(0) \neq 0$. By construction, $\dot{l}(0) = 0$ and $\ddot{l}(0) = \mathbf{t}'L_{\mathbf{cc}}\mathbf{t}$ so that $l(0)$ is a relative minimum of l in $(-\epsilon, \epsilon)$ only if $\mathbf{t}'L_{\mathbf{cc}}\mathbf{t} \geqslant 0$ for every $\mathbf{t} \in T(\mathbf{c})$.

An alternate form of this necessary condition is stated as

$$\mathbf{d}'(\mathbf{P}L_{\mathbf{cc}}\mathbf{P})\mathbf{d} \geqslant 0 \;\forall\, \mathbf{d} \in \mathscr{R}^n$$

using (2-22) and hence as $(\mathbf{P}L_{\mathbf{cc}}\mathbf{P})$ must be positive semi-definite.

A similar sufficiency theorem can also be given for strict relative minima which are constrained by equality constraints. This theorem is stated in terms of a condition that is analogous to the combined conditions $\mathbf{t}'L_{\mathbf{cc}}\mathbf{t} \geqslant 0 \;\forall\; \mathbf{t} \in \mathscr{R}^n$ and $L_{\mathbf{cc}}$ nonsingular, that is $L_{\mathbf{cc}}$ positive definite, for a strict relative minimum without constraints. Specifically, linear vector space methods are used to prove

Lemma 2-4 *Suppose* $\mathbf{G}$ *is a real matrix with linearly independent rows and* $\mathbf{F}$ *is a real symmetric matrix with* $\mathbf{t}'\mathbf{F}\mathbf{t} \geqslant 0$ *for all* $\mathbf{t} \in N(\mathbf{G})$. *Then* $\mathbf{t}'\mathbf{F}\mathbf{t} > 0$ *holds for all nonzero* $\mathbf{t} \in N(\mathbf{G})$ *if and only if matrix*

$$\mathbf{H} = \begin{bmatrix} \mathbf{F} & \mathbf{G}' \\ \mathbf{G} & \mathbf{0} \end{bmatrix}$$

is nonsingular.

Proof For $\mathbf{G}$ square, there is nothing to prove because $N(\mathbf{G}) = \{\mathbf{0}\}$. Therefore, let $\mathbf{G}$ have a smaller number of rows than columns so that the columns of $\mathbf{G}'$ are a basis set of $R(\mathbf{G}')$. Assume there exists a $\mathbf{t} \neq \mathbf{0}$ such that $\mathbf{t} \in N(\mathbf{G})$ and $\mathbf{t}'\mathbf{F}\mathbf{t} = 0$. Then $\mathbf{F}\mathbf{t} \in R(\mathbf{G}')$ because $\mathbf{t}$ and $\mathbf{F}\mathbf{t}$ are orthogonal, and hence there exists a unique vector $\boldsymbol{\tau}$ such that

$$\mathbf{F}\mathbf{t} + \mathbf{G}'\boldsymbol{\tau} = \mathbf{0}.$$

Therefore, matrix $\mathbf{H}$ is singular because

$$\begin{bmatrix} \mathbf{F} & \mathbf{G}' \\ \mathbf{G} & \mathbf{0} \end{bmatrix} \begin{bmatrix} \mathbf{t} \\ \boldsymbol{\tau} \end{bmatrix} = \mathbf{0} \text{ for some } \begin{bmatrix} \mathbf{t} \\ \boldsymbol{\tau} \end{bmatrix} \neq \mathbf{0}.$$

Conversely, assume $\mathbf{H}$ is singular so that there exists a vector $[\mathbf{t}' \ \ \tau']' \neq \mathbf{0}$ such that

$$\mathbf{F}\mathbf{t} + \mathbf{G}'\boldsymbol{\tau} = \mathbf{0} \quad \text{and} \quad \mathbf{G}\mathbf{t} = \mathbf{0}.$$

Then $\mathbf{t} \in N(\mathbf{G})$ and $\mathbf{t} \neq \mathbf{0}$ because $\mathbf{t} = \mathbf{0}$ implies $\boldsymbol{\tau} = \mathbf{0}$ in accordance with the assumption that $\mathbf{G}'$ has linearly independent columns. Moreover,

$$\mathbf{t}'(\mathbf{F}\mathbf{t} + \mathbf{G}'\boldsymbol{\tau}) = \mathbf{t}'\mathbf{F}\mathbf{t} = 0$$

because $\mathbf{t} \in N(\mathbf{G})$.

This sufficiency theorem for strict relative minima subject to equality constraints can now be stated as

Theorem 2-9 *Suppose*

i) M is the manifold of $\mathbf{g}$ *and $T(\mathbf{c})$ is the tangent space at* $\mathbf{c} \in M$
ii) f and $\mathbf{g}$ *are twice continuously differentiable at* $\mathbf{c}$
iii) $L_{\mathbf{c}} = \mathbf{0}$ *for some real* $\boldsymbol{\lambda}$
and
iv) $\mathbf{t}'L_{\mathbf{cc}}\mathbf{t} \geqslant 0$ *holds for all* $\mathbf{t} \in T(\mathbf{c})$ *and matrix*

$$\begin{bmatrix} L_{\mathbf{cc}} & \mathbf{g}_{\mathbf{c}}' \\ \mathbf{g}_{\mathbf{c}} & \mathbf{0} \end{bmatrix}$$

is nonsingular.
Then $f(\mathbf{c})$ is a strict relative minimum of f in M.

Proof The construction used in the proof of Theorem 2-8 reveals that $l(0)$ is a strict relative minimum of l in $(-\epsilon, \epsilon)$ when $\ddot{l}(0) > 0$ and hence $f(\mathbf{c})$ is a strict relative minimum of f in M when $\mathbf{t}'L_{\mathbf{cc}}\mathbf{t} > 0$ given any nonzero $\mathbf{t} \in T(\mathbf{c})$. The columns of $\mathbf{g}_{\mathbf{c}}'$ are linearly independent at all $\mathbf{c} \in M$ by definition of M so that Lemma 2-3 applies and the theorem is proved.

Condition iv) of Theorem 2-9 is of course satisfied whenever $L_{\mathbf{cc}}$ is positive definite. However, matrix $L_{\mathbf{cc}}$ need not even be positive semi-definite at strict relative minima which are constrained by one or more equality constraints. Condition iv) is always satisfied when $\mathbf{g_c}$ is square.

An elementary example of this sufficiency condition is provided by

$$f(\mathbf{c}) = c_1 c_2 \quad \text{and} \quad \mathbf{g}(\mathbf{c}) = [(c_1 - c_2 - 1)].$$

The Lagrangian for this example becomes

$$L = c_1 c_2 + \lambda_1 (c_1 - c_2 - 1),$$

and conditions $L_{\mathbf{c}} = \mathbf{0}$ and $L_{\boldsymbol{\lambda}} = \mathbf{0}$ yield

$$\mathbf{c} = \tfrac{1}{2}\begin{bmatrix} 1 \\ -1 \end{bmatrix} \quad \text{and } \boldsymbol{\lambda} = [\tfrac{1}{2}].$$

In addition, hessian

$$L_{\mathbf{cc}} = \begin{bmatrix} 0 & 1 \\ 1 & 0 \end{bmatrix}$$

has a negative eigenvalue. However, projection operator

$$\mathbf{P} = \tfrac{1}{2}\begin{bmatrix} 1 & 1 \\ 1 & 1 \end{bmatrix}$$

yields

$$\mathbf{P}L_{\mathbf{cc}}\mathbf{P} = \tfrac{1}{2}\begin{bmatrix} 1 & 1 \\ 1 & 1 \end{bmatrix}$$

which is positive semi-definite. Moreover, elementary calculations yield

$$\det \left[\begin{array}{cc|c} 0 & 1 & 1 \\ 1 & 0 & -1 \\ \hline 1 & -1 & 0 \end{array}\right] = -2$$

so that a strict relative minimum has been found. This result can, of course, be verified using $g_1(\mathbf{c}) = 0$ to eliminate say c_1.

Conditions for absolute minima constrained by equality constraints can also be stated conveniently in terms of Lagrangian $L = L(\mathbf{c},\boldsymbol{\lambda})$ as defined by (2-18). The main result is stated as

Theorem 2-10 *Suppose M is the manifold of* **g**, *and suppose f is differentiable and (strictly) convex on a convex set* $S \subset M$. *Then f has a (strict) absolute minimum on S at* **c** *if and only if* $L_{\mathbf{c}} = \mathbf{0}$ *for some real* $\boldsymbol{\lambda}$.

Proof Necessity of $L_{\mathbf{c}} = \mathbf{0}$ for some real $\boldsymbol{\lambda}$ has already been established in Theorem 2-7. In order to prove sufficiency, let $\boldsymbol{\lambda}$ be chosen so that $L_{\mathbf{c}} = \mathbf{0}$. Then, by definition of M,

$$L(\mathbf{c} + \mathbf{d}, \boldsymbol{\lambda}) = f(\mathbf{c} + \mathbf{d})$$

holds for any direction vector **d** chosen so that $\mathbf{c} + \mathbf{d} \in M$. Therefore, Lagrangian L is (strictly) convex on $S \subset M$ and hence

$$L(\mathbf{c} + \mathbf{d}, \boldsymbol{\lambda}) \geqslant L(\mathbf{c}, \boldsymbol{\lambda}) + \mathbf{d}' L_{\mathbf{c}} = L(\mathbf{c}, \boldsymbol{\lambda})$$

or rather

$$f(\mathbf{c} + \mathbf{d}) \geqslant f(\mathbf{c}) \text{ for any } \mathbf{c} + \mathbf{d} \in S.$$

Moreover, the above equality holds only for $\mathbf{d} = \mathbf{0}$ if f is strictly convex on S.

Linear constraint functions result in convex manifolds. Unfortunately, however, many other constraint functions do not result in manifolds that contain convex subsets, and hence this theorem is not applicable to many problems for which $L_{\mathbf{c}} = \mathbf{0}$ does imply $f(\mathbf{c})$ is an absolute minimum.

2.8 EXAMPLE OF EQUALITY CONSTRAINTS

A specialized form of equality constraint arises frequently in automated design problems associated with control. This form of constraint illustrates the elimination of variables that corresponds to equality constraints. Moreover, this form of constraint permits a slightly different and much more useful form of statements concerning sufficient conditions for strict relative minima. The usefulness of these alternate statements is primarily computational.

By way of introduction, consider the minimization of

$$F(\mathbf{a}) = f[\mathbf{a}, \mathbf{b}(\mathbf{a})] \tag{2-26}$$

with respect to **a** where f and **b** are twice continuously differentiable functions. Then conditions $F_{\mathbf{a}} = \mathbf{0}$ and $F_{\mathbf{aa}}$ positive definite are sufficient for a strict relative minimum at **a**. Moreover, the ith element of $F_{\mathbf{a}}$ is

$$\frac{\partial F}{\partial a_i} = \frac{\partial f}{\partial a_i} + \sum_{k=1}^{m} \frac{\partial f}{\partial b_k} \frac{\partial b_k}{\partial a_i}, \tag{2-27}$$

and the ij-element of $F_{\mathbf{aa}}$ is

$$\frac{\partial^2 F}{\partial a_i\, \partial a_j} = \frac{\partial^2 f}{\partial a_i\, \partial a_j} + \sum_{k=1}^{m} \frac{\partial^2 f}{\partial a_i\, \partial b_k}\frac{\partial b_k}{\partial a_j} + \frac{\partial^2 f}{\partial a_j\, \partial b_k}\frac{\partial b_k}{\partial a_i} + \frac{\partial f}{\partial b_k}\frac{\partial^2 b_k}{\partial a_i\, \partial a_j}. \tag{2-28}$$

Gradient $F_{\mathbf{a}}$ can also be written in vector-matrix form as

$$F_{\mathbf{a}} = f_{\mathbf{a}} + \mathbf{b}'_{\mathbf{a}} f_{\mathbf{b}}, \tag{2-29}$$

but hessian $F_{\mathbf{aa}}$ does not have a similar vector-matrix representation.

Now suppose that function $\mathbf{b}(\mathbf{a})$ is in fact defined implicitly by $\mathbf{h}(\mathbf{a}) = \mathbf{0}$ for all $\mathbf{a}$ where

$$\mathbf{h}(\mathbf{a}) = \mathbf{g}[\mathbf{a}, \mathbf{b}(\mathbf{a})]. \tag{2-30}$$

Function $\mathbf{g}$ is assumed to be twice continuously differentiable, and matrix $\mathbf{g}_{\mathbf{b}}$ is assumed to be nonsingular in accordance with the implicit function theorem. Differentiation then yields

$$\frac{dh_l}{da_i} = \frac{\partial g_l}{\partial a_i} + \sum_{k=1}^{m} \frac{\partial g_l}{\partial b_k}\frac{\partial b_k}{\partial a_i} = 0$$

for each l or rather

$$\mathbf{g}_{\mathbf{a}} + \mathbf{g}_{\mathbf{b}}\mathbf{b}_{\mathbf{a}} = \mathbf{0}. \tag{2-31}$$

Hence, matrix $\mathbf{b}_{\mathbf{a}}$ is given by

$$\mathbf{b}_{\mathbf{a}} = -\mathbf{g}_{\mathbf{b}}^{-1}\mathbf{g}_{\mathbf{a}} \tag{2-32}$$

so that (2-29) now becomes

$$F_{\mathbf{a}} = f_{\mathbf{a}} - \mathbf{g}'_{\mathbf{a}}(\mathbf{g}'_{\mathbf{b}})^{-1} f_{\mathbf{b}}. \tag{2-33}$$

Steps leading to this form of $F_{\mathbf{a}}$ illustrate the elimination of variables implied by equality constraints.

The form of gradient $F_{\mathbf{a}}$ appearing in (2-33) can be related to the Lagrange multiplier formulation of equality constraints by defining a function $\boldsymbol{\lambda} = \boldsymbol{\lambda}(\mathbf{a})$ as

$$\boldsymbol{\lambda} = -(\mathbf{g}'_{\mathbf{b}})^{-1} f_{\mathbf{b}}. \tag{2-34}$$

This expression for $\boldsymbol{\lambda}(\mathbf{a})$ can also be written as

$$f_{\mathbf{b}} + \mathbf{g}'_{\mathbf{b}}\boldsymbol{\lambda} = \mathbf{0}, \tag{2-35}$$

and gradient $F_{\mathbf{a}}$ now becomes

$$F_{\mathbf{a}} = f_{\mathbf{a}} + \mathbf{g}'_{\mathbf{a}}\boldsymbol{\lambda}. \tag{2-36}$$

These last two expressions combine to yield the familar condition $L_{\mathbf{c}} = \mathbf{0}$ when $F_{\mathbf{a}} = \mathbf{0}$ holds and definitions

$$\mathbf{c} = \begin{bmatrix} \mathbf{a} \\ \\ \mathbf{b} \end{bmatrix} \quad \text{and} \quad L = f + \boldsymbol{\lambda}'\mathbf{g} \tag{2-37}$$

are introduced.

Now suppose the Lagrangian defined in (2-37) is taken to be the function $L = L[\mathbf{a}, \mathbf{b}(\mathbf{a}), \boldsymbol{\lambda}(\mathbf{a})]$. Then condition $\mathbf{h}(\mathbf{a}) = \mathbf{0}$ becomes

$$L_{\lambda} = \mathbf{0}, \tag{2-38}$$

equation (2-35) becomes

$$L_{\mathbf{b}} = \mathbf{0}, \tag{2-39}$$

and gradient $F_{\mathbf{a}}$ given in (2-36) becomes

$$F_{\mathbf{a}} = L_{\mathbf{a}} \tag{2-40}$$

because $F = L$ for all $\mathbf{a}$ by definition of $\mathbf{b}(\mathbf{a})$. Moreover, differentiation of (2-38) with respect to $\mathbf{a}$ yields

$$L_{\lambda\mathbf{a}} + L_{\lambda\mathbf{b}}\mathbf{b}_{\mathbf{a}} = \mathbf{0}$$

because $L_{\lambda\lambda} = \mathbf{0}$, and hence an alternate expression for matrix $\mathbf{b}_{\mathbf{a}}$ is found to be

$$\mathbf{b}_{\mathbf{a}} = -L_{\lambda\mathbf{b}}^{-1}{}_{\lambda\mathbf{a}}. \tag{2-41}$$

In a similar fashion, differentiation of (2-39) with respect to $\mathbf{a}$ yields

$$L_{\mathbf{ba}} + L_{\mathbf{bb}}\mathbf{b}_{\mathbf{a}} + L_{\mathbf{b}\lambda}\boldsymbol{\lambda}_{\mathbf{a}} = \mathbf{0},$$

and hence an expression for matrix $\boldsymbol{\lambda}_{\mathbf{a}}$ is found to be

$$\boldsymbol{\lambda}_{\mathbf{a}} = -L_{\mathbf{b}\lambda}^{-1}(L_{\mathbf{ba}} + L_{\mathbf{bb}}\mathbf{b}_{\mathbf{a}}). \tag{2-42}$$

Finally, differentiation of $F_{\mathbf{a}}$ with respect to $\mathbf{a}$ yields

$$F_{\mathbf{aa}} = L_{\mathbf{aa}} + L_{\mathbf{ab}}\mathbf{b}_{\mathbf{a}} + L_{\mathbf{a}\lambda}\boldsymbol{\lambda}_{\mathbf{a}} \tag{2-43}$$

so that hessian $F_{\mathbf{aa}}$ can be computed easily.

This example is characterized by a specific set of dependent variables, namely $\mathbf{b}(\mathbf{a})$, that are uniquely defined by a corresponding set of algebraic constraint equations, namely $\mathbf{g}[\mathbf{a}, \mathbf{b}(\mathbf{a})] = \mathbf{0}$. In order to compute the gradient and hessian with respect to $\mathbf{a}$ in a convenient fashion, an additional set of dependent variables, namely $\boldsymbol{\lambda}(\mathbf{a})$, is introduced. These additional dependent variables are treated as independent Lagrange multipliers up through the computation of the gradient with respect to $\mathbf{a}$. However, these additional dependent variables are not treated as independent Lagrange multipliers in the computation of the hessian with respect to **a**. Therefore, this hessian not only is given by an expression that is different from those appearing in Theorems 2-8 and 2-9 but also has a different interpretation. The example presented in this section is typical of almost all automated design problems that are formulated in subsequent chapters.

2.9 INEQUALITY CONSTRAINTS

Numerous automated design problems which also arise in control involve one or more inequality constraint functions that must be satisfied in order to achieve an admissible solution. In this section, the problem of minimizing $f(\mathbf{c})$ subject to $\mathbf{g}(\mathbf{c}) \leqslant \mathbf{0}$ is considered. Unlike equality constraints which always reduce the number of independent elements of **c**, inequality constraints reduce the set of points on which a solution is defined but may not reduce the number of independent elements of **c** depending upon the point in question. Moreover, well-defined problems are often encountered where dim **g** > dim **c**.

In order to formulate notions concerning inequality constraints, the set

$$I_a = \{i : g_i(\mathbf{c}) = 0\} \tag{2-44}$$

is introduced and is called the index set of *active* constraints at **c**. Active constraints when arranged in vector form are denoted as $\bar{\mathbf{g}}$ and are defined by

$$\bar{\mathbf{g}} = [g_i] \quad \text{with} \quad i \in I_a. \tag{2-45}$$

Points at which necessary and sufficient conditions for a relative minimum are to be developed are defined by

Definition 2-10 Let **g** be continuously differentiable on some open set S, and let

$$F = \{\mathbf{c} : \mathbf{c} \in S, \mathbf{g}(\mathbf{c}) \leqslant \mathbf{0}, \quad \text{and} \quad \text{rank}\, \bar{\mathbf{g}}_{\mathbf{c}} = \dim \bar{\mathbf{g}}\}. \tag{2-46}$$

If F is not empty, then F is called the *feasible set* of **g**.

Feasible sets may include points at which dim $\bar{\mathbf{g}}$ = 0, 1, . . . , or dim **c**. At points where dim $\bar{\mathbf{g}}$ = 0, conditions for a relative minimum revert to those for unconstrained minima. At points where dim $\bar{\mathbf{g}}$ > 0, a feasible set contains a subset of the manifold of $\bar{\mathbf{g}}$ at **c**. Moreover, conditions for a relative minimum at points where dim $\bar{\mathbf{g}}$ > 0 involve some of the aspects of conditions for relative minima with equality constraints. A geometrical interpretation of active constraints and feasible sets is depicted in Figure 2-3.

Admissible direction vectors from points belonging to feasible sets are defined by

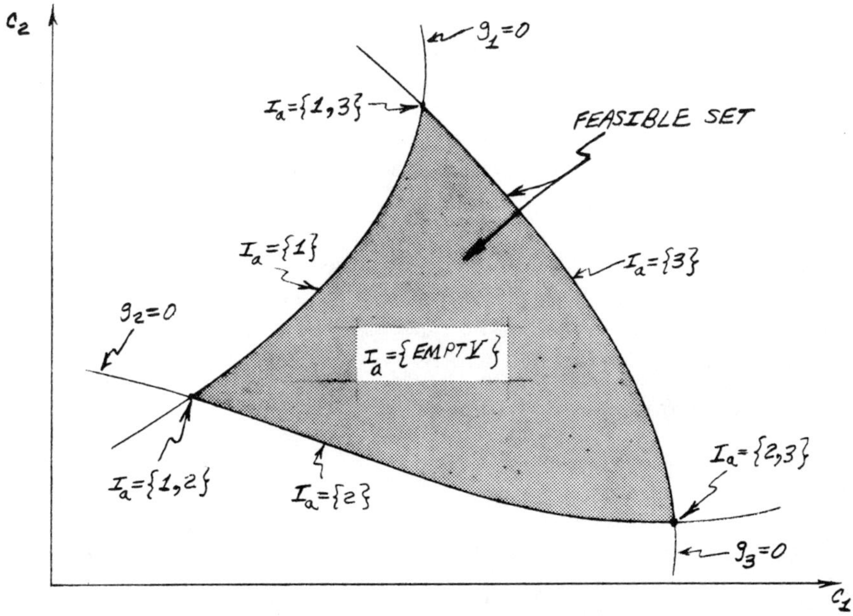

Figure 2-3 Geometry of a typical feasible set in two dimensions.

Definition 2-11 Let F be a feasible set and $\mathbf{c} \in F$. Then $\mathbf{d}$ is called an *admissible direction vector* from $\mathbf{c}$ if and only if there exists an $\epsilon > 0$ and a vector function $\boldsymbol{\gamma}(\alpha)$ which is defined on $(-\epsilon, \epsilon)$ such that $\alpha \in [0, \epsilon)$ implies $\boldsymbol{\gamma} \in F$, $\boldsymbol{\gamma}(0) = \mathbf{c}$, and $\dot{\boldsymbol{\gamma}}(0) = \mathbf{d}$.

Also, the set of all admissible direction vectors from $\mathbf{c}$ is called the *admissible cone* at $\mathbf{c}$ and is denoted as $A(\mathbf{c})$. Definition of $\boldsymbol{\gamma}$ on $(-\epsilon, 0)$, of course, poses no added restrictions on $A(\mathbf{c})$ due to the arbitrariness of $\boldsymbol{\gamma}$ on this open interval. The main result concerning relative minima constrained by inequality constraints can now be stated as

Theorem 2-11 *Suppose F is the feasible set of* $\mathbf{g}$, $A(\mathbf{c})$ *is the admissible cone at* $\mathbf{c} \in F$, *and f is differentiable at* $\mathbf{c}$. *Then $f(\mathbf{c})$ is a relative minimum of f in F only if* $\mathbf{d}'f_{\mathbf{c}} \geqslant 0$ *for all* $\mathbf{d} \in A(\mathbf{c})$.

Proof Let $\boldsymbol{\gamma}(\alpha)$ be specified as in Definition 2-11, and let

$$h(\alpha) = \begin{cases} \dfrac{f[\boldsymbol{\gamma}(\alpha)] - f(\mathbf{c})}{\alpha} & \text{for} \quad \alpha \in (-\epsilon, 0) \cup (0, \epsilon) \\ \\ \mathbf{d}'f_{\mathbf{c}} & \text{for} \quad \alpha = 0 \end{cases}$$

which is a continuous function at $\alpha = 0$ because f is differentiable at $\mathbf{c}$ and $\boldsymbol{\gamma}$ is continuous at $\alpha = 0$. Now suppose that $\mathbf{d}'f_{\mathbf{c}} < 0$ holds for some $\mathbf{d} \in A(\mathbf{c})$. Then $f[\boldsymbol{\gamma}(\alpha)] < f(\mathbf{c})$ holds for some $\alpha \in (0, \epsilon)$, and hence $f(\mathbf{c})$ is not a relative minimum of f in F.

This necessary condition for relative minima constrained by inequality constraints reverts to Theorem 2-1 for unconstrained relative minima when $\dim \bar{\mathbf{g}} = 0$ because $A(\mathbf{c}) = \mathscr{R}^n$.

The Lagrange multiplier rule for inequality constraints is not based directly on linear vector space methods as is the case for equality constraints. Instead, the Lagrange multiplier formulation for inequality constraints is based on the Farkas lemma [9] which is stated here as

Proposition 2-9 *If every real vector* $\mathbf{d}$ *which satisfies*

$$\mathbf{G}\mathbf{d} \leqslant \mathbf{0} \tag{2-47}$$

also satisfies

$$\mathbf{d}'\mathbf{f} \geqslant 0, \tag{2-48}$$

then there exists a real vector $\boldsymbol{\lambda} \geqslant \mathbf{0}$ *such that*

$$\mathbf{f} + \mathbf{G}'\boldsymbol{\lambda} = \mathbf{0}. \tag{2-49}$$

Notation $\boldsymbol{\lambda} \geqslant \mathbf{0}$ and $\boldsymbol{\lambda} > \mathbf{0}$ is used to denote respectively $\lambda_i \geqslant 0$ and $\lambda_i > 0$ for each i. This result due to Farkas, although somewhat difficult to prove, has a geometrical interpretation that is intuitively reasonable. The set defined as

$$C(-\mathbf{G}') = \{-\mathbf{G}'\boldsymbol{\lambda} : \boldsymbol{\lambda} \in \mathscr{R}^m \quad \text{and} \quad \boldsymbol{\lambda} \geqslant \mathbf{0}\} \tag{2-50}$$

is called the *convex polyhedral cone* of $(-\mathbf{G}')$. Thus, the condition given in (2-49) with $\boldsymbol{\lambda} \geqslant \mathbf{0}$ can be restated as $\mathbf{f} \in C(-\mathbf{G}')$. If the matrix $(-\mathbf{G}')$ is partitioned as

$$\mathbf{G}' = [\mathbf{g}_1 \ \mathbf{g}_2 \ \ldots],$$

then the Farkas lemma can be given the geometrical interpretation depicted in Figure 2-4.

The Lagrange multiplier rule for inequality constraints can now be stated and proved in the form

Theorem 2-12 *Let F be the feasible set of* $\mathbf{g}$, *f be differentiable at* $\mathbf{c} \in F$, *and Lagrangian L be defined by (2-18). Then* $f(\mathbf{c})$ *is a relative minimum of f in F only if* $L_{\mathbf{c}} = \mathbf{0}$ *for some real* $\boldsymbol{\lambda} \geqslant \mathbf{0}$ *with* $\lambda_i g_i(\mathbf{c}) = 0$ *for each i.*

Proof For each $i \notin I_a$, let $\lambda_i = 0$ so that the Lagrangian becomes

$$L = f + \bar{\boldsymbol{\lambda}}'\bar{\mathbf{g}}$$

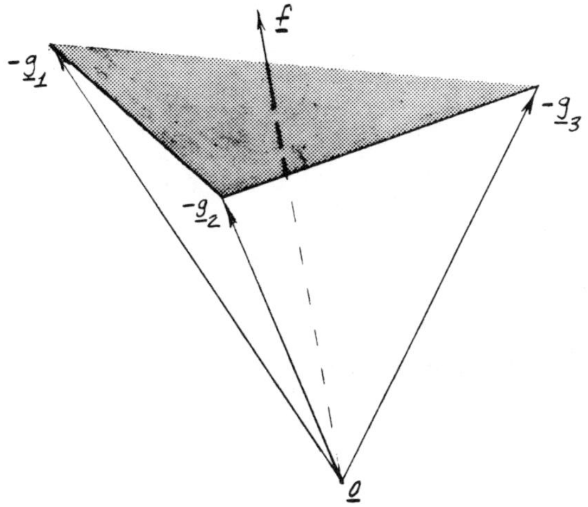

Figure 2-4 Geometrical interpretation of the Farkas lemma with dim $\boldsymbol{\lambda} = 3$.

and hence

$$L_{\mathbf{c}} = f_{\mathbf{c}} + \bar{\mathbf{g}}_{\mathbf{c}}'\bar{\boldsymbol{\lambda}}.$$

Now let curves $\boldsymbol{\gamma}(\alpha)$ be specified in accordance with Definition 2-11. Then functions

$$\bar{\mathbf{h}}(\alpha) = \begin{cases} \dfrac{1}{\alpha}\bar{\mathbf{g}}[\boldsymbol{\gamma}(\alpha)] & \text{for} \quad \alpha \in (-\epsilon, 0) \cup (0, \epsilon) \\ \bar{\mathbf{g}}_{\mathbf{c}}\mathbf{d} & \text{for} \quad \alpha = 0 \end{cases}$$

are continuous at $\alpha = 0$ because $\bar{\mathbf{g}}(\mathbf{c}) = \mathbf{0}$, $\bar{\mathbf{g}}$ is differentiable at $\mathbf{c}$, and $\boldsymbol{\gamma}$ is continuous at $\alpha = 0$. Moreover, curves $\boldsymbol{\gamma}(\alpha)$ are constructed on $(0, \epsilon)$ so that $\mathbf{g}[\boldsymbol{\gamma}(\alpha)] \leqslant \mathbf{0}$ and hence

$$\bar{\mathbf{g}}_{\mathbf{c}}\mathbf{d} \leqslant \mathbf{0} \tag{2-51}$$

holds for all $\mathbf{d} \in A(\mathbf{c})$. By Theorem 2-11, condition $\mathbf{d}'f_{\mathbf{c}} \geqslant 0$ for all $\mathbf{d} \in A(\mathbf{c})$ also holds at a relative minimum of f in F. Thus, by Proposition 2-9, there exists a real $\bar{\boldsymbol{\lambda}} \geqslant \mathbf{0}$ and hence $\boldsymbol{\lambda} \geqslant \mathbf{0}$ such that $L_{\mathbf{c}} = \mathbf{0}$ at a relative minimum.

In other words, the first necessary condition for a relative minimum which is constrained by inequality constraints is $f_{\mathbf{c}} \in C(-\bar{\mathbf{g}}_{\mathbf{c}})$. A geometrical interpretation of this form of the Lagrange multiplier rule for inequality constraints is given in Figure 2-5.

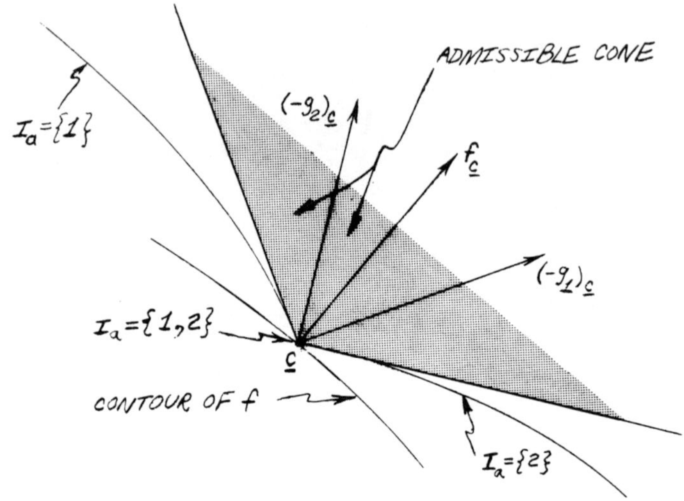

Figure 2-5 Geometrical interpretation of Lagrange multiplier rule for inequality constraints.

As in the case of equality constraints, perhaps the most useful aspects of the Lagrangian for inequality constraints concerns additional conditions for relative minima. For example, the second necessary condition for a constrained relative minimum becomes

Theorem 2-13 *Suppose*

i) F is the feasible set of $\mathbf{g}$ *and* $\overline{T}(\mathbf{c})$ *is the tangent space of* $\overline{\mathbf{g}}$ *at* $\mathbf{c} \in F$

ii) f and $\mathbf{g}$ *are twice continuously differentiable at* $\mathbf{c}$

and

iii) $L_{\mathbf{c}} = \mathbf{0}$ *for some real* $\boldsymbol{\lambda} \geqslant \mathbf{0}$ *with* $\lambda_i g_i(\mathbf{c}) = 0$ *for each i.*

Then $f(\mathbf{c})$ *is a relative minimum of f in F only if* $\mathbf{d}' L_{\mathbf{cc}} \mathbf{d} \geqslant 0$ *holds for all* $\mathbf{d} \in \overline{T}(\mathbf{c})$.

Proof The result follows from the same construction used in the proof of Theorem 2-8.

In addition, a sufficiency theorem for strict relative minima which are constrained by inequality constraints is stated in terms of *effective* constraints

$$\hat{\mathbf{g}} = [g_i] \quad \text{with} \quad i \in I_e \tag{2-52}$$

where the index set of effective constraints at $\mathbf{c}$ is defined as

$$I_e = \{i : \lambda_i > 0\}. \tag{2-53}$$

Specifically,

Theorem 2-14 *Suppose*

i) F *is the feasible set of* $\mathbf{g}$ *and* $\hat{T}(\mathbf{c})$ *is the tangent space of* $\hat{\mathbf{g}}$ *at* $\mathbf{c} \in F$

ii) f *and* $\mathbf{g}$ *are twice continuously differentiable at* $\mathbf{c}$

iii) $L_{\mathbf{c}} = \mathbf{0}$ *for some real* $\boldsymbol{\lambda} \geqslant \mathbf{0}$ *with* $\lambda_i g_i(\mathbf{c}) = 0$ *for each* i

and

iv) $\mathbf{d}' L_{\mathbf{cc}} \mathbf{d} \geqslant 0$ *holds for all* $\mathbf{d} \in \hat{T}(\mathbf{c})$ *and matrix*

$$\begin{bmatrix} L_{\mathbf{cc}} & \hat{\mathbf{g}}_{\mathbf{c}}' \\ \hat{\mathbf{g}}_{\mathbf{c}} & \mathbf{0} \end{bmatrix}$$

is nonsingular.

Then $f(\mathbf{c})$ *is a strict relative minimum of* f *in* F.

Proof Let the direction vector $\mathbf{d}$ be nonzero with $\mathbf{d} \in A(\mathbf{c})$ but $\mathbf{d} \notin \hat{T}(\mathbf{c})$. Then

$$\boldsymbol{\lambda}' \mathbf{g}_{\mathbf{c}} \mathbf{d} = \hat{\boldsymbol{\lambda}}' \hat{\mathbf{g}}_{\mathbf{c}} \mathbf{d} < 0$$

so that $\mathbf{d}' L_{\mathbf{c}} = 0$ implies

$$\mathbf{d}' f_{\mathbf{c}} > 0.$$

Thus, given any such direction vector, there exists corresponding curves $\boldsymbol{\gamma}(\alpha) \in F$ and a $\delta > 0$ such that

$$f[\boldsymbol{\gamma}(\alpha)] > f(\mathbf{c})$$

for all $\alpha \in (0, \delta)$. Let curves $\boldsymbol{\gamma}(\alpha)$ again be specified as in Definition 2-11 but now with nonzero direction vectors $\mathbf{d} \in \hat{T}(\mathbf{c})$, and let

$$l(\alpha) = \begin{Bmatrix} \dfrac{L[\boldsymbol{\gamma}(\alpha), \boldsymbol{\lambda}] - L(\mathbf{c}, \boldsymbol{\lambda})}{\alpha^2} & \text{for} \quad \alpha \in (-\delta, 0) \cup (0, \delta) \\ \tfrac{1}{2} \mathbf{d}' L_{\mathbf{cc}} \mathbf{d} & \text{for} \quad \alpha = 0. \end{Bmatrix}.$$

Then $l(\alpha)$ is continuous at $\alpha = 0$ because L is twice continuously differentiable at $\mathbf{c}$ and $L_{\mathbf{c}} = \mathbf{0}$. Also, by Lemma 2-3, $\mathbf{d}' L_{\mathbf{cc}} \mathbf{d} > 0$ so that

$$L[\boldsymbol{\gamma}(\alpha), \boldsymbol{\lambda}] > L(\mathbf{c}, \boldsymbol{\lambda})$$

and hence

$$f[\boldsymbol{\gamma}(\alpha)] > f(\mathbf{c})$$

for all $\alpha \neq 0$ on some neighborhood of 0.

Effective constraints are of significance in this sufficiency condition for inequality constraints whereas active constraints are of significance in the corresponding sufficiency condition given for equality constraints.

An example of these necessary and sufficient conditions is provided by

$$f(\mathbf{c}) = \frac{c_1^2}{2} + \frac{c_1 c_2}{4} + \frac{c_2^2}{8} \quad \text{and} \quad \mathbf{g}(\mathbf{c}) = \begin{bmatrix} (1 - c_1) \\ \left(1 - c_1 - \frac{c_2}{4}\right) \end{bmatrix}.$$

The Lagrangian for this example becomes

$$L = \frac{c_1^2}{2} + \frac{c_1 c_2}{4} + \frac{c_2^2}{8} + \lambda_1(1 - c_1) + \lambda_2\left(1 - c_1 - \frac{c_2}{4}\right),$$

and conditions $L_\mathbf{c} = \mathbf{0}$, $\mathbf{g}(\mathbf{c}) \leqslant \mathbf{0}$, and $\boldsymbol{\lambda} \geqslant \mathbf{0}$ yield

$$\mathbf{c} = \begin{bmatrix} 1 \\ 0 \end{bmatrix} \quad \text{and} \quad \boldsymbol{\lambda} = \begin{bmatrix} 0 \\ 1 \end{bmatrix}.$$

In addition, tangent space $\overline{T}(\mathbf{c}) = \{\mathbf{0}\}$ so that the necessary condition given in Theorem 2-13 is satisfied trivially. However, tangent space $\hat{T}(\mathbf{c})$ has dimension one so that additional computations must be performed in order to verify that Theorem 2-14 applies. Hessian

$$L_{\mathbf{cc}} = \begin{bmatrix} 1 & \frac{1}{4} \\ \frac{1}{4} & \frac{1}{4} \end{bmatrix}$$

is positive semi-definite and elementary calculations yield

$$\det \left[\begin{array}{cc|c} 1 & \frac{1}{4} & -1 \\ \frac{1}{4} & \frac{1}{4} & -\frac{1}{4} \\ \hline -1 & -\frac{1}{4} & 0 \end{array}\right] = -\tfrac{3}{16}$$

so that a strict relative minimum has been found.

Conditions for absolute minima constrained by inequality constraints can also be stated conveniently in terms of Lagrangian $L = L(\mathbf{c}, \boldsymbol{\lambda})$ as defined in (2-18). The main result is stated as

Theorem 2-15 *Suppose F is the feasible set of* $\mathbf{g}$, *and suppose f and* $\mathbf{g}$ *are differentiable and convex on a convex set $S \subset F$. Then $f(\mathbf{c})$ is an absolute minimum on S if and only if $L_\mathbf{c} = \mathbf{0}$ for some real $\boldsymbol{\lambda} \geqslant \mathbf{0}$ with $\lambda_i g_i(\mathbf{c}) = 0$ for* *each i. If in addition f is strictly convex on S, then the absolute minimum is strict.*

Proof Necessity of $L_{\mathbf{c}} = \mathbf{0}$ for some real $\boldsymbol{\lambda} \geqslant \mathbf{0}$ with $\lambda_i g_i(\mathbf{c}) = 0$ for each i has already been established in Theorem 2-12. In order to prove sufficiency, note that

$$f(\boldsymbol{\gamma}) \geqslant f(\mathbf{c}) + f_{\mathbf{c}}'(\boldsymbol{\gamma} - \mathbf{c})$$

and

$$\mathbf{g}(\boldsymbol{\gamma}) \geqslant \mathbf{g}(\mathbf{c}) + \mathbf{g}_{\mathbf{c}}(\boldsymbol{\gamma} - \mathbf{c})$$

hold for all $\boldsymbol{\gamma} \in S$. Because $\boldsymbol{\lambda} \geqslant \mathbf{0}$, condition

$$L(\boldsymbol{\gamma}, \boldsymbol{\lambda}) \geqslant L(\mathbf{c}, \boldsymbol{\lambda}) + (\boldsymbol{\gamma} - \mathbf{c})' L_{\mathbf{c}}$$

also holds for all $\boldsymbol{\gamma} \in S$. Because $L_{\mathbf{c}} = \mathbf{0}$ and $\lambda_i g_i(\mathbf{c}) = 0$ for each i, the condition

$$f(\boldsymbol{\gamma}) \geqslant f(\mathbf{c}) - \boldsymbol{\lambda}' \mathbf{g}(\boldsymbol{\gamma}) \geqslant f(\mathbf{c})$$

holds for all $\boldsymbol{\gamma} \in S$ and hence $f(\mathbf{c})$ is an absolute minimum of f in S. Moreover, the equality holds only if $\boldsymbol{\gamma} = \mathbf{c}$ when f is strictly convex on S.

This theorem differs from Theorem 2-10 for equality constraints in that constraint functions $\mathbf{g}$ not only must result in a convex subset of F but now must also be convex.

2.10 SUMMARY

Conditions for relative and absolute minima have been selected and presented here in a form that is most useful for subsequent computational purposes. Stationarity and convexity are found to be necessary conditions for unconstrained relative minima. Moreover, stationarity and convexity are found to be sufficient conditions at nondegenerate points for unconstrained relative minima.

These basic conditions for unconstrained relative minima are extended to equality and inequality constraints by the introduction of appropriate Lagrangians. The Lagrange multiplier formulation and the concepts of active constraints and effective constraints permit a unified approach to relative minima of problems simultaneously having both equality and inequality constraints. This unified treatment of constraints is introduced in the context of computational methods given in the next chapter.

Conditions for absolute minima have also been given in this chapter. In this context, global properties are introduced in the form of convex functions. These conditions for absolute minima are included primarily for the purposes of identifying the structure of automated design problems which occur frequently in control.

BIBLIOGRAPHY

1. Fleming, W. H.: *Functions of Several Variables*, Addison-Wesley Publishing Company, Inc., Reading, 1965.
2. Apostol, T. M.: *Calculus*, Vol. I, Blaisdell Publishing Company, Waltham, 1967.
3. Ostrowski, A. M.: *Solution of Linear Equations and Systems of Equations*, Academic Press, New York, 1966.
4. Eggleston, H. G.: *Convexity*, Cambridge University Press, New York, 1958.
5. Hoffman, K. and R. Kunze: *Linear Algebra*, Prentice-Hall, Inc., Englewood Cliffs, 1961.
6. Rudin, W.: *Principles of Mathematical Analysis*, McGraw-Hill Book Company, Inc., New York, 1964.
7. Kuhn, H. W. and A. W. Tucker: *Nonlinear Programming*, Proceedings of the Second Berkeley Symposium on Mathematical Statistics and Probability, University of California Press, Berkeley, 1951.
8. McCormick, G. P.: "Second order conditions for constrained minima", *SIAM J. Appl. Math.*, **15**, No. 3, May, 1967.
9. Gale, D.: *The Theory of Linear Economic Models*, McGraw-Hill Book Company, Inc., New York, 1960.

PROBLEMS

2-1 Let f be defined on $\mathscr{R}^2$ as

$$f(\mathbf{c}) = (1 - c_1)^2 + (c_1^2 - c_2)^2.$$

a) Sketch contours of f.
b) Locate all stationary points of f.
c) Locate all relative minima if there are any.
d) Locate the absolute minimum if there is one.

2-2 Repeat Problem 2-1 for

$$f(\mathbf{c}) = 2c_1^2 - c_2(c_2 - 1)^2.$$

2-3 Repeat Problem 2-1 for

$$f(\mathbf{c}) = (c_1 + c_2)^2 - 2c_1.$$

2-4 Repeat Problem 2-1 for

$$f(\mathbf{c}) = c_1(c_1 - 1)c_2^2.$$

2-5

a) Give appropriate definitions of absolute and relative maxima.
b) Restate Theorems 2-1, 2-2, and 2-3 for relative maxima.

2-6 Locate all stationary points of

$$f(c) = c_1^2 + c_2^2 - c_3^2$$

in $\mathscr{R}^3$ and locate all relative minima and maxima.

2-7 Repeat Problem 2-6 for

$$f(c) = c_1 + c_1^3 - 4c_1c_2 - 2c_2^2 + c_3^2.$$

2-8 Let

$$f(\mathbf{c}) = f_0 + \mathbf{c}'\mathbf{f} + \tfrac{1}{2}\mathbf{c}'\mathbf{F}\mathbf{c}$$

be defined on $\mathscr{R}^n$ with F symmetric.
a) Show that every relative minima of f in $\mathscr{R}^n$ is also the absolute minimum of f on $\mathscr{R}^n$.
b) Show that the absolute minimum of f in $\mathscr{R}^n$ is strict when the point is not degenerate and locate all such minima.

2-9 Given points $\mathbf{c}_1, \mathbf{c}_2, \ldots, \mathbf{c}_m$, locate and evaluate the minimum value of

$$f(\mathbf{c}) = \sum_{i=1}^{m} (\mathbf{c} - \mathbf{c}_i)'(\mathbf{c} - \mathbf{c}_i).$$

2-10 Given time functions $g(t)$ and $\mathbf{f}(t)$, locate and evaluate the minimum value of

$$f(\mathbf{c}) = \int_0^1 [g(t) - \mathbf{c}'\mathbf{f}(t)]^2 \, dt$$

under the assumption that elements of $\mathbf{f}(t)$ are continuous and linearly independent on [0, 1].

2-11 Given time functions $\mathbf{z}(t)$ and constant matrices F, G, and H, locate and evaluate the minimum value of

$$f(\mathbf{c}) = \int_0^1 [\mathbf{z}(t) - \mathbf{y}(t)]'[\mathbf{z}(t) - \mathbf{y}(t)] \, dt,$$

where

$$\mathbf{y} = \mathbf{H}\mathbf{x} \quad \text{and} \quad \dot{\mathbf{x}} = \mathbf{F}\mathbf{x}; \; \mathbf{x}(0) = \mathbf{G}\mathbf{c}.$$

Assume the pair {H, F} is completely observable, and assume rank G = dim c.

2-12 Show that the intersection of two convex sets is convex but that the union of two convex sets may not be convex.

2-13 Consider norms

$$\|\mathbf{f}\|_p = \left[\sum_{i=1}^{n} |f_i|^p\right]^{1/p}$$

which are defined on $\mathscr{R}^n$ with $p \geqslant 1$. Sketch sets

$$\{\mathbf{f} : \|\mathbf{f}\|_p \leqslant \delta\}$$

for $p = 1, 2$, and ∞ with $n = 2$.

2-14 Prove that the neighborhood defined in (2-1) is convex.

2-15 Prove that the set $\{\mathbf{c} : \mathbf{G}\mathbf{c} + \mathbf{g} = \mathbf{0}\}$ is convex.

2-16 Prove that a norm defined on $\mathscr{R}^n$ is convex on $\mathscr{R}^n$.

2-17 Suppose every element of g(c) is convex on a convex set S and every element of $\boldsymbol{\lambda}$ is nonnegative. Prove that $\boldsymbol{\lambda}'\mathbf{g}(\mathbf{c})$ is convex on S.

2-18 Suppose $g(\mathbf{c})$ is convex on a convex set S and $h(g)$ is convex and nondecreasing on an interval I where

$$g(\mathbf{c}) \in I \Leftrightarrow \mathbf{c} \in S.$$

Prove that

$$f(\mathbf{c}) = h[g(\mathbf{c})]$$

is convex on S.

2-19 Determine the largest convex set S on which

$$f(\mathbf{c}) = -e^{-(\mathbf{c}'\mathbf{c})/2}$$

is convex.

2-20 Repeat Problem 2-19 for

$$f(\mathbf{c}) = (1 + \mathbf{c}'\mathbf{c})^{(\mathbf{c}'\mathbf{c})}.$$

2-21 Suppose

$$f(\mathbf{a}, \mathbf{b}) = \tfrac{1}{2}\mathbf{a}'\mathbf{P}\mathbf{a} + \mathbf{b}'\mathbf{Q}\mathbf{a} + \tfrac{1}{2}\mathbf{b}'\mathbf{R}\mathbf{b}$$

with **P** and **R** symmetric is known to be strictly convex with respect to **b** and also nonnegative.

a) What restrictions have been imposed on **P**, **Q**, and **R** by these conditions?

b) Must the function f be (strictly) convex with respect to **a** under these conditions?

2-22 Suppose f is differentiable on a convex set $S \subset \mathscr{R}^1$. Prove that

a) f is convex on S if and only if $\dot{f}$ is nondecreasing on S

b) f is strictly convex on S if and only if $\dot{f}$ is increasing.

2-23 Prove Proposition 2-6.

2-24 Locate and evaluate all relative minima of

$$f(\mathbf{c}) = c_1^2 + 4c_2^2 - 4c_1$$

subject to equality constraint

$$\mathbf{g}(\mathbf{c}) = [(2c_2 - c_1 - 12)].$$

2-25 Repeat Problem 2-24 for

$$f(\mathbf{c}) = \tfrac{1}{2}\left[\left(\frac{c_1}{a_1}\right)^2 + \left(\frac{c_2}{a_2}\right)^2\right]$$

and

$$\mathbf{g}(\mathbf{c}) = [(c_1 + a_3c_2 - a_4)].$$

2-26 Repeat Problem 2-24 for

$$f(\mathbf{c}) = c_1^2 + c_2^2$$

and

$$\mathbf{g}(\mathbf{c}) = [(5 - c_1 - c_2^2)].$$

2-27 Repeat Problem 2-24 for

$$f(\mathbf{c}) = -(c_1 + c_2)$$

and

$$\mathbf{g}(\mathbf{c}) = [(c_1^2 + c_2^2 - 2)].$$

2-28 Repeat Problem 2-24 for

$$f(\mathbf{c}) = \frac{1}{2}\left[\left(\frac{c_1}{a_1}\right)^2 + \left(\frac{c_2}{a_2}\right)^2\right]$$

and

$$\mathbf{g}(\mathbf{c}) = [(a_3 - c_1 c_2)].$$

2-29 Repeat Problem 2-24 for

$$f(\mathbf{c}) = 8c_1c_2c_3$$

and

$$\mathbf{g}(\mathbf{c}) = [(a_1c_1^2 + a_2c_2^2 + a_3c_3^2 - 1)].$$

2-30 Repeat Problem 2-24 for

$$f(\mathbf{c}) = c_1^2 + c_2^2 + c_3^2$$

and

$$\mathbf{g}(\mathbf{c}) = \begin{bmatrix} (c_1 + 2c_2 - 3c_3 - 10) \\ (c_1 - c_2 + 2c_3 - 1) \end{bmatrix}.$$

2-31 Restate Theorem's 2-7, 2-8, and 2-9 for relative maxima.

2-32 Suppose

$$f(\mathbf{c}) = \tfrac{1}{2}\mathbf{a}'\mathbf{F}^{-1}\mathbf{a} + \mathbf{c}'\mathbf{a} + \tfrac{1}{2}\mathbf{c}'\mathbf{F}\mathbf{c}$$

is to be minimized subject to equality constraints

$$\mathbf{G}\mathbf{c} + \mathbf{b} = \mathbf{0}.$$

Assume $\mathbf{F}$ is positive definite and det $(\mathbf{G}\mathbf{G}') \neq 0$. Let f_{min} denote the minimum value of f without constraints, and let f_* denote the minimum value of f achievable with constraints.
a) Find f_{min} and the corresponding value of $\mathbf{c}$ in terms of $\mathbf{a}$.
b) Find f_* and the corresponding values of $\mathbf{c}$ and Lagrange multipliers $\boldsymbol{\lambda}$ in terms of $\mathbf{a}$ and $\mathbf{b}$.
c) Give a condition for $f_* > f_{min}$ and give a significance to the case where $f_* = f_{min}$.

2-33 Under the conditions of Lemma 2-3, show that $\mathbf{H}$ is nonsingular if $\mathbf{F}$ is nonsingular.

2-34 Show that matrix $\mathbf{H}$ defined in Lemma 2-3 is nonsingular only if $(\mathbf{F}^2 + \mathbf{G}'\mathbf{G})$ and $(\mathbf{G}\mathbf{G}')$ are nonsingular.

2-35 Summarize the results of Section 2.8 for problems where $\mathbf{b}(\mathbf{a})$ is defined by constraint functions

$$\mathbf{g}(\mathbf{a}, \mathbf{b}) = \mathbf{F}(\mathbf{a})\mathbf{b} + \mathbf{f}(\mathbf{a}).$$

2-36 Sketch the feasible set for inequality constraints

$$\mathbf{g}(\mathbf{c}) = \begin{bmatrix} -c_1 \\ -c_2 \\ c_2 - (1 - c_1)^3 \end{bmatrix}.$$

2-37 Minimize

$$f(\mathbf{c}) = 2c_1 + c_1^2 - c_2^2$$

subject to inequality constraints

$$\mathbf{g}(\mathbf{c}) = [(c_1^2 + c_2^2 - 1)].$$

2-38 Repeat Problem 2-37 for

$$f(\mathbf{c}) = 4c_1c_2 - 10c_1^2 - 5c_2c_3 - 3c_2^2$$

and

$$\mathbf{g}(\mathbf{c}) = \begin{bmatrix} (c_1 + 2c_2 - 3) \\ (2 - c_2 + c_3) \\ (1 - c_1) \end{bmatrix}.$$

2-39 Repeat Problem 2-37 for

$$f(\mathbf{c}) = 3c_1 + c_2$$

and

$$\mathbf{g}(\mathbf{c}) = \begin{bmatrix} (14 - 3c_1 - c_2) \\ (4 + 2c_2 - 5c_1) \\ -c_1 \\ -c_2 \end{bmatrix}.$$

2-40 Restate Theorems 2-12, 2-13, and 2-14 for relative maxima.

2-41 Find values of $a > 0$ for which $\mathbf{c} = \mathbf{0}$ yields a relative minimum and a relative maximum of

$$f(\mathbf{c}) = (x_1 - 1)^2 + x_2^2$$

subject to inequality constraint

$$\mathbf{g}(\mathbf{c}) = [(ac_1 - c_2^2)].$$

2-42 Show that stationarity conditions for minimizing

$$f(\mathbf{c}) = \mathbf{a}'\mathbf{c}$$

subject to

$$\mathbf{F}'\mathbf{c} + \mathbf{f} \leqslant \mathbf{0}$$

are equivalent to stationarity conditions for maximizing

$$g(\boldsymbol{\lambda}) = \mathbf{f}'\boldsymbol{\lambda}$$

subject to

$$\mathbf{F}\boldsymbol{\lambda} + \mathbf{a} = \mathbf{0} \quad \text{and} \quad \boldsymbol{\lambda} \geqslant \mathbf{0}.$$

CHAPTER 3

Gradient methods

Even relatively simple parameter optimization problems arising in control are tedious if not intractable when solved by purely analytical methods. In addition, the usefulness of optimization techniques is primarily attributed to applications of considerable complexity which are computational by their very nature. Practicability of automated design techniques thus depends directly upon the efficiency and flexibility of available computational methods for locating relative minima.

This chapter is intended to be an introduction to computational methods for locating relative minima by successive approximations. However, only methods based on gradient calculations are discussed here because performance functionals and required partial derivatives are given analytically for most problems arising in control. Gradient methods first are developed for unconstrained minima and then are extended to equality and inequality constraints. These methods are, in one form or another, computer implementations of the sufficiency conditions given in Chapter 2 for relative minima.

3.1 SUCCESSIVE APPROXIMATIONS FOR UNCONSTRAINED MINIMA

The problem of locating relative minima of a function $f(\mathbf{c})$ by computational means is introduced in this section. The intent here is to identify the basic ingredients of gradient methods. These basic ingredients are found to be a procedure for selecting direction vectors along which one-dimensional searches are performed and a procedure for searches in one dimension using gradient information.

By way of introduction, suppose parameter vector $\mathbf{c}$ is replaced by a parameterized curve $\boldsymbol{\Upsilon}(t)$ so that a composite function can be defined as

$$h(t) = f[\boldsymbol{\Upsilon}(t)]. \tag{3-1}$$

If f and $\boldsymbol{\Upsilon}$ are differentiable, then derivative

$$\dot{h}(t) = f'_{\boldsymbol{\Upsilon}}\dot{\boldsymbol{\Upsilon}} \tag{3-2}$$

also is defined. Now suppose that $\boldsymbol{\gamma}(t)$ has been selected so that $\dot{h}(t) < 0$ for all $f_\gamma \neq \mathbf{0}$ and $\dot{h}(t) = 0$ for $f_\gamma = \mathbf{0}$. Then such selections of $\boldsymbol{\gamma}(t)$ constitute a search procedure for locating a relative minimum of f. For example, the curve $\boldsymbol{\gamma}(t)$ defined by

$$\dot{\boldsymbol{\gamma}} = -\boldsymbol{\Gamma} f_\gamma ; \boldsymbol{\gamma}(0) = \boldsymbol{\gamma}_0 \tag{3-3}$$

yields

$$\dot{h}(t) = -f_\gamma' \boldsymbol{\Gamma} f_\gamma \tag{3-4}$$

and hence has the prescribed properties given any positive definite matrix $\boldsymbol{\Gamma} = \boldsymbol{\Gamma}(\boldsymbol{\gamma})$. Selection $\boldsymbol{\Gamma} = \mathbf{I}$ corresponds to the *steepest-descent* method, whereas selection $\boldsymbol{\Gamma} = f_{\gamma\gamma}^{-1}$ for some strictly convex and twice continuously differentiable functions corresponds to the *Newton–Raphson* procedure.

Analytical or analog computer solutions of (3-3) can, of course, be used to locate relative minima of f. For instance, suppose f is given by

$$f(\boldsymbol{\gamma}) = f_0 + \boldsymbol{\gamma}'\mathbf{f} + \tfrac{1}{2}\boldsymbol{\gamma}'\mathbf{F}\boldsymbol{\gamma} \tag{3-5}$$

with $\mathbf{F}$ positive definite. Then differential equations given in (3-4) become

$$\dot{\boldsymbol{\gamma}} = -\boldsymbol{\Gamma}\mathbf{F}\boldsymbol{\gamma} - \boldsymbol{\Gamma}\mathbf{f} ; \boldsymbol{\gamma}(0) = \boldsymbol{\gamma}_0 . \tag{3-6}$$

Moreover, solution of these equations in the form of (1-47) is found for constant $\boldsymbol{\Gamma}$ to be

$$\boldsymbol{\gamma} - \boldsymbol{\gamma}_e = e^{-\boldsymbol{\Gamma}\mathbf{F}t}(\boldsymbol{\gamma}_0 - \boldsymbol{\gamma}_e) \tag{3-7}$$

where equilibrium point $\boldsymbol{\gamma}_e$ is given by

$$\boldsymbol{\gamma}_e = -\mathbf{F}^{-1}\mathbf{f}. \tag{3-8}$$

Matrix $(\boldsymbol{\Gamma}\mathbf{F})$ has only positive real eigenvalues, and hence this system of differential equations is asymptotically stable so that norm bounding in the form of (1-33) yields

$$\| \boldsymbol{\gamma} - \boldsymbol{\gamma}_e \| \leqslant \Gamma e^{-\gamma t} \| \boldsymbol{\gamma}_0 - \boldsymbol{\gamma}_e \| . \tag{3-9}$$

In other words,

$$\lim_{t \to \infty} \boldsymbol{\gamma}(t) = \boldsymbol{\gamma}_e \;\; \forall \boldsymbol{\gamma}_0$$

holds so that the equilibrium solution of (3-3) locates the minimum of f.

Even though all solutions of (3-6) share the equilibrium point $\boldsymbol{\gamma}_e$, considerable difficulty may be encountered in obtaining an analog computer solution. Such difficulties arise when there is a large spread in the eigenvalues of $(\boldsymbol{\Gamma}\mathbf{F})$. For example, suppose $\boldsymbol{\gamma}_0$ occurs so that

$$(\boldsymbol{\Gamma}\mathbf{F})(\boldsymbol{\gamma}_0 - \boldsymbol{\gamma}_e) = \lambda_{\max}(\boldsymbol{\gamma}_0 - \boldsymbol{\gamma}_e)$$

where $\lambda_{\max}$ denotes the maximum eigenvalue of $(\mathbf{\Gamma F})$. Then (3-7) yields

$$\boldsymbol{\gamma} - \boldsymbol{\gamma}_e = e^{-\lambda_{\max} t}(\boldsymbol{\gamma}_0 - \boldsymbol{\gamma}_e).$$

On the other hand, suppose $\boldsymbol{\gamma}_0$ occurs so that

$$(\mathbf{\Gamma F})(\boldsymbol{\gamma}_0 - \boldsymbol{\gamma}_e) = \lambda_{\min}(\boldsymbol{\gamma}_0 - \boldsymbol{\gamma}_e)$$

where $\lambda_{\min}$ denotes the minimum eigenvalue of $(\mathbf{\Gamma F})$. Then (3-7) now yields

$$\boldsymbol{\gamma} - \boldsymbol{\gamma}_e = e^{-\lambda_{\min} t}(\boldsymbol{\gamma}_0 - \boldsymbol{\gamma}_e).$$

When $\lambda_{\min}$ and $\lambda_{\max}$ differ by sufficient orders of magnitude, time scaling for analog computer solutions becomes impossible. Thus, judicious choices of $\mathbf{\Gamma}$ are an important consideration in using (3-3) to locate relative minima. For instance, the selection of $\mathbf{\Gamma} = \mathbf{F}^{-1}$ for strictly convex functions yields $\lambda_{\min} = \lambda_{\max} = 1$ and

$$\boldsymbol{\gamma} - \boldsymbol{\gamma}_e = e^{-t}(\boldsymbol{\gamma}_0 - \boldsymbol{\gamma}_e)$$

holds for all $\boldsymbol{\gamma}_0$.

Recourse to the digital computer inevitably will be required in order to locate relative minima of f when f and $f_{\boldsymbol{\gamma}}$ are complicated functions of $\boldsymbol{\gamma}$. One approach using the digital computer is the numerical integration of (3-3). Suppose

$$\boldsymbol{\gamma}(t) = \mathbf{c} \quad \text{and} \quad \boldsymbol{\gamma}(t + \alpha) = \mathbf{c} + \alpha\mathbf{d} \tag{3-10}$$

are used to define $\mathbf{c}$, $\mathbf{d}$, and α. Then the first difference approximation to $\dot{\boldsymbol{\gamma}}$ yields

$$\frac{\boldsymbol{\gamma}(t + \alpha) - \boldsymbol{\gamma}(t)}{\alpha} = \mathbf{d}$$

and hence (3-3) is replaced by

$$\mathbf{d} = -\mathbf{\Gamma} f_{\mathbf{c}}. \tag{3-11}$$

Repeated use of this first difference approximation presumably produces an estimate of $\boldsymbol{\gamma}_e$ starting from an initial estimate $\boldsymbol{\gamma}_0$. In order to achieve computational efficiency, parameter $\alpha > 0$ is chosen to be as large as possible subject to restrictions such as

$$f(\mathbf{c} + \alpha\mathbf{d}) < f(\mathbf{c}). \tag{3-12}$$

Suitable values of α are in fact dependent upon selections of $\mathbf{\Gamma}$, and considerable difficulty may be encountered in finding such values of α.

By way of example, suppose the function given in (3-5) is evaluated along direction vector $\mathbf{d}$ as

$$h(\alpha) = f(\mathbf{c} + \alpha\mathbf{d}) \tag{3-13}$$

and α_* is defined as a value of α for which h has a relative minimum. Specifically, value α_* is defined by

$$\dot{h}(\alpha_*) = 0 \quad \text{and} \quad \ddot{h}(\alpha_*) \geqslant 0. \tag{3-14}$$

Then elementary manipulations for this example yield

$$h(\alpha) = f(\mathbf{c}) + \alpha \mathbf{d}' f_{\mathbf{c}} + \frac{\alpha^2}{2} \mathbf{d}'\mathbf{F}\mathbf{d} \tag{3-15}$$

and hence

$$\alpha_* = \frac{-\mathbf{d}' f_{\mathbf{c}}}{\mathbf{d}'\mathbf{F}\mathbf{d}}. \tag{3-16}$$

In addition, the direction vector given in (3-11) yields

$$\alpha_* = \frac{f_{\mathbf{c}}' \mathbf{\Gamma} f_{\mathbf{c}}}{f_{\mathbf{c}}'(\mathbf{\Gamma}\mathbf{F}\mathbf{\Gamma}) f_{\mathbf{c}}}. \tag{3-17}$$

If the Newton–Raphson method is used with $\mathbf{\Gamma} = \mathbf{F}^{-1}$, then $\alpha_* = 1$ and $\mathbf{c} + \alpha_* \mathbf{d} = \boldsymbol{\gamma}_e$ for all $\mathbf{c}$. On the other hand, if the steepest-descent method is used with $\mathbf{\Gamma} = \mathbf{I}$, then upper and lower bounds on α_* can be determined in terms of inequalities of the form

$$\lambda_{\min}\{\mathbf{F}\} \leqslant \frac{\mathbf{f}'\mathbf{F}\mathbf{f}}{\mathbf{f}'\mathbf{f}} \leqslant \lambda_{\max}\{\mathbf{F}\} \tag{3-18}$$

which hold for all $\mathbf{f} \in \mathscr{R}^n$ given any positive semi-definite matrix $\mathbf{F}$. In particular, the minimizing value of α along direction vector $\mathbf{d}$ is bounded as

$$\frac{1}{\lambda_{\max}\{\mathbf{F}\}} \leqslant \alpha_* \leqslant \frac{1}{\lambda_{\min}\{\mathbf{F}\}} \tag{3-19}$$

for steepest–descent. Trial-and-error determination of α_*, such as required for nonquadratic functions, thus may require a search over a very large range of α unless $\mathbf{\Gamma}$ is chosen judiciously.

The approach introduced here for the minimization of a function by successive approximations can be summarized with the aid of Figure 3-1 which defines an *iteration* of the procedure. Corresponding to any point $\mathbf{c}$, a direction vector $\mathbf{d}$ is computed on each iteration and is used to form the function h of a single variable α that is defined in (3-13). Methods for computing direction vectors, such as steepest-descent and Newton–Raphson, and properties of these methods are one of the two major considerations in the construction of efficient algorithms. Also, a one-dimensional search is performed on each iteration in order to locate a relative minimum of h, and the minimizing value of step-size α must satisfy (3-14). Methods for computing step-size is the other major consideration in the construction of efficient algorithms. Although these two major considerations are somewhat interrelated, they can be and are discussed independently in the sequel.

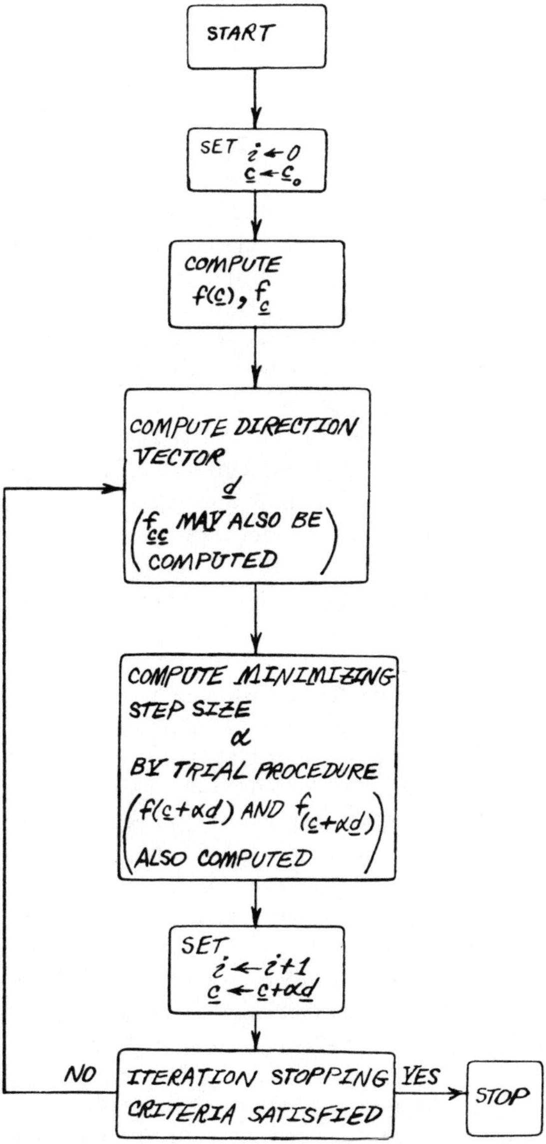

Figure 3-1 Computational flow chart for general iteration procedure.

3.2 GRADIENT SEARCH PROCEDURES FOR FUNCTIONS OF ONE VARIABLE

Perhaps the most difficult aspect of constructing an efficient algorithm arises with one-dimensional search procedures for $h(\alpha)$ which is defined in (3-13). These search procedures are intended to find a value of α, namely α_*, that corresponds to a relative minimum of h and hence satisfies (3-14). There are, of course, many approaches to search in one dimension [1, 2], and these approaches differ primarily in accordance with assumptions made about $h(\alpha)$.

Assumptions that are appropriate for most control problems discussed in this book are summarized briefly as follows. Function h is assumed to be twice continuously differentiable. Moreover, relative minima of h are assumed not to be degenerate so that $\ddot{h}(\alpha_*) > 0$ and hence h is strictly convex on some neighborhood of α_*. Selections of direction vector **d** are always made so that $\dot{h}(0) < 0$ and hence only searches on interval $(0, \infty)$ need be performed in order to satisfy (3-12). Moreover, selections of direction vector **d** are always made so that $h(\alpha) \to \infty$ as $\alpha \to \infty$ and hence interval $(0, \infty)$ contains at least one point at which h has a relati[ve] minimum. However, no finite upper bound on α_* can be assigned *a priori* so that the search cannot be confined to a definite interval at the outset. On each trial, function $f(\boldsymbol{\gamma})$ and gradient $f_{\boldsymbol{\gamma}}$ can be computed with relative economy but $f_{\boldsymbol{\gamma}\boldsymbol{\gamma}}$ cannot be computed due to excessive costs.

These assumptions suggest an approach to one-dimensional search problems that can be summarized with the aid of Figure 3-2. This flow chart defines a *trial* of the procedure used to determine an estimate α_+ of α_* defined by (3-14). This trial procedure is initialized with an estimate α_0 and is organized so that $\dot{h}(\beta) < 0$ on every trial. In keeping with the assumptions of h being well-defined on $(0, \infty)$ and strictly convex on some suitably large neighborhood of α_*, the procedure continues the search for some $\alpha_+ > \beta$. Each new trial point α then results in one of three possibilities. Estimate $\alpha_+ = \alpha$ corresponds to condition $\dot{h}(\alpha) = 0$ and trials would then be terminated. On the other hand, an estimate $\alpha_+ < \alpha$ corresponds to condition $\dot{h}(\alpha) > 0$, and α_+ would be generated by some method of interpolation. Finally, an estimate $\alpha_+ > \alpha$ corresponds to condition $\dot{h}(\alpha) < 0$, and α_+ would be generated by some method of extrapolation. In order to account for the fact that h is not always well-defined on $(0, \infty)$ such as arises with control system instability, interval length $(\alpha - \beta)$ is decreased by some factor $r_{\min} \in (0, 1)$ whenever $f(\boldsymbol{\gamma}) > f_{\max}$ and then an additional trial is performed.

The major topics of extrapolation, interpolation, and initialization which are indicated in Figure 3-2 are discussed subsequently in this section. The reader should bear in mind, however, that the assumptions which pertain to this

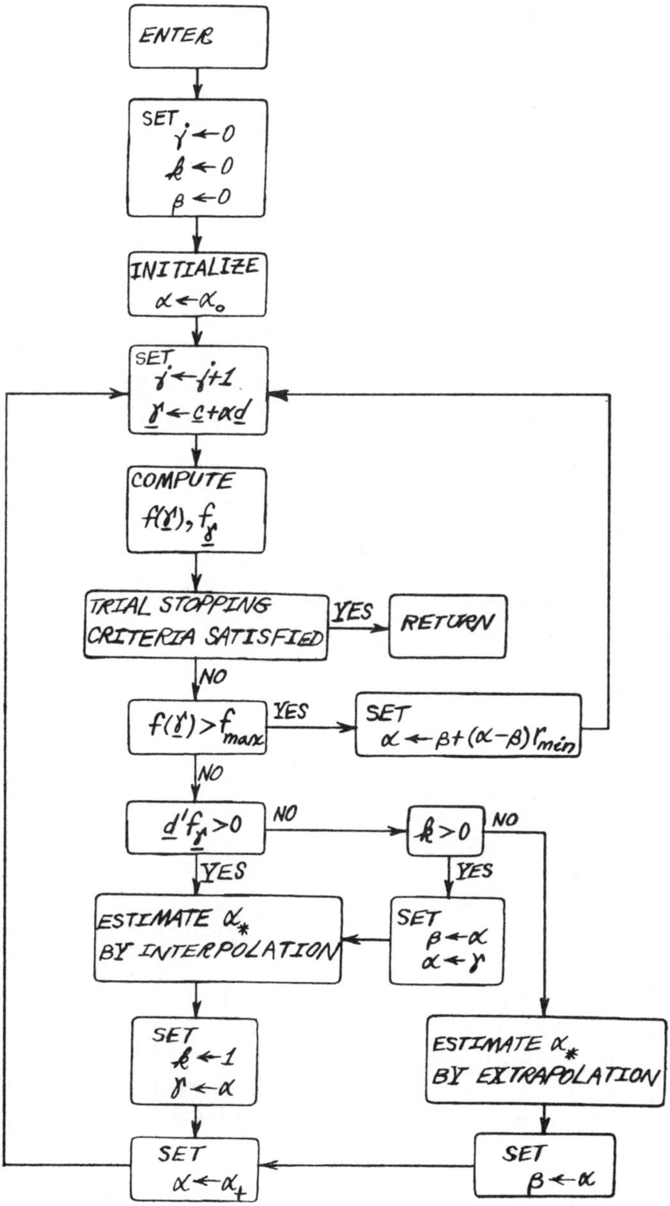

Figure 3-2 Computational flow chart for general trial procedure.

procedure do not apply to all functions encountered in control system optimization. In particular, function h may not be strictly convex on any given subinterval of $(0, \infty)$. Therefore, any complete implementation of a trial procedure involves additional diagnostic and stopping considerations which are not explicitly discussed in this book.

A number of extrapolation methods for estimating α_* can be employed with the trial procedure depicted in Figure 3-2. An elementary method involves increasing interval length $(\alpha - \beta)$ by some factor $r_{\max} \in (1, \infty)$. In particular, parameter

$$\alpha_+ = \beta + (\alpha - \beta) r_{\max}$$

can be used as a crudely constructed estimate of α_*. Somewhat more sophisticated methods of extrapolation can also be constructed using data that has already been computed for the function $h(a) = f(\mathbf{c} + a\mathbf{d})$. For instance, the quadratic extrapolating function [3]

$$g(a) = h(\beta) + \dot{h}(\beta)(a - \beta) + \frac{\dot{h}(\alpha) - \dot{h}(\beta)}{2(\alpha - \beta)}(a - \beta)^2 \tag{3-20}$$

can be used for typical geometry depicted in Figure 3-3 under the assumption $\dot{h}(\beta) < \dot{h}(\alpha) \leqslant 0$. The condition $\dot{g}(a) = 0$ yields

$$\alpha_+ = \alpha + (\alpha - \beta)\frac{\dot{h}(\beta)}{\dot{h}(\beta) - \dot{h}(\alpha)}.$$

An extrapolation method which combines both of these approaches has been found to be effective. Specifically, the extrapolated estimate of α_* is formed as

$$\alpha_+ = \beta + (\alpha - \beta) r, \tag{3-21}$$

where

$$r = \begin{cases} \dfrac{\dot{h}(\beta)}{\dot{h}(\beta) - \dot{h}(\alpha)} & \text{for} \quad \dot{h}(\alpha) \geqslant \left(1 - \dfrac{1}{r_{\max}}\right)\dot{h}(\beta) \\[2ex] r_{\max} & \text{for} \quad \dot{h}(\alpha) < \left(1 - \dfrac{1}{r_{\max}}\right)\dot{h}(\beta) \end{cases} \tag{3-22}$$

and $r_{\max} \in (1, \infty)$.

A number of interpolating methods for estimating α_* can also be employed with the trial procedure depicted in Figure 3-2. For instance, the cubic inter-

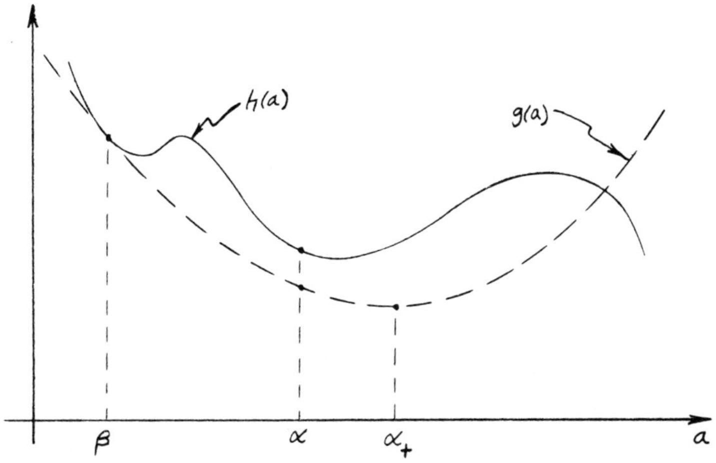

Figure 3-3 Geometry of α_* estimation by quadratic extrapolation.

polating function [4]

$$g(a) = h(\beta) + \dot{h}(\beta)(a - \beta) - \frac{z + \dot{h}(\beta)}{(\alpha - \beta)}(a - \beta)^2 + \frac{\dot{h}(\beta) + \dot{h}(\alpha) + 2z}{3(\alpha - \beta)^2}(a - \beta)^3 \tag{3-23}$$

where

$$z = 3\,\frac{h(\beta) - h(\alpha)}{(\alpha - \beta)} + \dot{h}(\beta) + \dot{h}(\alpha) \tag{3-24}$$

can be used for typical geometry depicted in Figure 3-4 under assumptions $\dot{h}(\beta) < 0$ and $\dot{h}(\alpha) > 0$. Condition $\dot{g}(a) = 0$ then yields estimate α_+ in the form of (3-21) with $r \in (0, 1)$ being given by

$$r = \begin{cases} \dfrac{\dot{h}(\beta) + z + q}{\dot{h}(\beta) + \dot{h}(\alpha) + 2z} & \text{for} \quad \dot{h}(\beta) + \dot{h}(\alpha) + 2z \neq 0 \\[2ex] \dfrac{\dot{h}(\beta)}{\dot{h}(\beta) - \dot{h}(\alpha)} & \text{for} \quad \dot{h}(\beta) + \dot{h}(\alpha) + 2z = 0 \end{cases} \tag{3-25}$$

where

$$q = [z^2 - \dot{h}(\beta)\dot{h}(\alpha)]^{1/2}. \tag{3-26}$$

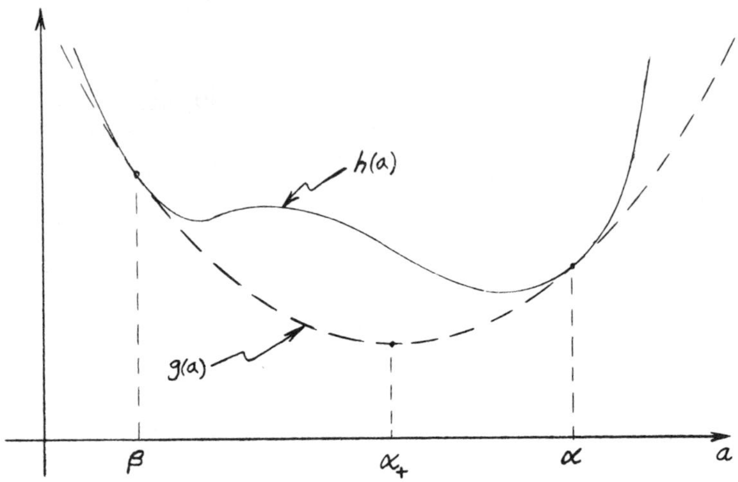

Figure 3-4 Geometry of α_* estimation by cubic interpolation.

This estimate has been found to be effective and corresponds to the relative minimum of $g(a)$ occurring in (β, α) because

$$\ddot{g}(\alpha_+) = \begin{cases} \dfrac{2q}{(\alpha - \beta)} & \text{for} \quad \dot{h}(\beta) + \dot{h}(\alpha) + 2z \neq 0 \\[2ex] \dfrac{\dot{h}(\alpha) - \dot{h}(\beta)}{(\alpha - \beta)} & \text{for} \quad \dot{h}(\beta) + \dot{h}(\alpha) + 2z = 0 \end{cases} \tag{3-27}$$

demonstrates that $\ddot{g}(\alpha_+) > 0$.

Initialization of parameter α perhaps is the most troublesome aspect of the trial procedure depicted in Figure 3-2. Initialization primarily is intended to establish the order of magnitude of α_*. Hence, initial estimate α_0 can be taken as

$$\alpha_0 = \begin{cases} \alpha_{max} & \text{for} \quad \alpha_{max} \leqslant \alpha_+ \\ \alpha_+ & \text{for} \quad \alpha_{min} < \alpha_+ < \alpha_{max} \\ \alpha_{min} & \text{for} \quad \alpha_+ \leqslant \alpha_{min} \end{cases}, \tag{3-28}$$

where α_+ is again some estimate of α_*. For instance, quadratic extrapolating function

$$g(a) = f(\mathbf{c}) + (\mathbf{d}'f_\mathbf{c})a + \tfrac{1}{2}(\mathbf{d}'\mathbf{\Gamma}^{-1}\mathbf{d})a^2 \tag{3-29}$$

can be used and is reasonably effective when $\mathbf{\Gamma}^{-1} \cong f_{\mathbf{cc}}$. Condition $\dot{g}(a) = 0$ then yields

$$\alpha_+ = \frac{-\mathbf{d}' f_{\mathbf{c}}}{\mathbf{d}' \mathbf{\Gamma}^{-1} \mathbf{d}} \tag{3-30}$$

so that

$$g(\alpha_+) = f(\mathbf{c}) + \frac{\mathbf{d}' f_{\mathbf{c}}}{2} \alpha_+ .$$

Alternatively, estimate α_+ can also be computed from the expression [4]

$$\alpha_+ = 2 \frac{g(\alpha_+) - f(\mathbf{c})}{\mathbf{d}' f_{\mathbf{c}}} \tag{3-31}$$

where now $g(\alpha_+)$ is taken from the previous iteration as determined from the previous extrapolation or interpolation.

Normal termination of the trial procedure depicted in Figure 3-2 is usually based on achieving the condition

$$\left| \frac{\mathbf{d}' f_{\Upsilon}}{\mathbf{d}' f_{\mathbf{c}}} \right| < R \tag{3-32}$$

where R is chosen to be a suitably small number. Whether or not a normal termination will be achieved, however, cannot be guaranteed *a priori* for this procedure or any other known procedure under the assumptions invoked here.

3.3 STEEPEST-DESCENT METHOD

The steepest-descent method is based on using direction vector $\mathbf{d} = -f_{\mathbf{c}}$ and, by selection of α, establishing condition $\mathbf{d}' f_{(\mathbf{c}+\alpha\mathbf{d})} = 0$ on each iteration. This method has the inherent advantages of having minimal computer memory requirements and hessian $f_{\mathbf{cc}}$ need not be computed. However, even assuming the trial procedure being used converges to a relative minimum of $f(\mathbf{c} + \alpha\mathbf{d})$ and $f(\mathbf{c} + \alpha\mathbf{d}) \leqslant f(\mathbf{c})$ occurs on every iteration, questions arise concerning what can be said about convergence.

By way of introduction, consider the quadratic function

$$f(\mathbf{c}) = \tfrac{1}{2} \mathbf{c}' \mathbf{F} \mathbf{c}, \tag{3-33}$$

where $\mathbf{F}$ is positive definite. The direction vector for steepest-descent then becomes $\mathbf{d} = -\mathbf{F}\mathbf{c}$, and substitution into (3-33) yields

$$f(\mathbf{c} + \alpha\mathbf{d}) = f(\mathbf{c}) - \alpha \mathbf{d}' \mathbf{d} + \frac{\alpha^2}{2} \mathbf{d}' \mathbf{F} \mathbf{d}.$$

Condition

$$\frac{\partial f(\mathbf{c} + \alpha\mathbf{d})}{\partial \alpha} = 0$$

results in

$$\alpha = \frac{\mathbf{d}'\mathbf{d}}{\mathbf{d}'\mathbf{F}\mathbf{d}}$$

and $f(\mathbf{c} + \alpha\mathbf{d})$ also has been minimized with respect to α. Moreover, manipulation of these relationships yields

$$\frac{f(\mathbf{c}) - f(\mathbf{c} + \alpha\mathbf{d})}{f(\mathbf{c})} = \frac{(\mathbf{d}'\mathbf{d})^2}{(\mathbf{d}'\mathbf{F}\mathbf{d})(\mathbf{d}'\mathbf{F}^{-1}\mathbf{d})}$$

for steepest-descent. Inequalities given in (3-18) for positive definite matrices then yield

$$\frac{f(\mathbf{c}) - f(\mathbf{c} + \alpha\mathbf{d})}{f(\mathbf{c})} \geqslant \frac{\lambda_{\min}\{\mathbf{F}\}}{\lambda_{\max}\{\mathbf{F}\}}$$

or rather

$$\frac{f(\mathbf{c} + \alpha\mathbf{d})}{f(\mathbf{c})} \leqslant 1 - \frac{\lambda_{\min}\{\mathbf{F}\}}{\lambda_{\max}\{\mathbf{F}\}}.$$

Sequence $\{f(\mathbf{c}_i)\}$ corresponding to the steepest-descent method for the function given in (3-33) thus satisfies

$$f(\mathbf{c}_i) \leqslant \left[1 - \frac{\lambda_{\min}\{\mathbf{F}\}}{\lambda_{\max}\{\mathbf{F}\}}\right]^i f(\mathbf{c}_0) \tag{3-34}$$

for $i = 0, 1, \ldots$, given any point $\mathbf{c}_0 \in \mathscr{R}^n$. Moreover, inequality

$$|\mathbf{c}_i|^2 \leqslant \frac{f(\mathbf{c}_i)}{\lambda_{\min}\{\mathbf{F}\}} \tag{3-35}$$

is obtained by use of (3-18) and (3-33). Therefore, sequence $\{\mathbf{c}_i\}$ converges to $\mathbf{0}$ and the steepest-descent method is seen to be convergent for strictly convex quadratic functions.

The steepest-descent method also converges for a rather general class of nonquadratic functions. The main result for steepest-descent is stated as

Theorem 3-1 *Let S be any open convex set containing* $\{\mathbf{c} : f(\mathbf{c}) \leqslant f(\mathbf{c}_0)\}$ *on which*

i) f is bounded from below and is twice continuously differentiable and

ii) there exist constants m, M > 0 *such that* $\lambda_{\min}\{f_{\mathbf{cc}}\} \geqslant m$ *and* $\lambda_{\max}\{f_{\mathbf{cc}}\} \leqslant M$. *Then sequence* $\{\mathbf{c}_i\}$ *generated by the steepest-descent method converges to a point* $\mathbf{c}_\infty \in S$ *with* $|\mathbf{c}_\infty| < \infty$, *and* $f(\mathbf{c}_\infty)$ *is the strict absolute minimum of f on S.*

Proof Let h be defined as

$$h(\alpha) = f(\mathbf{c} - \alpha f_{\mathbf{c}})$$

with $\mathbf{c}$, $(\mathbf{c} - \alpha f_{\mathbf{c}}) \in S$ so that Taylor's theorem applied to $\dot{h}(\alpha)$ yields

$$\dot{h}(\alpha) = -f_{\mathbf{c}}' f_{\mathbf{c}} + \alpha f_{\mathbf{c}}' f_{\boldsymbol{\gamma}\boldsymbol{\gamma}} f_{\mathbf{c}},$$

where $\boldsymbol{\gamma} = \mathbf{c} - \theta f_{\mathbf{c}}$ and $\theta \in (0, \alpha)$. Then condition $\dot{h}(\alpha_*) = 0$ yields

$$\alpha_* = \frac{f_{\mathbf{c}}' f_{\mathbf{c}}}{f_{\mathbf{c}}' f_{\boldsymbol{\gamma}\boldsymbol{\gamma}} f_{\mathbf{c}}}$$

and hence

$$\frac{1}{M} \leqslant \alpha_* \leqslant \frac{1}{m}$$

holds. Moreover, Taylor's theorem applied to $h(\alpha)$ yields

$$f(\mathbf{c} - \alpha_* f_{\mathbf{c}}) \leqslant f(\mathbf{c}) - \frac{\alpha_*^2}{2} f_{\mathbf{c}}' f_{\mathbf{c}}$$

so that steepest-descent results in nonincreasing sequences $\{f(\mathbf{c}_i)\}$ with $\mathbf{c}_i \in \{\mathbf{c}: f(\mathbf{c}) \leqslant f(\mathbf{c}_0)\}$ for each i. Also, each member of $\{f(\mathbf{c}_i)\}$ is bounded from below so that this sequence converges to some value f_∞. Now suppose that the corresponding sequence $\{f_{\mathbf{c}_i}\}$ does not converge to $\mathbf{0}$. Then there is a number $\epsilon > 0$ such that, given any $I > 0$,

$$|f_{\mathbf{c}_i}| \geqslant \epsilon$$

holds for some $i > I$. Therefore, I can always be chosen suitably large so that

$$|f(\mathbf{c}_i) - f_\infty| \leqslant \frac{\epsilon^2}{4M} \quad \forall\, i > I.$$

Use of Taylor's theorem now yields

$$f(\mathbf{c} - \alpha_* f_{\mathbf{c}}) \leqslant f(\mathbf{c} - \frac{1}{M} f_{\mathbf{c}}) \leqslant f(\mathbf{c}) - \frac{1}{2M} |f_{\mathbf{c}}|^2$$

so that

$$f(\mathbf{c}_{i+1}) \leqslant f(\mathbf{c}_i) - \frac{\epsilon^2}{2M}$$

and hence

$$f(\mathbf{c}_{i+1}) \leqslant f_\infty - \frac{\epsilon^2}{4M}.$$

In other words, $f(\mathbf{c}_{i+1}) < f_\infty$ for some i which is impossible by construction. Sequence $\{f_{\mathbf{c}i}\}$ thus converges to $\mathbf{0}$, and sequence $\{\mathbf{c}_i\}$ converges to the point $\mathbf{c}_\infty \in \{\mathbf{c}: f(\mathbf{c}) \leqslant f(\mathbf{c}_0)\}$ at which $f_{\mathbf{c}_\infty} = \mathbf{0}$ by Theorem 2-5 because f is strictly convex on S. Moreover, set $\{\mathbf{c} : f(\mathbf{c}) \leqslant f(\mathbf{c}_0)\}$ is bounded because

$$f(\mathbf{c}) \geqslant f(\mathbf{c}_0) + (\mathbf{c} - \mathbf{c}_0)' f_{\mathbf{c}_0} + m \mid \mathbf{c} - \mathbf{c}_0 \mid^2$$

and hence $| \mathbf{c}_\infty | < \infty$. Finally, by Theorem 2-5, $f(\mathbf{c}_\infty)$ is the absolute minimum of f on S.

For many functions in addition to those to which Theorem 3-1 applies, steepest-descent is a surprisingly effective method of successive approximations. Steepest-descent has the primary advantages that computations required to determine direction vector $\mathbf{d}$ are minimal and hence computer memory requirements are minimal. In particular, hessian $f_{\mathbf{cc}}$ is not required. Trial procedures required for successful use with steepest-descent, however, must be relatively complex and hence entail a relatively complicated and lengthy computer program. Moreover, convergence rates achieved in $\mathbf{c}_n \to \mathbf{c}_\infty$ as $n \to \infty$ may be relatively slow and hence many iterations may be required in order to achieve suitable accuracy. Various alternatives to steepest-descent are examined in the next three sections in regard to these disadvantages of steepest-descent.

3.4 SUCCESSIVE SUBSTITUTIONS AND THE NEWTON-RAPHSON METHOD

In order to investigate the possibility of eliminating the rather complicated trial procedure discussed in Section 3.2, vector function $\mathbf{h}(\mathbf{c})$ is introduced as

$$\mathbf{h}(\mathbf{c}) = \mathbf{c} + \alpha \mathbf{d}, \tag{3-36}$$

where parameter α is some fixed constant in $(0, \infty)$ and direction vector $\mathbf{d}$ is some appropriate function of $\mathbf{c}$ such as given in (3-11) with $\mathbf{\Gamma} = \mathbf{\Gamma}(\mathbf{c})$. Computation of the sequence $\{\mathbf{c}_i\}$ that is now formed by *successive substitutions* as

$$\mathbf{c}_{i+1} = \mathbf{h}(\mathbf{c}_i) \quad \text{for} \quad i = 0, 1, \ldots \tag{3-37}$$

can be considered to be an alternate method of locating a relative minimum under some conditions.

Convergence properties of sequences generated by successive substitutions are summarized in very general form by the well-known contraction mapping theorem which is given here as

Theorem 3-2 *Let S be any subset of* $\mathscr{R}^n$ *for which*

i) $\mathbf{c} \in S \Rightarrow \mathbf{h}(\mathbf{c}) \in S$

and

ii) $\bar{\mathbf{c}}, \hat{\mathbf{c}} \in S \Rightarrow \| \mathbf{h}(\bar{\mathbf{c}}) - \mathbf{h}(\hat{\mathbf{c}}) \| \leqslant \beta \| \bar{\mathbf{c}} - \hat{\mathbf{c}} \|$ *with* $\beta \in [0, 1)$.

Then there exists a unique point $\mathbf{c}_\infty \in S$ *such that* $\mathbf{c}_\infty = \mathbf{h}(\mathbf{c}_\infty)$, *and sequence* $\{\mathbf{c}_i\}$ *corresponding to (3-37) converges to the point* $\mathbf{c}_\infty$ *given any* $\mathbf{c}_0 \in S$.

Proof Let $\mathbf{c}_i$ correspond to any $\mathbf{c}_0 \in S$ in accordance with (3-37). Then, given any integer $j \geqslant 1$, norm bounding and assumptions of the theorem yield

$$\| \mathbf{c}_{i+j} - \mathbf{c}_i \| \leqslant \sum_{l=1}^{j} \| \mathbf{c}_{i+l} - \mathbf{c}_{i+l-1} \| \leqslant \left[\sum_{l=1}^{j} \beta^{(i+l-1)} \right] \| \mathbf{c}_1 - \mathbf{c}_0 \|.$$

Thus, inequality

$$\| \mathbf{c}_{i+j} - \mathbf{c}_i \| \leqslant \frac{\beta^i}{1-\beta} \| \mathbf{c}_1 - \mathbf{c}_0 \|$$

is obtained for $i = 0, 1, \ldots$ with $\beta \in [0, 1)$ so that $\{\mathbf{c}_i\}$ is a Cauchy sequence and hence is convergent to some point $\mathbf{c}_\infty \in S$. Moreover, similar manipulations yield

$$\begin{aligned} \| \mathbf{c}_\infty - \mathbf{h}(\mathbf{c}_\infty) \| &\leqslant \| \mathbf{c}_\infty - \mathbf{c}_i \| + \| \mathbf{h}(\mathbf{c}_{i-1}) - \mathbf{h}(\mathbf{c}_\infty) \| \\ &\leqslant \| \mathbf{c}_\infty - \mathbf{c}_i \| + \beta \| \mathbf{c}_{i-1} - \mathbf{c}_\infty \| \end{aligned}$$

and hence $\| \mathbf{c}_\infty - \mathbf{h}(\mathbf{c}_\infty) \| = 0$ or rather $\mathbf{c}_\infty = \mathbf{h}(\mathbf{c}_\infty)$ because $\mathbf{c}_i, \mathbf{c}_{i-1} \to \mathbf{c}_\infty$ as $n \to \infty$. Now suppose that $\bar{\mathbf{c}}_\infty$ and $\hat{\mathbf{c}}_\infty$ are distinct values of $\mathbf{c}_\infty$. Then inequalities

$$\| \bar{\mathbf{c}}_\infty - \hat{\mathbf{c}}_\infty \| \leqslant \beta \| \bar{\mathbf{c}}_\infty - \hat{\mathbf{c}}_\infty \| < \| \bar{\mathbf{c}}_\infty - \hat{\mathbf{c}}_\infty \|$$

must be satisfied which is impossible.

Sequences which converge in accordance with Theorem 3-2 are said to converge *linearly* on S because

$$\frac{\|\mathbf{c}_{i+1} - \mathbf{c}_\infty\|}{\|\mathbf{c}_i - \mathbf{c}_\infty\|} \leqslant \beta < 1 \tag{3-38}$$

holds for all $i \geqslant 0$ and all $\mathbf{c}_0 \in S$.

Preconditions i) and ii) of Theorem 3-2 generally are very difficult to verify in advance of actually computing sequences $\{\mathbf{c}_i\}$. For instance, consider the constant β in the following context. Let scalar function $e(a)$ be defined on $[0, 1]$ as

$$e(a) = \mathbf{f}'[\mathbf{h}(\mathbf{c} + a\boldsymbol{\delta}) - \mathbf{h}(\mathbf{c})] \tag{3-39}$$

where $\mathbf{c}$, $\mathbf{c} + \mathbf{d} \in S$ and $\mathbf{f}$ is any vector belonging to $\mathscr{R}^n$ for which

$$\mathbf{f}'[\mathbf{h}(\mathbf{c} + \boldsymbol{\delta}) - \mathbf{h}(\mathbf{c})] = \|\mathbf{f}\|\,\mathbf{h}(\mathbf{c} + \boldsymbol{\delta}) - \mathbf{h}(\mathbf{c})\|.$$

Then, assuming $\mathbf{h}$ is continuously differentiable on S and S is convex, the first degree Taylor series of $e(1)$ about $e(0)$ yields

$$\mathbf{f}'[\mathbf{h}(\mathbf{c} + \boldsymbol{\delta}) - \mathbf{h}(\mathbf{c})] = \mathbf{f}'\mathbf{h}_{\boldsymbol{\gamma}}\boldsymbol{\delta}$$

where $\boldsymbol{\gamma} \in (\mathbf{c}, \mathbf{c} + \boldsymbol{\delta})$. Moreover, by construction and norm bounding, inequality

$$\|\mathbf{f}\|\;\|\mathbf{h}(\mathbf{c} + \boldsymbol{\delta}) - \mathbf{h}(\mathbf{c})\| \leqslant \|\mathbf{f}\|\;\|\mathbf{h}_{\boldsymbol{\gamma}}\|\;\|\boldsymbol{\delta}\|$$

holds for each $\mathbf{c}$, $\mathbf{c} + \boldsymbol{\delta} \in S$, and hence inequality

$$\|\mathbf{h}(\mathbf{c} + \boldsymbol{\delta}) - \mathbf{h}(\mathbf{c})\| \leqslant \beta \|\boldsymbol{\delta}\|$$

holds for all $\mathbf{c}$, $\mathbf{c} + \boldsymbol{\delta} \in S$ when

$$\beta = \sup_{\mathbf{c} \in S} \|\mathbf{h}_{\mathbf{c}}\|. \tag{3-40}$$

Verification that $\beta < 1$ for some S on which precondition i) of Theorem 3-2 also holds is thus seen to be very difficult indeed.

Perhaps the most frequently used method of successive substitutions is Newton–Raphson. This method is defined for twice continuously differentiable functions f with a nonsingular hessian $f_{\mathbf{cc}}$ as

$$\mathbf{h}(\mathbf{c}) = \mathbf{c} - f_{\mathbf{cc}}^{-1} f_{\mathbf{c}}. \tag{3-41}$$

This method has some obvious advantages for locating a stationary point of f in any set S to which Theorem 3-2 applies. First, limit point $\mathbf{c}_\infty$ of course yields $f_{\mathbf{c}_\infty} = \mathbf{0}$ because $\mathbf{c}_\infty = \mathbf{h}(\mathbf{c}_\infty)$. In addition, direct computation yields $\mathbf{h}_{\mathbf{c}_\infty} = \mathbf{0}$ so that there is a suitably small neighborhood N of any nondegenerate stationary point of f for which (3-40) yields $\beta < 1$. If in addition $\mathbf{c} \in N \Rightarrow \mathbf{h}(\mathbf{c}) \in N$, then linear convergence is obtained in accordance with Theorem 3-2.

Perhaps the most attractive feature of the Newton–Raphson method of successive substitution concerns rate of convergence which is in fact much faster than linear under some conditions. In particular, suppose $\mathbf{h}(\mathbf{c})$ defined in (3-41) is twice continuously differentiable and converges in accordance with Theorem 3-2 on some closed, bounded, and convex set S. The second degree Taylor series expansion of $e(1)$ about $e(0)$, as defined by (3-39) with $\mathbf{c} = \mathbf{c}_\infty$, then yields

$$\|\mathbf{h}(\mathbf{c}) - \mathbf{c}_\infty\| \leqslant \gamma \|\mathbf{c} - \mathbf{c}_\infty\|^2$$

because $\mathbf{h}_{\mathbf{c}_\infty} = \mathbf{0}$. In other words, *quadratic convergence* on S, as defined by

$$\frac{\|\mathbf{c}_{i+1} - \mathbf{c}_\infty\|}{\|\mathbf{c}_i - \mathbf{c}_\infty\|^2} \leqslant \gamma < \infty, \tag{3-42}$$

is obtained. Of course, the price paid for quadratic convergence is the necessity of computing hessian $f_{\mathbf{cc}}$ on each iteration.

Convergence in accordance with Theorem 3-2 unfortunately does not guarantee that the stationary point which is located by successive substitutions is also a relative minimum of f on S. On the other hand, if f is known *a priori* to be convex on some open convex set containing S, the $f(\mathbf{c}_\infty)$ corresponding to convergence in accordance with Theorem 3-2 is the absolute minimum of f on this open set. If, in addition, f is four times continuously differentiable on this set and the Newton–Raphson method is used, successive substitution without the need of any trial procedure yields quadratic convergence.

3.5 CONJUGATE GRADIENT METHOD OF CONJUGATE DIRECTIONS

Steepest-descent, with α adjusted so that $f_\mathbf{c}' f_{(\mathbf{c}-\alpha f_\mathbf{c})} = 0$ holds on each iteration, yields linear convergence for strictly convex quadratic functions. A number of modifications to steepest-descent, which do not require direct computation of hessian $f_{\mathbf{cc}}$, can be introduced for the purpose of obtaining finite step convergence with strictly convex quadratic functions. These methods are also employed, as a means of successive approximations using the trial procedure discussed in Section 4.2, for nonquadratic functions to which Proposition 2-6 applies. These methods presumably yield convergence rates that are much faster than linear even though not quadratic once $\mathbf{c}$ belongs to some suitably small neighborhood of $\mathbf{c}_\infty$. Two examples of these methods are the conjugate gradient method of conjugate directions discussed in this section and the Fletcher–Powell method of conjugate directions discussed in the next section.

The method of conjugate gradients for function minimization is based on relationships derived explicitly for the quadratic function

$$f(\mathbf{c}) = f(\mathbf{b}) + \tfrac{1}{2}(\mathbf{c} - \mathbf{b})'\mathbf{F}(\mathbf{c} - \mathbf{b}), \tag{3-43}$$

where $\mathbf{F}$ is positive definite and $\mathbf{b}$ is an arbitrary element of $\mathscr{R}^n$. The derivation is facilitated by introduction of error vector

$$\mathbf{e} = -f_{\mathbf{c}}. \tag{3-44}$$

Iterations are performed in accordance with

$$\mathbf{c}_{i+1} = \mathbf{c}_i + \alpha_i \mathbf{d}_i \quad \text{for} \quad i = 0, 1, \ldots \tag{3-45}$$

given any $\mathbf{c}_0 \in \mathscr{R}^n$. Parameter α_i satisfies

$$\frac{\partial f(\mathbf{c}_i + \alpha_i \mathbf{d}_i)}{\partial \alpha_i} = 0 \tag{3-46}$$

on each iteration. This condition yields

$$\mathbf{d}_i'\mathbf{F}(\mathbf{c}_i + \alpha_i \mathbf{d}_i - \mathbf{b}) = -\mathbf{d}_i'\mathbf{e}_i + \alpha_i \mathbf{d}_i'\mathbf{F}\mathbf{d}_i = 0$$

for the quadratic function given in (3-43) where

$$\mathbf{e}_i = -\mathbf{F}(\mathbf{c}_i - \mathbf{b}).$$

Parameter

$$\alpha_i = \frac{\mathbf{d}_i'\mathbf{e}_i}{\mathbf{d}_i'\mathbf{F}\mathbf{d}_i} \tag{3-47}$$

hence is obtained for a quadratic function. Correspondingly, error vector

$$\mathbf{e}_{i+1} = \mathbf{e}_i - \alpha_i \mathbf{F}\mathbf{d}_i \tag{3-48}$$

is obtained from (3-44) for this quadratic function.

The notion of conjugacy is an extension of orthogonality for vectors. This notion is defined in the following fundamental result for conjugate directions; namely

Lemma 3-1 *Let* $\{\mathbf{d}_0, \mathbf{d}_1, \ldots\}$ *be a set of nonzero vectors for which*

$$i \neq j \Rightarrow \mathbf{d}_i'\mathbf{F}\mathbf{d}_j = 0$$

given some positive definite matrix $\mathbf{F}$. *Then this set of vectors, which is said to be* conjugate *with respect to* $\mathbf{F}$, *is linearly independent.*

Proof Suppose $\{\mathbf{d}_0, \mathbf{d}_1, \ldots\}$ is conjugate with respect to $\mathbf{F}$ and also linearly dependent. Then there exists a corresponding set of constants $\{\beta_0, \beta_1, \ldots\}$, not all of which are zero, such that

$$\mathbf{0} = \beta_0 \mathbf{d}_0 + \beta_1 \mathbf{d}_1 + \ldots.$$

Premultiplication by $\mathbf{d}_i'\mathbf{F}$ then yields

$$0 = \beta_i \mathbf{d}_i' \mathbf{F} \mathbf{d}_i$$

so that $\beta_i = 0$ for each i. This result contradicts the assumption of linear dependence.

The significance of conjugate directions for the minimization of quadratic functions is stated as

Lemma 3-2 *Suppose* $\{\mathbf{d}_0, \mathbf{d}_1, \ldots, \mathbf{d}_{n-1}\}$ *is conjugate with respect to* $\mathbf{F}$. *Then (3-45) and (3-47) yield* $\mathbf{c}_m = \mathbf{b}$ *for some* $m \leqslant n$ *given any* $\mathbf{c}_0 \in \mathscr{R}^n$.

Proof Let m be the smallest integer for which elements of $\{\mathbf{d}_0, \mathbf{d}_1, \ldots, \mathbf{d}_{m-1}\}$ span a subspace containing $(\mathbf{c}_0 - \mathbf{b})$. By linear independence of $\{\mathbf{d}_i\}$, such an integer exists. There also exist constants $\{\beta_0, \beta_1, \ldots, \beta_{m-1}\}$ such that

$$\mathbf{0} = (\mathbf{c}_0 - \mathbf{b}) + \sum_{j=0}^{m-1} \beta_j \mathbf{d}_j.$$

Premultiplication by $\mathbf{d}_i'\mathbf{F}$ now yields

$$0 = \mathbf{d}_i'\mathbf{F}(\mathbf{c}_0 - \mathbf{b}) + \beta_i \mathbf{d}_i'\mathbf{F}\mathbf{d}_i$$

and hence

$$\beta_i = \frac{-\mathbf{d}_i'\mathbf{F}(\mathbf{c}_0 - \mathbf{b})}{\mathbf{d}_i'\mathbf{F}\mathbf{d}_i} = \frac{-\mathbf{d}_i'\mathbf{F}(\mathbf{c}_i - \mathbf{b})}{\mathbf{d}_i'\mathbf{F}\mathbf{d}_i} = \frac{\mathbf{d}_i'\mathbf{e}_i}{\mathbf{d}_i'\mathbf{F}\mathbf{d}_i}.$$

Thus, $\alpha_i = \beta_i$ and hence $\mathbf{c}_m = \mathbf{b}$ for some $m \leqslant n$. In other words, gradient methods based on direction vectors that are conjugate with respect to $f_{\mathbf{cc}}$ result in n-step convergence for quadratic functions with $n = \dim \mathbf{c}$.

A method of generating conjugate directions with respect to $f_{\mathbf{cc}}$ for quadratic functions [5] is given in

Lemma 3-3 *Nonzero direction vectors*

$$\mathbf{d}_0 = \mathbf{e}_0 \quad \textit{and} \quad \mathbf{d}_{i+1} = \mathbf{e}_{i+1} + \beta_i \mathbf{d}_i \; \textit{for} \; i = 0, 1, \ldots \tag{3-49}$$

are conjugate with respect to **F** *when (3-45), (3-47), and*

$$\beta_i = \frac{\mathbf{e}_{i+1}'\mathbf{e}_{i+1}}{\mathbf{e}_i'\mathbf{e}_i} \tag{3-50}$$

are used.

Proof First note that (3-49) and (3-50) combine to yield

$$\mathbf{d}_i = \mathbf{e}_i'\mathbf{e}_i \left[\frac{\mathbf{e}_i}{\mathbf{e}_i'\mathbf{e}_i} + \frac{\mathbf{d}_{i-1}}{\mathbf{e}_{i-1}'\mathbf{e}_{i-1}}\right]$$

so that successive substitutions yield

$$\mathbf{d}_i = \mathbf{e}_i'\mathbf{e}_i \sum_{j=0}^{i} \frac{\mathbf{e}_j}{\mathbf{e}_j'\mathbf{e}_j}. \tag{3-51}$$

Proof of the lemma then is accomplished by showing that

$$\mathbf{d}_i'\mathbf{e}_j = \mathbf{e}_i'\mathbf{e}_i \quad \text{for} \quad i \geqslant j, \tag{3-52}$$

$$\mathbf{d}_i'\mathbf{F}\mathbf{d}_j = 0 \quad \text{for} \quad i \neq j, \tag{3-53}$$

$$\mathbf{e}_i'\mathbf{F}\mathbf{d}_i = \mathbf{d}_i'\mathbf{F}\mathbf{d}_i, \tag{3-54}$$

$$\mathbf{e}_i'\mathbf{F}\mathbf{d}_j = 0 \quad \text{for} \quad i \neq j, j+1, \tag{3-55}$$

$$\mathbf{e}_i'\mathbf{e}_j = 0 \quad \text{for} \quad i \neq j, \tag{3-56}$$

and

$$\mathbf{d}_i'\mathbf{e}_j = 0 \quad \text{for} \quad i < j \tag{3-57}$$

hold when (3-45), (3-47), (3-49), and (3-50) are used. These relationships are shown to hold by induction noting that (3-51) and (3-56) in conjunction with (3-47) yield

$$\alpha_i = \frac{\mathbf{e}_i'\mathbf{e}_i}{\mathbf{d}_i'\mathbf{F}\mathbf{d}_i}. \tag{3-58}$$

First, consider sets of nonzero vectors $\{\mathbf{d}_0\}$ and $\{\mathbf{e}_0, \mathbf{e}_1\}$. Equations (3-53) and (3-55) are clearly valid, equations (3-52) and (3-54) hold because of (3-49), and thus equations (3-56) and (3-57) hold because use of (3-48) and then (3-49) yield

$$\mathbf{e}_0'\mathbf{e}_1 = \mathbf{e}_0'\mathbf{e}_0 - \alpha_0\mathbf{e}_0'\mathbf{F}\mathbf{d}_0 = 0.$$

Second, suppose (3-52) through (3-57) hold for sets of nonzero vectors $\{\mathbf{d}_0, \mathbf{d}_1, \ldots, \mathbf{d}_{k-1}\}$ and $\{\mathbf{e}_0, \mathbf{e}_1, \ldots, \mathbf{e}_k\}$, and suppose also that nonzero vector $\mathbf{d}_k$ is added to $\{\mathbf{d}_0, \mathbf{d}_1, \ldots, \mathbf{d}_{k-1}\}$. Then equations (3-56) and (3-57) are not altered. In order to show that (3-52) and (3-53) still hold, equations

$$\mathbf{e}_i'\mathbf{d}_k = \mathbf{e}_k'\mathbf{e}_k \quad \text{for} \quad i \leqslant k \tag{3-59}$$

and

$$\mathbf{d}_i'\mathbf{F}\mathbf{d}_k = 0 \quad \text{for} \quad i < k \tag{3-60}$$

respectively must be shown to hold. In order to show that (3-54) for $i = k$ and (3-55) for $i < k - 1$ in accordance with (3-60) still hold, equation

$$\mathbf{e}_k'\mathbf{F}\mathbf{d}_i = \mathbf{d}_k'\mathbf{F}\mathbf{d}_i \quad \text{for} \quad i \leqslant k \quad \text{and} \quad i \neq k-1 \tag{3-61}$$

must be shown to hold. Equation (3-59) follows directly from (3-51) and (3-56). Equation (3-60) is established by first using (3-48) to obtain

$$\mathbf{e}_{i+1}'\mathbf{d}_k = \mathbf{e}_i'\mathbf{d}_k - \alpha_i\mathbf{d}_i'\mathbf{F}\mathbf{d}_k.$$

Then (3-59) is used twice to obtain

$$\mathbf{e}_k'\mathbf{e}_k = \mathbf{e}_k'\mathbf{e}_k - \alpha_i\mathbf{d}_i'\mathbf{F}\mathbf{d}_k$$

for $i \leqslant k - 1$. Finally, equation (3-60) is established because $\alpha_i \neq 0$ as determined from (3-58). Equation (3-61) is established by first substituting (3-49) and then using (3-60) to obtain

$$\mathbf{d}_k'\mathbf{F}\mathbf{d}_i = \mathbf{e}_k'\mathbf{F}\mathbf{d}_i + \beta_{k-1}\mathbf{d}_{k-1}'\mathbf{F}\mathbf{d}_i = \mathbf{e}_k'\mathbf{F}\mathbf{d}_i \quad \text{for} \quad i \neq k-1.$$

Third, now suppose nonzero vector $\mathbf{e}_{k+1}$ is added to $\{\mathbf{e}_0, \mathbf{e}_1, \ldots, \mathbf{e}_k\}$. Then equations (3-52), (3-53), and (3-54) are not altered. In order to show that (3-56), (3-55), and (3-57) still hold, equations

$$\mathbf{e}_i'\mathbf{e}_{k+1} = 0 \quad \text{for} \quad i \leqslant k, \tag{3-62}$$

$$\mathbf{d}_i'\mathbf{F}\mathbf{e}_{k+1} = 0 \quad \text{for} \quad i < k, \tag{3-63}$$

and

$$\mathbf{d}_i'\mathbf{e}_{k+1} = 0 \quad \text{for} \quad i \leqslant k \tag{3-64}$$

must be shown to hold. Use of (3-48) yields

$$\mathbf{e}_i'\mathbf{e}_{k+1} = \mathbf{e}_i'\mathbf{e}_k - \alpha_k e_i'\mathbf{F}\mathbf{d}_k.$$

If $i < k$, then (3-62) is seen to hold from (3-56) and (3-55) which have already been established for these indices by assumption. If $i = k$, then (3-62) is seen to hold by (3-54) and (3-58). Use of (3-62) with $i < k$ and substitution of (3-48) now yield

$$0 = \mathbf{e}_{k+1}'\mathbf{e}_{i+1} = \mathbf{e}_{k+1}'\mathbf{e}_i - \alpha_i\mathbf{e}_{k+1}'\mathbf{F}\mathbf{d}_i = -\alpha_i\mathbf{e}_{k+1}'\mathbf{F}\mathbf{d}_i.$$

Because $\alpha_i \neq 0$, equation (3-63) hence results. Finally, use of (3-51) and (3-62) yields (3-64) and the induction is complete.

The conjugate gradient method of minimizing the quadratic function given in (3-43) is stated as

Theorem 3-3 *Equations (3-45), (3-47), (3-49), and (3-50) yield* $\mathbf{c}_m = \mathbf{b}$ *for some* $m \leqslant n$ *given any* $\mathbf{c}_0 \in \mathscr{R}^n$.

Proof This result follows directly from Lemmas 3-2 and 3-3 noting that $\mathbf{d}_i = \mathbf{0}$ if and only if $\mathbf{c}_i = \mathbf{b}$ so that a sufficient number of nonzero direction vectors is always obtained.

For quadratic functions, the conjugate gradient method therefore not only results in conjugate direction vectors by virtue of (3-53) but also in orthogonal error vectors by virtue of (3-56).

An elementary example of the conjugate gradient method for quadratic functions is provided by

$$f(\mathbf{c}) = 1 - \mathbf{c}' \begin{bmatrix} 0 \\ 1 \\ 2 \end{bmatrix} + \tfrac{1}{2}\mathbf{c}' \begin{bmatrix} 1 & 1 & 1 \\ 1 & 2 & 2 \\ 1 & 2 & 3 \end{bmatrix} \mathbf{c}. \tag{3-65}$$

Typical convergence characteristics of the steepest-descent method are illustrated by Table 3-1. The steepest-descent method, of course, corresponds to the conjugate gradient method when $\beta_i = 0$ is used on each iteration. By way of comparison, typical convergence characteristics of the conjugate gradient method are illustrated in Table 3-2. Convergence is obtained in three steps for initial point $\mathbf{c}_0' = [1 \;\; 0 \;\; 0]$.

For nonquadratic functions, equations (3-45), (3-46), (3-49), and (3-50) define the *conjugate gradient method* [6] with the understanding that

$$-\frac{\partial^2 f(\mathbf{c}_i + \alpha_i \mathbf{d}_i)}{\partial \alpha_i^2} > 0 \tag{3-66}$$

holds on each iteration for the value of α_i found from (3-46). This method is used on the presumption that the twice continuously differentiable function approaches an appropriate quadratic function in a suitably small neighborhood of a relative minimum at which hessian $f_{\mathbf{cc}}$ is positive definite. However, sufficient conditions for convergence of the method with nonquadratic functions have never been derived. For all practical purposes, this method has the same computer memory requirements and computer program complexities as steepest-descent.

Table 3-1–Convergence for (3-65) using the steepest-descent method

Iteration	c_1	c_2	c_3	$f(c)$	e_1	e_2	e_3	β	d_1	d_2	d_3	α
0	1	0	0	$\frac{3}{2}$	-1	0	1	–	-1	0	1	1
1	0	0	1	$\frac{1}{2}$	-1	-1	-1	0	-1	-1	-1	$\frac{3}{14}$
2	$-\frac{3}{14}$	$-\frac{3}{14}$	$\frac{11}{14}$	$\frac{5}{28}$	$-\frac{5}{14}$	$\frac{1}{14}$	$\frac{4}{14}$	0	$-\frac{5}{14}$	$\frac{1}{14}$	$\frac{4}{14}$	$\frac{42}{41}$
3	$-\frac{333}{574}$	$-\frac{81}{574}$	$\frac{619}{574}$	$\frac{553}{8036}$	$-\frac{205}{574}$	$-\frac{169}{574}$	$-\frac{214}{574}$	–	–	–	–	–

Table 3-2–Convergence for (3-65) using the conjugate gradient method

Iteration	c_1	c_2	c_3	$f(c)$	e_1	e_2	e_3	β	d_1	d_2	d_3	α
0	1	0	0	1.5	-1	0	1	–	-1	0	1	1
1	0	0	1	$\frac{1}{2}$	-1	-1	-1	$\frac{3}{2}$	$-\frac{5}{2}$	-1	$\frac{1}{2}$	$\frac{6}{19}$
2	$-\frac{15}{19}$	$-\frac{6}{19}$	$\frac{22}{19}$	$\frac{1}{38}$	$-\frac{1}{19}$	$\frac{2}{19}$	$-\frac{1}{19}$	$\frac{2}{361}$	$-\frac{24}{361}$	$\frac{36}{361}$	$-\frac{18}{361}$	$\frac{19}{6}$
3	-1	0	1	0	0	0	0	–	–	–	–	–

3.6 FLETCHER–POWELL METHOD OF CONJUGATE DIRECTIONS

An alternate method of generating conjugate direction vectors is presented here. This method [7] is based on an indirect computation of $f_{\mathbf{cc}}^{-1}$ for the quadratic function given in (3-43) and is based on the result

Lemma 3-4 *Suppose (3-45), (3-47), and*

$$\mathbf{d}_i = \mathbf{\Gamma}_i \mathbf{e}_i \tag{3-67}$$

are used where $\mathbf{\Gamma}_i$ *is positive definite for each i and*

$$\mathbf{\Gamma}_i \mathbf{F} \mathbf{d}_j = \mathbf{d}_j \quad \text{for} \quad j < i. \tag{3-68}$$

Then nonzero vectors $\{\mathbf{d}_0, \mathbf{d}_1, \ldots\}$ *are conjugate with respect to* $\mathbf{F}$.

Proof Conjugacy of nonzero direction vectors is established by induction as follows. Equation (3-53) is clearly valid for nonzero vector $\{\mathbf{d}_0\}$. Now suppose this equation holds for nonzero vectors $\{\mathbf{d}_0, \mathbf{d}_1, \ldots, \mathbf{d}_{k-1}\}$ and $\mathbf{d}_k$ is also a nonzero vector. Then use of (3-48) yields

$$\mathbf{e}_k = \mathbf{e}_{i+1} - \mathbf{F} \sum_{j=i+1}^{k-1} \alpha_j \mathbf{d}_j$$

so that

$$\mathbf{d}_i' \mathbf{e}_k = \mathbf{d}_i' \mathbf{e}_{i+1} - \sum_{j=i+1}^{k-1} \alpha_j \mathbf{d}_i' \mathbf{F} \mathbf{d}_j.$$

Use of (3-46) and (3-53) now yields

$$\mathbf{d}_i' \mathbf{e}_k = 0 \quad \text{for} \quad i < k. \tag{3-69}$$

Moreover, substitution of (3-68) and then (3-67) into (3-69) yields (3-60). Therefore, equation (3-53) also holds for nonzero direction vectors $\{\mathbf{d}_0, \mathbf{d}_1, \ldots, \mathbf{d}_k\}$ and the induction is complete.

A desirable property of the Fletcher–Powell method is $\mathbf{\Gamma}_n = \mathbf{F}^{-1}$ so that $\alpha_n = 1$. This property results from conjugacy of direction vectors and (3-68) which imply that $\{\mathbf{d}_0, \mathbf{d}_1, \ldots, \mathbf{d}_{n-1}\}$ is a set of linearly independent eigenvectors of $(\mathbf{\Gamma}_n \mathbf{F})$ corresponding to unit eigenvalues.

A suitable structure for $\mathbf{\Gamma}_i$ can be deduced easily by introducing the relationship

$$\mathbf{\Gamma}_{i+1} = \mathbf{\Gamma}_i + \mathbf{A}_i + \mathbf{B}_i \quad \text{for} \quad i = 0, 1, \ldots. \tag{3-70}$$

In order to satisfy (3-68), the relationship

$$\mathbf{\Gamma}_{i+1} \mathbf{F} \mathbf{d}_i = \mathbf{d}_i$$

and hence

$$\mathbf{\Gamma}_i\mathbf{F}\mathbf{d}_i + \mathbf{A}_i\mathbf{F}\mathbf{d}_i + \mathbf{B}_i\mathbf{F}\mathbf{d}_i = \mathbf{d}_i$$

must hold for each i. Matrix $\mathbf{A}_i$ can always be chosen so that

$$\mathbf{A}_i\mathbf{F}\mathbf{d}_i = \mathbf{d}_i$$

which is satisfied by the positive semi-definite matrix

$$\mathbf{A}_i = \frac{\mathbf{d}_i\mathbf{d}_i'}{\mathbf{d}_i'\mathbf{F}\mathbf{d}_i}. \tag{3-71}$$

Then matrix $\mathbf{B}_i$ must satisfy

$$\mathbf{B}_i\mathbf{F}\mathbf{d}_i = -\mathbf{\Gamma}_i\mathbf{F}\mathbf{d}_i$$

which in turn is satisfied by the negative semi-definite matrix

$$\mathbf{B}_i = -\frac{(\mathbf{\Gamma}_i\mathbf{F}\mathbf{d}_i)(\mathbf{\Gamma}_i\mathbf{F}\mathbf{d}_i)'}{(\mathbf{\Gamma}_i\mathbf{F}\mathbf{d}_i)'\mathbf{F}\mathbf{d}_i}. \tag{3-72}$$

An additional structural property desired of $\mathbf{\Gamma}_i$ is established in

Lemma 3-5 *Suppose* $\{\mathbf{d}_0, \mathbf{d}_1, \ldots\}$ *is a set of nonzero direction vectors. Then corresponding matrices* $\mathbf{\Gamma}_i$ *are positive definite if* $\mathbf{\Gamma}_0$ *is positive definite.*

Proof It suffices to show that $\mathbf{\Gamma}_{i+1}$ is positive definite when $\mathbf{\Gamma}_i$ is positive definite so assume $\mathbf{\Gamma}_i$ is in fact positive definite. Then there exists a nonsingular matrix $\mathbf{\Delta}_i$ such that $\mathbf{\Gamma}_i = \mathbf{\Delta}_i^2$. Given any nonzero $\gamma \in \mathscr{R}^n$, let

$$\mathbf{p} = \mathbf{\Delta}_i\mathbf{f} \quad \text{and} \quad \mathbf{q} = \mathbf{\Delta}_i\mathbf{F}\mathbf{d}_i.$$

Use of (3-70), (3-71), and (3-72) now yields

$$\mathbf{f}'\mathbf{\Gamma}_{i+1}\mathbf{f} = \mathbf{f}'\mathbf{\Gamma}_i\mathbf{f} + \frac{(\mathbf{f}'\mathbf{d}_i)^2}{\mathbf{d}_i'\mathbf{F}\mathbf{d}_i} - \frac{(\mathbf{f}'\mathbf{\Gamma}_i\mathbf{F}\mathbf{d}_i)^2}{(\mathbf{F}\mathbf{d}_i)'\mathbf{\Gamma}_i\mathbf{F}\mathbf{d}_i}.$$

and substitution of $\mathbf{p}$ and $\mathbf{q}$ results in

$$\mathbf{f}'\mathbf{\Gamma}_{i+1}\mathbf{f} = \frac{(\mathbf{p}'\mathbf{p})(\mathbf{q}'\mathbf{q}) - (\mathbf{p}'\mathbf{q})^2}{\mathbf{q}'\mathbf{q}} + \frac{(\mathbf{f}'\mathbf{d}_i)^2}{\mathbf{d}_i'\mathbf{F}\mathbf{d}_i}.$$

By Schwartz's inequality, the relationship

$$\mathbf{f}'\mathbf{\Gamma}_{i+1}\mathbf{f} \geqslant \frac{(\mathbf{f}'\mathbf{d}_i)^2}{\mathbf{d}_i'\mathbf{F}\mathbf{d}_i}$$

is obtained where the equality holds only if $\mathbf{p} = \beta\mathbf{q}$. Because $\mathbf{f} \neq \mathbf{0}$ implies $\beta \neq 0$ when the equality holds as seen from the definition of $\mathbf{p}$ and $\mathbf{q}$, inequality

$$\frac{(\mathbf{f}'\mathbf{d}_i)^2}{\mathbf{d}_i'\mathbf{F}\mathbf{d}_i} = \beta^2 \mathbf{d}_i'\mathbf{F}\mathbf{d}_i > 0$$

is obtained. Thus, inequality $\mathbf{f}'\mathbf{\Gamma}_{i+1}\mathbf{f} > 0$ is established for all nonzero $\mathbf{f} \in \mathscr{R}^n$.

The final structural property desired of $\mathbf{\Gamma}_i$ for quadratic functions is established in

Lemma 3-6 *Matrices* $\mathbf{\Gamma}_i$ *corresponding to (3-70), (3-71), (3-72), and a positive definite* $\mathbf{\Gamma}_0$ *satisfy (3-68) when (3-45), (3-47), and (3-67) are used.*

Proof Equation (3-68) is valid for $i = 0$ so assume that this equation is also valid for some $i > 0$. Then the set of nonzero direction vectors $\{\mathbf{d}_0, \mathbf{d}_1, \ldots, \mathbf{d}_i\}$ is conjugate with respect to $\mathbf{F}$ as determined from the construction used to prove Lemma 3-4. Use of (3-68) yields

$$(\mathbf{\Gamma}_i\mathbf{F}\mathbf{d}_i)'\mathbf{F}\mathbf{d}_j = \mathbf{d}_i'\mathbf{F}\mathbf{d}_j = 0 \quad \text{for} \quad j < i,$$

and use of (3-70), (3-71), and (3-72) then yields

$$\mathbf{\Gamma}_{i+1}\mathbf{F}\mathbf{d}_j = \mathbf{d}_j \quad \text{for} \quad j < i.$$

The induction hence is complete.

The Fletcher–Powell method of minimizing the quadratic function given in (3-43) is stated as

Theorem 3-4 *Equations (3-45), (3-47), (3-67), (3-70), (3-71), and (3-72) yield* $\mathbf{c}_m = \mathbf{b}$ *for some* $m \leqslant n$ *given any* $\mathbf{c}_0 \in \mathscr{R}^n$ *and positive definite* $\mathbf{\Gamma}_0$.

Proof This result follows directly from Lemmas 3-2, 3-5, and 3-6 noting again that $\mathbf{d}_i = \mathbf{0}$ if and only if $\mathbf{c}_i = \mathbf{b}$ so that a sufficient number of nonzero direction vectors is always obtained.

An elementary example of the Fletcher–Powell method for quadratic functions is provided by (3-65). Typical convergence characteristics of this method are illustrated in Table 3-3. Convergence is obtained in three steps using the initialization $\mathbf{\Gamma}_0 = \mathbf{I}$ and $\mathbf{c}_0' = [1 \ \ 0 \ \ 0]$. This initialization results in the same sequence as obtained with the conjugate gradient method summarized in Table 3-2. The data given in Table 3-3 also illustrates typical convergence of $\mathbf{\Gamma}_i$ to $\mathbf{F}^{-1}$. In general, matrix $\mathbf{\Gamma}_i$ for $i < n$ is not a very satisfactory estimate of $\mathbf{F}^{-1}$.

In order to put the Fletcher–Powell method into a useful form for non-quadratic functions, matrix $\mathbf{F}$ must be eliminated from (3-71) and (3-72). This goal is accomplished by the introduction of

$$\boldsymbol{\delta}_i = \alpha_i\mathbf{d}_i \quad \text{and} \quad \boldsymbol{\epsilon}_i = \mathbf{e}_i - \mathbf{e}_{i+1} \tag{3-73}$$

Table 3-3–Convergence for (3-65) using the Fletcher–Powell method

Iteration	γ_{11}	γ_{12}	γ_{13}	γ_{22}	γ_{23}	γ_{33}	c_1	c_2	c_3	d_1	d_2	d_3	α
0	1.00	0.00	0.00	1.00	0.00	1.00	1	0	0	−1	0	1	1
1	1.50	0.00	−0.50	0.80	−0.40	0.70	0	0	1	−1	$-\frac{2}{5}$	$\frac{1}{5}$	$\frac{15}{19}$
2	1.42	−0.13	−0.44	0.70	−0.35	0.67	$-\frac{15}{19}$	$-\frac{6}{19}$	$\frac{22}{19}$	$-\frac{4}{61}$	$\frac{6}{61}$	$-\frac{3}{61}$	$\frac{61}{19}$
3	2.00	−1.00	0.00	2.00	−1.00	1.00	−1	0	1	–	–	–	–

and use of (3-48). Multiplication of numerator and denominator by α_i^2 and substitution then yields

$$\mathbf{A}_i = \frac{\boldsymbol{\delta}_i \boldsymbol{\delta}_i'}{\boldsymbol{\delta}_i' \boldsymbol{\epsilon}_i} \tag{3-74}$$

and

$$\mathbf{B}_i = -\frac{(\boldsymbol{\Gamma}_i \boldsymbol{\epsilon}_i)(\boldsymbol{\Gamma}_i \boldsymbol{\epsilon}_i)'}{(\boldsymbol{\Gamma}_i \boldsymbol{\epsilon}_i)' \boldsymbol{\epsilon}_i}. \tag{3-75}$$

The Fletcher–Powell method for nonquadratic functions is the computation of $\boldsymbol{\Gamma}_i$ by (3-70), (3-74), and (3-75), and the computation of $\mathbf{d}_i$ by (3-67). As for other methods requiring the trial procedure discussed in Section 3.2, parameter α_i is computed in accordance with (3-46) and (3-66).

As in the case of the conjugate gradient method, the Fletcher–Powell method is used on the presumption that the twice continuously differentiable function approaches an appropriate quadratic function in a suitably small neighborhood of a relative minimum at which hessian $f_{\mathbf{cc}}$ is positive definite. Moreover, this presumption is not accompanied with an adequate theory of error. The Fletcher–Powell method does not involve a significantly more complicated computer program than the steepest-descent and conjugate gradient methods. However, matrix $\boldsymbol{\Gamma}_i$ must be stored and hence there may be a significant increase in computer memory requirements for the Fletcher–Powell method. The chief advantage of the Fletcher–Powell method over the conjugate gradient method is that the number of trials required to determine α_i may be decreased by judicious selections of $\boldsymbol{\Gamma}_0$.

3.7 SUCCESSIVE APPROXIMATIONS FOR CONSTRAINED MINIMA

The problem of locating constrained relative minima by computational means is introduced in this section. The intent here is to identify additional basic ingredients required to use gradient methods for problems with constraints. Constraints generally are troublesome from a computational point of view so that attempts to avoid the problem by approximation are introduced first.

Penalty functions are often suggested as a method of computing approximations to the solution of minimization problems with equality constraints. For instance, the problem of locating a relative minimum of $f(\mathbf{c})$ subject to $\mathbf{g}(\mathbf{c}) = \mathbf{0}$ is replaced by the problem of simply locating a relative minimum of

$$h(\mathbf{c}) = f(\mathbf{c}) + \tfrac{1}{2}\mathbf{g}'(\mathbf{c})\mathbf{E}\mathbf{g}(\mathbf{c}), \tag{3-76}$$

where $\mathbf{E}$ is some positive definite matrix. The quadratic form appearing in (3-76) is one example of a penalty function for equality constraints. Presumably, as $\lambda_{min}\{\mathbf{E}\} \to \infty$, the location of a relative minimum of $h(\mathbf{c})$ approaches the location of the corresponding relative minimum of $f(\mathbf{c})$ which is constrained by $\mathbf{g}(\mathbf{c}) = \mathbf{0}$. In fact, a number of suggestions [2] have been made for iteratively altering penalty functions during successive approximations such as increasing $\lambda_{min}\{\mathbf{E}\}$.

In order to illustrate penalty function approximations, consider minimization of the quadratic function given in (3-43) subject to linear equality constraints

$$\mathbf{g}(\mathbf{c}) = \mathbf{G}(\mathbf{c} - \mathbf{a}). \tag{3-77}$$

Also assume $\mathbf{F}$ and $(\mathbf{GG}')$ are positive definite. Solutions to the minimization of $h(\mathbf{c})$ given in (3-76) are now investigated as a function of $\mathbf{E}$ by minimizing function

$$f(\mathbf{c}) + \tfrac{1}{2}\mathbf{e}'\mathbf{E}\mathbf{e}$$

with respect to $\mathbf{c}$, $\mathbf{e}$ and subject to equality constraints

$$\mathbf{g} - \mathbf{e} = \mathbf{0}.$$

This auxiliary problem is solved readily in terms of a Lagrangian formulation using

$$L = f(\mathbf{b}) + \tfrac{1}{2}(\mathbf{c} - \mathbf{b})'\mathbf{F}(\mathbf{c} - \mathbf{b}) + \tfrac{1}{2}\mathbf{e}'\mathbf{E}\mathbf{e} + \boldsymbol{\lambda}'[\mathbf{G}(\mathbf{c} - \mathbf{a}) - \mathbf{e}].$$

Stationarity conditions for L with respect to $\mathbf{c}$, $\mathbf{e}$, and $\boldsymbol{\lambda}$ yield

$$\mathbf{F}(\mathbf{c} - \mathbf{b}) + \mathbf{G}'\boldsymbol{\lambda} = \mathbf{0},$$

$$\mathbf{E}\mathbf{e} - \boldsymbol{\lambda} = \mathbf{0},$$

and

$$\mathbf{G}(\mathbf{c} - \mathbf{a}) - \mathbf{e} = \mathbf{0}$$

respectively. These three equations are then solved in order, thereby obtaining

$$\mathbf{c} = \mathbf{b} - \mathbf{F}^{-1}\mathbf{G}'\boldsymbol{\lambda}, \tag{3-78}$$

$$\mathbf{e} = \mathbf{E}^{-1}\boldsymbol{\lambda}, \tag{3-79}$$

and

$$\boldsymbol{\lambda} = (\mathbf{G}\mathbf{F}^{-1}\mathbf{G}' + \mathbf{E}^{-1})^{-1}\mathbf{G}(\mathbf{b} - \mathbf{a}) \tag{3-80}$$

respectively. Behavior of solution $\mathbf{c}$ as a function of penalty function matrix $\mathbf{E}$ is identified easily from these equations. In particular, Lagrange multipliers $\boldsymbol{\lambda}$ approach those of the original equality constraint problem as $\lambda_{min}\{\mathbf{E}\} \to \infty$ and hence solution $\mathbf{c}$ also approaches that of the original problem. Moreover, auxiliary variables $\mathbf{e} \to \mathbf{0}$ and hence equality constraints $\mathbf{g}(\mathbf{c}) \to \mathbf{0}$ as $\lambda_{\min}\{\mathbf{E}\} \to \infty$.

Iterative selection of penalty function matrix $\mathbf{E}$ so that a suitably large value of $\lambda_{\min}\{\mathbf{E}\}$ is attained may be very difficult. An alternate approach is based on the Lagrangian formulation of the original problem. Both parameters $\mathbf{c}$ and Lagrange multiplers $\boldsymbol{\lambda}$ are found by successive approximations when this approach is taken. By way of introduction, suppose parameters $\mathbf{c}$ and Lagrange multipliers $\boldsymbol{\lambda}$ are replaced by parameterized curves $\boldsymbol{\gamma}(t)$ and $\mathbf{l}(t)$ respectively. Then the Lagrangian defined in (2-18) becomes

$$h(t) = L(\boldsymbol{\gamma}, \mathbf{l}) = f[\boldsymbol{\gamma}(t)] + \mathbf{l}'(t)\mathbf{g}[\boldsymbol{\gamma}(t)]. \tag{3-81}$$

If differentiability of all functions is assumed, then derivative

$$\dot{h}(t) = \begin{bmatrix} L_{\gamma} \\ \\ L_{l} \end{bmatrix}' \begin{bmatrix} \dot{\boldsymbol{\gamma}} \\ \\ \dot{\mathbf{l}} \end{bmatrix} \tag{3-82}$$

also is defined. As is the case without constraints, a vector $\boldsymbol{\gamma}$ is being sought that minimizes h amongst those which satisfy $\mathbf{g} = \mathbf{0}$. Derivative $\dot{\boldsymbol{\gamma}}$ presumably should then be chosen so that $\dot{\boldsymbol{\gamma}}' L_{\gamma} \leqslant 0$. On the other hand, given that $\mathbf{g} \neq \mathbf{0}$, a vector $\mathbf{l}$ is being sought that maximizes penalty function $\mathbf{l}'\mathbf{g}$ and thus also maximizes h. Derivative $\dot{\mathbf{l}}$ presumably should be chosen so that $\dot{\mathbf{l}}' L_{l} \geqslant 0$. These considerations suggest that curves $\boldsymbol{\gamma}(t)$ and $\mathbf{l}(t)$ can be defined by

$$\begin{bmatrix} \dot{\boldsymbol{\gamma}} \\ \\ \dot{\mathbf{l}} \end{bmatrix} = \boldsymbol{\Gamma} \begin{bmatrix} -L_{\gamma} \\ \\ L_{l} \end{bmatrix}; \quad \begin{bmatrix} \boldsymbol{\gamma}(0) \\ \\ \mathbf{l}(0) \end{bmatrix} = \begin{bmatrix} \boldsymbol{\gamma}_0 \\ \\ \mathbf{l}_0 \end{bmatrix} \tag{3-83}$$

when $\boldsymbol{\Gamma} = \boldsymbol{\Gamma}(\boldsymbol{\gamma}, \mathbf{l})$ is suitably selected. For instance, selection $\boldsymbol{\Gamma} = \mathbf{I}$ is analogous to steepest-descent, and selection

$$\boldsymbol{\Gamma} = \begin{bmatrix} L_{\gamma\gamma} & L_{\gamma l} \\ \\ -L_{l\gamma} & \mathbf{0} \end{bmatrix}^{-1}$$

is analogous to Newton–Raphson.

An example of the continuous version of successive approximations for equality constraints is provided by (3-43) and (3-77) with $\mathbf{F}$ and $(\mathbf{GG}')$ assumed to be positive definite. For any constant matrix $\boldsymbol{\Gamma}$, the differential equations appearing in (3-83) are now linear and can be rewritten as

$$\begin{bmatrix} \dot{\boldsymbol{\gamma}} \\ \\ \dot{\mathbf{l}} \end{bmatrix} = \mathbf{A} \begin{bmatrix} (\boldsymbol{\gamma} - \boldsymbol{\gamma}_e) \\ \\ (\mathbf{l} - \mathbf{l}_e) \end{bmatrix}; \quad \begin{bmatrix} \boldsymbol{\gamma}(0) \\ \\ \mathbf{l}(0) \end{bmatrix} = \begin{bmatrix} \boldsymbol{\gamma}_0 \\ \\ \mathbf{l}_0 \end{bmatrix} \tag{3-84}$$

where

$$\begin{bmatrix} \boldsymbol{\gamma}_e \\ \mathbf{l}_e \end{bmatrix} = \begin{bmatrix} \mathbf{F} & \mathbf{G}' \\ \mathbf{G} & \mathbf{0} \end{bmatrix}^{-1} \begin{bmatrix} \mathbf{Fb} \\ \mathbf{Ga} \end{bmatrix} \tag{3-85}$$

and

$$\mathbf{A} = \boldsymbol{\Gamma} \begin{bmatrix} -\mathbf{F} & -\mathbf{G}' \\ \mathbf{G} & \mathbf{0} \end{bmatrix}. \tag{3.86}$$

Curves $\boldsymbol{\gamma}(t)$ and $\mathbf{l}(t)$ thus are given by

$$\begin{bmatrix} (\boldsymbol{\gamma} - \boldsymbol{\gamma}_e) \\ (\mathbf{l} - \mathbf{l}_e) \end{bmatrix} = e^{\mathbf{A}t} \begin{bmatrix} (\boldsymbol{\gamma}_0 - \boldsymbol{\gamma}_e) \\ (\mathbf{l}_0 - \mathbf{l}_e) \end{bmatrix}. \tag{3-87}$$

Matrix $\boldsymbol{\Gamma}$ can have a variety of structures which nevertheless result in asymptotic stability. Use of

$$\boldsymbol{\Gamma} = \begin{bmatrix} \mathbf{F} & \mathbf{G}' \\ -\mathbf{G} & \mathbf{0} \end{bmatrix}$$

results in $\mathbf{A} = -\mathbf{I}$, and hence the Newton-Raphson approach is asymptotically stable. Alternatively, use of $\boldsymbol{\Gamma} = \mathbf{I}$ results in

$$\mathbf{A} = \begin{bmatrix} -\mathbf{F} & -\mathbf{G}' \\ \mathbf{G} & \mathbf{0} \end{bmatrix}$$

which happens to be asymptotically stable as seen from what follows. Let $[\mathbf{m}' \;\; \mathbf{n}']'$ be an eigenvector of $\mathbf{A}$ and λ be the corresponding eigenvalue. Then, by definition,

$$\lambda \begin{bmatrix} \mathbf{m} \\ \mathbf{n} \end{bmatrix} = \begin{bmatrix} -(\mathbf{Fm} + \mathbf{G}'\mathbf{n}) \\ \mathbf{Gm} \end{bmatrix}$$

so that premultiplication by $[\mathbf{m}^{\dagger} \;\; \mathbf{n}^{\dagger}]$ can be used to obtain

$$\lambda = \frac{-\mathbf{m}^{\dagger}\mathbf{Fm} - \mathbf{m}^{\dagger}\mathbf{G}'\mathbf{n} + \mathbf{n}^{\dagger}\mathbf{Gm}}{\mathbf{m}^{\dagger}\mathbf{m} + \mathbf{n}^{\dagger}\mathbf{n}}.$$

In addition, the complex conjugate of λ is given by

$$\lambda^* = \frac{-\mathbf{m}^\dagger \mathbf{F}\mathbf{m} - \mathbf{n}^\dagger \mathbf{G}\mathbf{m} + \mathbf{m}^\dagger \mathbf{G}'\mathbf{n}}{\mathbf{m}^\dagger \mathbf{m} + \mathbf{n}^\dagger \mathbf{n}}$$

and hence

$$\mathrm{Re}\{\boldsymbol{\lambda}\} = -\frac{\mathbf{m}^\dagger \mathbf{F}\mathbf{m}}{\mathbf{m}^\dagger \mathbf{m} + \mathbf{n}^\dagger \mathbf{n}}.$$

Thus, eigenvalues of **A** for the steepest-descent approach have the property $\mathrm{Re}\{\lambda\} \leqslant 0$. However, vector **m** cannot vanish unless the corresponding eigenvalue is zero because otherwise **n** would also have to vanish. Matrix **A** is nonsingular because **F** and $(\mathbf{GG}')$ are positive definite and hence $\mathrm{Re}\{\lambda\} < 0$ for the steepest-descent approach.

Considerations similar to those pointed out for problems without constraints suggest that equilibrium solutions of (3-83) can be computed by successive approximations. Similarly, successive approximations would now be performed

$$\mathbf{d} = \boldsymbol{\Gamma} \begin{bmatrix} -L_\mathbf{c} \\ \\ L_\lambda \end{bmatrix} \tag{3-88}$$

with direction vector and a one-dimensional search along this direction vector. Unfortunately, even when $\boldsymbol{\Gamma}$ is selected analogously to the Newton-Raphson method, search in one dimension does not just involve relative minima of the parameterized Lagrangian. In fact, depending upon **c** and $\boldsymbol{\lambda}$, both inflection points and relative maxima must also be located by any one-dimensional search procedure. This difficulty arises from the saddle-value structure of the Lagrangian for equality constraints and greatly complicates trial procedures. Moreover, reliable criteria have not been established for determining whether a relative minimum, inflection point, or a relative maximum of the Lagrangian is being sought on a given iteration. Thus, without the introduction of additional concepts for equality constraints, the only approach introduced for unconstrained problems that retains some usefulness is successive substitutions. Analogous to the Newton–Raphson method given in (3-41), a method of successive substitutions

$$\mathbf{h}\left(\begin{bmatrix} \mathbf{c} \\ \\ \boldsymbol{\lambda} \end{bmatrix}\right) = \begin{bmatrix} \mathbf{c} \\ \\ \boldsymbol{\lambda} \end{bmatrix} - \begin{bmatrix} L_{\mathbf{cc}} & L_{\mathbf{c}\lambda} \\ \\ L_{\lambda\mathbf{c}} & \mathbf{0} \end{bmatrix}^{-1} \begin{bmatrix} L_\mathbf{c} \\ \\ L_\lambda \end{bmatrix} \tag{3-89}$$

for equality constraints is assuming Lagrangian L has a nonsingular hessian with respect to $[\mathbf{c}' \;\; \lambda']'$.

Difficulties encountered with constrained minima are compounded when constraints are of the inequality type. In order to extend penalty-function methods to inequality constraints, slack variables are often introduced. Suppose $f(\mathbf{c})$ is to be minimized subject to $\mathbf{g}(\mathbf{c}) \leqslant \mathbf{0}$. Then slack vector $\mathbf{s}$ can be introduced so that equivalent equality constraints

$$\mathbf{g}(\mathbf{c}) + \boldsymbol{\sigma}(\mathbf{s}) = \mathbf{0} \tag{3-90}$$

are constructed with $\boldsymbol{\sigma}(\mathbf{s}) \geqslant \mathbf{0}$. For instance, slack function $\boldsymbol{\sigma}$ can be selected with elements

$$\sigma_i = \tfrac{1}{2}s_i^2. \tag{3-91}$$

Moreover, a penalty function, such as appearing in (3-76), can be introduced so that the equivalent problem becomes the minimization of

$$h(\mathbf{c}, \mathbf{s}) = f(\mathbf{c}) + \tfrac{1}{2}[\mathbf{g}(\mathbf{c}) + \boldsymbol{\sigma}(\mathbf{s})]'\mathbf{E}[\mathbf{g}(\mathbf{c}) + \boldsymbol{\sigma}(\mathbf{s})] \tag{3-92}$$

with respect to $\mathbf{c}$ and $\mathbf{s}$. Added difficulties experienced by the penalty-function method with inequality constraints stem primarily from added dimensionality and a somewhat more complicated structure of the hessian of h with respect to $[\mathbf{c}' \ \mathbf{s}']'$.

Slack vectors for inequality constraints can also be introduced for use in an equivalent Lagrangian with equality constraints. Specifically, Lagrangian

$$L = f(\mathbf{c}) + \boldsymbol{\lambda}'[\mathbf{g}(\mathbf{c}) + \boldsymbol{\sigma}(\mathbf{s})] \tag{3-93}$$

is introduced for equality constraints given in (3-90) so that restriction $\boldsymbol{\lambda} \geqslant \mathbf{0}$ presumably is no longer required. Gradient $L_{\mathbf{c}}$ is unaltered by the addition of a slack function, and gradient $L_{\boldsymbol{\lambda}}$ is, of course, equal to the equivalent equality constraint function. When the slack function defined by (3-91) is used, the ith element of gradient $L_{\mathbf{s}}$ is equal to $\lambda_i s_i$, and hence $\lambda_i s_i = 0$ must hold for each i at a solution. This result is compatible with the Lagrangian formulation for inequality constraints. Moreover, the hessian of L, defined by (3-91) and (3-93), with respect to $[\mathbf{c}' \ \mathbf{s}' \mid \boldsymbol{\lambda}']'$ becomes

$$\mathbf{H} = \left[\begin{array}{cc|c} \mathbf{F} & \mathbf{0} & \mathbf{G}' \\ \mathbf{0} & \boldsymbol{\Lambda} & \mathbf{S}' \\ \hline \mathbf{G} & \mathbf{S} & \mathbf{0} \end{array}\right] \tag{3-94}$$

where

$$\mathbf{F} = (f + \boldsymbol{\lambda}'\mathbf{g})_{\mathbf{cc}}, \quad \mathbf{G} = \mathbf{g}_{\mathbf{c}} \tag{3-95}$$

and

$$\boldsymbol{\Lambda} = \begin{bmatrix} \lambda_1 & 0 & \cdots \\ 0 & \lambda_2 & \cdots \\ \cdots & \cdots & \cdots \end{bmatrix}, \mathbf{S} = \begin{bmatrix} s_1 & 0 & \cdots \\ 0 & s_2 & \cdots \\ \cdots & \cdots & \cdots \end{bmatrix}. \tag{3-96}$$

Assuming that rank $\mathbf{G} = \dim \mathbf{g}$, no restrictions are placed by the null space of matrix $[\mathbf{G}\ \ \mathbf{S}]$ on vector elements corresponding to s. Thus, by Theorem 2-8, matrix $\boldsymbol{\Lambda}$ must be positive semi-definite when rank $\mathbf{G} = \dim \mathbf{g}$, and restriction $\boldsymbol{\lambda} \geqslant \mathbf{0}$ in fact must hold at a solution.

Slack variables are particularly justified from a computational point of view when successive substitutions can be used with the equality constraint version of the Newton–Raphson method. This application of slack variables is possible when matrix $\mathbf{H}$ given in (3-94) is nonsingular on some neighborhood of a solution. If matrix

$$\begin{bmatrix} \mathbf{F} & \mathbf{G}' \\ \mathbf{G} & \mathbf{0} \end{bmatrix}$$

is nonsingular on this neighborhood, then elementary determinant manipulations yield

$$\det \mathbf{H} = \det \begin{bmatrix} \mathbf{F} & \mathbf{G}' \\ \mathbf{G} & \mathbf{0} \end{bmatrix} \times \det\left(\boldsymbol{\Lambda} - \begin{bmatrix} \mathbf{0} \\ \mathbf{S} \end{bmatrix}' \begin{bmatrix} \mathbf{F} & \mathbf{G}' \\ \mathbf{G} & \mathbf{0} \end{bmatrix}^{-1} \begin{bmatrix} \mathbf{0} \\ \mathbf{S} \end{bmatrix}\right). \tag{3-97}$$

If in addition matrix $\mathbf{F}$ is nonsingular and rank $\mathbf{G} = \dim \mathbf{g}$, then this expression simplifies to

$$\det \mathbf{H} = \det \begin{bmatrix} \mathbf{F} & \mathbf{G}' \\ \mathbf{G} & \mathbf{0} \end{bmatrix} \times \det[\boldsymbol{\Lambda} + \mathbf{S}'(\mathbf{G}\mathbf{F}^{-1}\mathbf{G}')^{-1}\mathbf{S}]. \tag{3-98}$$

The matrix appearing in the second determinant of (3-98) is positive semi-definite at a constrained relative minimum when $\mathbf{F}$ is positive semi-definite. However, matrix $\mathbf{H}$ is singular whenever an active but ineffective constraint occurs at the solution because $\lambda_i = s_i = 0$ occurs for some i at the solution.

3.8 MIN–MAX METHOD

The method of successive substitutions defined by (3-89) unfortunately requires direct computation of the hessian of the Lagrangian with respect to $[\mathbf{c}'\ \ \boldsymbol{\lambda}']'$, and hence this method is impractical for many problems arising in control. This method also requires initialization in a suitably small neighborhood of a solution in order to obtain convergence. In addition, use of slack variables for inequality constraints accentuates the difficulties associated with initialization. For the purposes of circumventing these problems, a new method is developed in this section. This new method is based on the peculiar structure of Lagrangians which in general precludes the use of one-dimensional search techniques. This method is founded on the assumption that the Lagrangian, given Lagrange multipliers $\boldsymbol{\lambda}$, is strongly convex with respect to parameters $\mathbf{c}$ on some neighborhood of a solution.

By way of introduction to the case with equality constraints, consider the quadratic function

$$V(\boldsymbol{\gamma}, \mathbf{l}) = \alpha \begin{bmatrix} \boldsymbol{\gamma} \\ \mathbf{l} \end{bmatrix}' \begin{bmatrix} \mathbf{p} \\ \mathbf{q} \end{bmatrix} + \tfrac{1}{2} \begin{bmatrix} \boldsymbol{\gamma} \\ \mathbf{l} \end{bmatrix}' \begin{bmatrix} \mathbf{P} & \mathbf{Q}' \\ \mathbf{Q} & \mathbf{0} \end{bmatrix} \begin{bmatrix} \boldsymbol{\gamma} \\ \mathbf{l} \end{bmatrix} \tag{3-99}$$

where $\alpha > 0$, $\mathbf{P}$ is positive definite, and rank $\mathbf{Q}$ = dim $\mathbf{l}$. This quadratic function is also constructed from the suitably differentiable Lagrangian of the problem under consideration with

$$\mathbf{p} = L_{\mathbf{c}} \quad \text{and} \quad \mathbf{q} = L_{\boldsymbol{\lambda}}. \tag{3-100}$$

In other words, the linear terms appearing in (3-99) with $\alpha = 1$ are also the first degree terms of a Taylor series expansion of Lagrangian $L(\mathbf{c} + \boldsymbol{\gamma}, \boldsymbol{\lambda} + \mathbf{l})$ about the point $\mathbf{c}, \boldsymbol{\lambda}$. If, in addition, the quadratic terms appearing in (3-99) are specialized to

$$\mathbf{P} = L_{\mathbf{cc}} \quad \text{and} \quad \mathbf{Q} = L_{\boldsymbol{\lambda}\mathbf{c}}, \tag{3-101}$$

then these terms are also the second degree terms of a Taylor series expansion of $L(\mathbf{c} + \boldsymbol{\gamma}, \boldsymbol{\lambda} + \mathbf{l})$ about the point $\mathbf{c}, \boldsymbol{\lambda}$. Thus, the quadratic function defined in (3-99) can be used flexibly to model the Lagrangian of the equality-constraint problem under consideration on a neighborhood of the point $\mathbf{c}, \boldsymbol{\lambda}$. Moreover, suitable selections of $[\boldsymbol{\gamma}' \;\; \mathbf{l}']'$ can also be used as the increment vector in a one-dimensional search of the function $L(\mathbf{c} + \boldsymbol{\gamma}, \boldsymbol{\lambda} + \mathbf{l})$.

The first step in the min–max procedure is the minimization of V with respect to $\boldsymbol{\gamma}$ given some fixed value of $\mathbf{l}$. This function is minimized with respect to $\boldsymbol{\gamma}$ when $V_{\boldsymbol{\gamma}} = \mathbf{0}$ because $\mathbf{P}$ is assumed to be positive definite. Therefore, minimization yields

$$\boldsymbol{\gamma} = -\mathbf{P}^{-1}(\alpha\mathbf{p} + \mathbf{Q}'\mathbf{l}) \tag{3-102}$$

given any $\mathbf{l}$, and the corresponding value of V is

$$V_{\min} = -\frac{\alpha^2}{2}\mathbf{p}'\mathbf{P}^{-1}\mathbf{p} + \alpha\mathbf{l}'(\mathbf{q} - \mathbf{Q}\mathbf{P}^{-1}\mathbf{p}) - \tfrac{1}{2}\mathbf{l}'(\mathbf{Q}\mathbf{P}^{-1}\mathbf{Q}')\mathbf{l}. \tag{3-103}$$

The second step in the min–max procedure is the maximization of $V_{\min}$ with respect to $\mathbf{l}$. This function is maximized when the gradient with respect to $\mathbf{l}$

vanishes because $(\mathbf{QP}^{-1}\mathbf{Q}')$ is positive definite by assumption. Therefore, maximization yields

$$\mathbf{l} = \alpha(\mathbf{QP}^{-1}\mathbf{Q}')^{-1}(\mathbf{q} - \mathbf{QP}^{-1}\mathbf{p}), \tag{3-104}$$

and the corresponding value of $V_{\min}$ is

$$V_{\text{min-max}} = -\frac{\alpha^2}{2}\mathbf{p}'\mathbf{P}^{-1}\mathbf{p} + \frac{\alpha^2}{2}(\mathbf{q} - \mathbf{QP}^{-1}\mathbf{p})'(\mathbf{QP}^{-1}\mathbf{Q}')^{-1}(\mathbf{q} - \mathbf{QP}^{-1}\mathbf{p}). \tag{3-105}$$

Finally, elimination of $\mathbf{l}$ in (3-102) using (3-104) results in

$$\boldsymbol{\gamma} = -\alpha\mathbf{P}^{-1}[\mathbf{p} + \mathbf{Q}'(\mathbf{QP}^{-1}\mathbf{Q}')^{-1}(\mathbf{q} - \mathbf{QP}^{-1}\mathbf{p})]. \tag{3-106}$$

Expressions for $\mathbf{l}$ and $\boldsymbol{\gamma}$ given in (3-104) and (3-106) respectively identify the direction vector which corresponds to the min–max method for the function given in (3-99). In this particular case, the min–max method simply amounts to matrix inversion by partitions of the matrix appearing in (3-99).

For application to problems with equality constraints, the min–max method is performed in a sequence of two steps on each iteration as follows. The first step is the minimization of Lagrangian $L(\mathbf{c}, \boldsymbol{\lambda})$ with respect to $\mathbf{c}$ given some fixed value of $\boldsymbol{\lambda}$. Any one of the methods presented previously for locating relative minima without constraints can of course be used for this step. If matrix $L_{\mathbf{cc}}$ cannot be computed directly in a practical fashion, however, then the Fletcher–Powell method of conjugate directions must be used. In particular, the inverse of the $n \times n$ matrix appearing in (3-99) is taken to be

$$\mathbf{P}^{-1} = \boldsymbol{\Gamma}_n \tag{3-107}$$

where $\boldsymbol{\Gamma}_n$ is computed from (3-70), (3-74), and (3-75). Matrix $\mathbf{P}$ thus approximates $L_{\mathbf{cc}}$ in a suitably small neighborhood of a solution. If matrix $L_{\boldsymbol{\lambda}\mathbf{c}}$ cannot be computed directly in a practical fashion, then either the conjugate gradient or the Fletcher–Powell method of conjugate directions must be used. In particular, matrix $\mathbf{Q}$ appearing in (3-99) is constructed to satisfy

$$\boldsymbol{\psi}_i = \mathbf{Q}\boldsymbol{\delta}_i \quad \text{for} \quad i = 0, 1, \ldots, n \tag{3-108}$$

where $\{\boldsymbol{\delta}_0, \boldsymbol{\delta}_1, \ldots, \boldsymbol{\delta}_{n-1}\}$ is a set of linearly independent increments in $\mathbf{c} \in \mathscr{R}^n$ and

$$\boldsymbol{\psi}_i = \mathbf{g}(\mathbf{c}_i + \boldsymbol{\delta}_i) - \mathbf{g}(\mathbf{c}_i). \tag{3-109}$$

Matrix $\mathbf{Q}$ thus can be computed as

$$\mathbf{Q} = \boldsymbol{\Psi}\boldsymbol{\Delta}^{-1} \tag{3-110}$$

where

$$\boldsymbol{\Psi} = [\boldsymbol{\psi}_0\ \boldsymbol{\psi}_1 \ldots \boldsymbol{\psi}_{n-1}] \quad \text{and} \quad \boldsymbol{\Delta} = [\boldsymbol{\delta}_0\ \boldsymbol{\delta}_1 \ldots \boldsymbol{\delta}_{n-1}]. \tag{3-111}$$

Therefore, matrix $\mathbf{Q}$ computed in this manner approximates $g_c = L_{\lambda \mathbf{c}}$ in a suitably small neighborhood of a solution and hence $L_{\lambda \mathbf{c}}$ need not be computed directly.

The first step of the sequential min–max method is not only used to minimize $L(\mathbf{c}, \boldsymbol{\lambda})$ with respect to $\mathbf{c}$ but in general is also used to compute $\mathbf{P}$ and $\mathbf{Q}$ indirectly. Moreover, the quadratic function given in (3-99) with $\alpha = 1$ includes first degree and approximate second degree terms of a Taylor series expansion of $L(\mathbf{c}, \boldsymbol{\lambda})$. Minimization has established the condition $\mathbf{p} = \mathbf{0}$ so that (3-102) becomes

$$\boldsymbol{\gamma} = -\mathbf{P}^{-1}\mathbf{Q}'\mathbf{l} \tag{3-112}$$

and (3-103) becomes

$$V_{\min} = \alpha \mathbf{l}'\mathbf{q} - \tfrac{1}{2}\mathbf{l}'(\mathbf{Q}\mathbf{P}^{-1}\mathbf{Q}')\mathbf{l}. \tag{3-113}$$

Therefore, maximization of $V_{\min}$ now yields $V_{\text{min-max}} > 0$ assuming $\mathbf{q} \neq \mathbf{0}$.

The second step of the sequential min–max method is the maximization of $L[(\mathbf{c} - \mathbf{P}^{-1}\mathbf{Q}'\mathbf{l}), (\boldsymbol{\lambda} + \mathbf{l})]$ with respect to $\mathbf{l}$. Maximization can of course be performed by any one of the methods presented previously for locating relative minima. Alternatively, a single one-dimensional search for a relative maximum can be performed using

$$\mathbf{l} = \alpha(\mathbf{Q}\mathbf{P}^{-1}\mathbf{Q}')^{-1}\mathbf{q} \tag{3-114}$$

which is found from (3-104) with $\mathbf{p} = \mathbf{0}$. In either case, maximization completes one iteration of the sequential min–max method for equality constraints.

The sequential min–max method is extended to inequality constraints by first extending the notion of effective constraints to points $\mathbf{c}, \boldsymbol{\lambda}$ which are not a solution point. Specifically, the index set of effective inequality constraints is redefined as

$$I_e = \{i : g_i(\mathbf{c}) > 0 \quad \text{or} \quad \lambda_i > 0\}. \tag{3-115}$$

This definition is consistent with (2-53) at a solution point because there $g_i(\mathbf{c}) \leqslant 0$ and $\lambda_i \geqslant 0$ hold for each i. As in Chapter 2, constraint functions and Lagrange multipliers corresponding to effective inequality constraints are denoted by $\hat{\mathbf{g}}$ and $\hat{\boldsymbol{\lambda}}$ respectively. Vectors $\boldsymbol{\gamma}$ and $\mathbf{l}$ appearing in (3-99) now are associated with $\mathbf{c}$ and $\hat{\boldsymbol{\lambda}}$ respectively so that vectors $\mathbf{p}$ and $\mathbf{q}$ are taken to be

$$\mathbf{p} = L_{\mathbf{c}} \quad \text{and} \quad \mathbf{q} = L_{\hat{\lambda}} \tag{3-116}$$

for inequality constraints. Moreover, matrices $\mathbf{P}$ and $\mathbf{Q}$ are now taken to be approximations of $L_{\mathbf{cc}}$ and $L_{\hat{\lambda}\mathbf{c}}$ respectively.

Minimization with respect to $\boldsymbol{\gamma}$ of V defined in (3-99) for inequality constraints again results in (3-102) and (3-103). However, maximization of $V_{\min}$ with respect to $\mathbf{l}$ now is complicated by the restriction that $(\hat{\boldsymbol{\lambda}}+\mathbf{l}) \geqslant \mathbf{0}$. This restriction is required in order to preserve the structure of the Lagrangian formulation for inequality constraints. A crude approximation [8] to the vector found from this maximization generally is useful for successive approximations; namely, elements of $\mathbf{l}$ given in (3-104) are altered in accordance with the corresponding element of $-\hat{\boldsymbol{\lambda}}$ such that

$$\mathbf{l} = \begin{cases} \mathbf{l}_s & \text{for} \quad \mathbf{l}_s > -\hat{\boldsymbol{\lambda}} \\ -\hat{\boldsymbol{\lambda}} & \text{for} \quad \mathbf{l}_s \leqslant -\hat{\boldsymbol{\lambda}} \end{cases} \tag{3-117}$$

where

$$\mathbf{l}_s = \alpha(\mathbf{Q}\mathbf{P}^{-1}\mathbf{Q}')^{-1}(\mathbf{q} - \mathbf{Q}\mathbf{P}^{-1}\mathbf{p}). \tag{3-118}$$

Alternately, although somewhat inconvenient, slack variables can be introduced so that a maximum constrained only by equality constraints need be sought.

The first step of the sequential min–max method for inequality constraints again is the minimization of $L(\mathbf{c},\boldsymbol{\lambda})$ with respect to $\mathbf{c}$ for some value of $\boldsymbol{\lambda} \geqslant \mathbf{0}$ as well as perhaps the indirect computation of $\mathbf{P}$ and $\mathbf{Q}$. At the end of this step, the index set of effective constraints, as defined in (3-115), also must be determined. The second step of the method now becomes the maximization of $L[(\mathbf{c} - \mathbf{P}^{-1}\mathbf{Q}'\mathbf{l}), (\hat{\boldsymbol{\lambda}} + \mathbf{l})]$ with respect to $\mathbf{l}$ and subject to $\mathbf{l} \geqslant -\hat{\boldsymbol{\lambda}}$. Frequently, maximization is simplified to a one-dimensional search by use of (3-117) and (3-118) with $\mathbf{p} = \mathbf{0}$.

A simple example is introduced here for the purposes of illustrating effective inequality constraints and Lagrange multiplier thresholding given in (3-117). The function

$$f = c_1^2 + 2c_2^2 + 3c_3^2$$

is to be minimized subject to inequality constraint functions

$$g_1 = c_1 + c_2 + c_3 + 1$$

and

$$g_2 = c_1 + 2c_2 + c_3 + 2.$$

Data given first for this example are based on using (3-102), (3-117), and (3-118) with $\alpha = 1$, $\mathbf{P} = L_{\mathbf{cc}}$ and $\mathbf{Q} = L_{\hat{\boldsymbol{\lambda}}\mathbf{c}}$ as a method of successive substitutions. Table 3-4 illustrates convergence of the approximate min–max method for a quadratic function and linear inequality constraint functions. These iterations illustrate the definition of effective constraints, and the threshold $\lambda_1 + l_1 \geqslant 0$ is used when

Table 3-4—Example convergence of the approximate min-max method

Iteration	c_1	c_2	c_3	λ_1	λ_2	f	g_1	g_2	I_e
0	3.00	−3.00	0.00	0.00	0.00	27.00	1.00	−1.00	{1}
1	−0.55	−0.27	−0.18	1.09	0.00	0.55	0.00	0.73	{1, 2}
2	−2.00	−2.00	−0.67	0.00	4.00	13.33	−3.67	−4.67	{2}
3	−0.60	−0.60	−0.20	0.00	1.20	1.20	−0.40	0.00	{2}

Table 3-5—Example convergence of the approximate min-max method

Iteration	c_1	c_2	c_3	λ_1	λ_2	f	g_1	g_2	I_e
0	−1.00	−1.00	0.00	0.00	0.00	3.00	−1.00	−1.00	{−}
1	0.00	0.00	0.00	0.00	0.00	0.00	1.00	2.00	{1, 2}
2	−2.00	−2.00	−0.67	0.00	4.00	13.33	−3.67	−4.67	{2}
3	−0.60	−0.60	−0.20	0.00	1.20	1.20	−0.40	0.00	{2}

constraint 1 becomes ineffective on iteration 2. Table 3-5 illustrates similar convergence and also illustrates that the solution point is achieved independently of the initial point.

Data given second for this example are based on (3-102) and on (3-94) for the maximization of $V_{\min}$ subject to $(\hat{\boldsymbol{\lambda}}+\mathbf{l}) \geqslant \mathbf{0}$. These data are also given for $\alpha = 1$, $\mathbf{P} = \boldsymbol{\Gamma}_n^{-1}$, and $\mathbf{Q} = L_{\hat{\lambda}\mathbf{c}}$. Table 3-6 illustrates convergence of the sequential min-max method with $\boldsymbol{\Gamma}_0 = L_{\mathbf{cc}}^{-1}$ for a quadratic function and linear inequality constraints. The sequential min-max method is a sequential Newton-Raphson method under these conditions and hence two-step convergence results. Table 3-7, on the other hand, illustrates convergence of the sequential min-max method with $\boldsymbol{\Gamma}_0 = \mathbf{I}$. Under these conditions, three iterations are required to obtain $\mathbf{P} = L_{\mathbf{cc}}$ by the Fletcher-Powell method and hence five-step convergence results.

Table 3-6—Example convergence of the sequential min-max method with $\boldsymbol{\Gamma}_0 = L_{\mathbf{cc}}^{-1}$

Iteration	c_1	c_2	c_3	λ_1	λ_2	f	g_1	g_2	I_e
0	3.00	−3.00	0.00	0.00	0.00	27.00	1.00	−1.00	{1}
1	0.00	0.00	0.00	0.00	1.20	0.00	1.00	2.00	{1, 2}
2	−0.60	−0.60	−0.20	0.00	1.20	1.20	−0.40	0.00	{2}

Table 3-7—Example convergence of the sequential min–max method with $\mathbf{\Gamma}_0 = \mathbf{I}$

Iteration	c_1	c_2	c_3	λ_1	λ_2	f	g_1	g_2	I_e
0	3.00	−3.00	0.00	0.00	0.00	27.00	1.00	−1.00	{1}
1	1.33	0.33	0.00	0.00	0.00	2.00	2.67	4.00	{1, 2}
2	0.00	0.00	0.00	0.00	0.80	0.00	0.00	2.00	{1, 2}
3	−0.13	−0.13	−0.27	0.00	1.09	0.27	0.27	1.33	{1, 2}
4	−0.54	−0.54	−0.16	0.00	1.20	0.96	0.96	0.21	{2}
5	−0.60	−0.60	−0.20	0.00	1.20	1.20	1.20	0.00	{2}

3.9 SUMMARY

Gradient methods for locating relative minima are found to be either successive substitution procedures typified by Newton–Raphson or successive approximation procedures typified by steepest-descent. Successive approximations involve both a selection procedure for direction vectors and a trial procedure for one-dimensional searches. In addition to steepest-descent, two other procedures for selecting direction vectors are presented; namely, conjugate gradient and Fletcher–Powell. These two methods share the property of conjugate directions and hence n-step convergence for $\mathbf{c} \in \mathscr{R}^n$ when applied to quadratic functions. The Fletcher–Powell method has the additional useful property of generating an approximation to $f_{\mathbf{cc}}^{-1}$. Convergence of all methods presented is restricted to stationary points that satisfy Theorem 2-3 and hence are not degenerate.

Gradient methods of successive substitution are extended directly to relative minima that are constrained by equality constraints using the Lagrangian formulation. However, gradient methods of successive approximations require the two step min–max procedure in order to be applicable to the Lagrangian formulation for equality constraints. Minimization with respect to parameter $\mathbf{c}$ generally is performed by the Fletcher–Powell method in order to indirectly compute $L_{\mathbf{cc}}$ and $L_{\boldsymbol{\lambda}\mathbf{c}}$. Maximization with respect to Lagrange multipliers $\boldsymbol{\lambda}$ can then be performed by any gradient method presented for unconstrained relative minima. Convergence of all methods presented for equality constraints is restricted to stationary points which satisfy Theorem 2-9. The min–max method of successive approximations, however, is restricted to stationary points at which $L_{\mathbf{cc}}$ positive definite also applies.

Gradient methods for equality constraints are also applicable to problems with inequality constraints when slack variables are introduced. However, successful initialization of these procedures in essence requires identification of the set of constraints that is effective at a solution point. Moreover, convergence is restricted to stationary points which satisfy Theorem 2-14 and perhaps at points where L_{cc} positive definite also applies. Alternately, the min–max method of successive approximations can be applied directly to inequality constraints by extending the notion of effective constraints to points other than a solution point. When this method is applied to inequality constraints, however, maximization generally is performed on the quadratic function given in (3-113) instead of on the function $L[(\mathbf{c} - \mathbf{P}^{-1}\mathbf{Q}'\mathbf{l}), \hat{\boldsymbol{\lambda}} + \mathbf{l}]$. The maximization step often is further approximated by the use of (3-117) and (3-118) so that the min–max method can be used directly with an arbitrary combination of equality and effective inequality constraints.

Although this chapter summarizes the computational methods that are perhaps the most useful for control problems, the reader should be cognizant that many other computational methods exist [9]. These additional methods are used in situations where, for instance, gradients cannot be computed directly. Similarly, assumptions, such as parameters can always be initialized so that all constraints are satisfied, permit the use of additional methods [10, 11]. A more broadly based treatment than presented here of computational methods for minimizing functions is a book in itself.

BIBLIOGRAPHY

1. Wild. D. J.: *Optimum Seeking Methods*, Prentice-Hall, Inc., Englewood Cliffs, 1964.
2. Beveridge, G. S. C. and R. S. Schechter: *Optimization: Theory and Practice*, McGraw-Hill Book Company, New York, 1970.
3. McDaniel, J. G.: "The human operator model: A comparison of parameter optimization methods", M. S. Thesis, Cornell University, Ithaca, 1970.
4. Davidon, W. C.: "Variable metric method for minimization", Argonne National Laboratory, ANL-5990 Rev., University of Chicago, Chicago, 1959.
5. Hestenes, M. R. and E. Stiefel: "Methods of conjugate gradients for solving linear systems", *Journal of Research of the National Bureau of Standards*, **49**, No. 6, December, 1952.
6. Fletcher, R. and C. M. Reeves: "Function minimization by conjugate gradients", *The Computer Journal*, **7**, 1964.
7. Fletcher, R. and M. J. D. Powell: "A rapidly convergent descent method for minimization", *The Computer Journal*, **6**, 1963.
8. Merriam, C. W. III and D. Jordan: "A computational method for parameter optimization problems arising in control", *International Journal of Control*, **14**, 385–397.
9. Wilde, D. J. and C. S. Beightler: *Foundations of Optimization*, Prentice-Hall, Inc., Englewood Cliffs, 1967.
10. Rosen, J. B.: "The gradient projection method for non-linear programming, Part I, Linear constraints", *J. Soc. Ind. Appl. Math.*, **8**, 1960.
11. Rosen, J. B.: "The gradient projection method for non-linear programming, Part II, Non-linear constraints", *J. Soc. Ind. Appl. Math.*, **9**, 1961.

PROBLEMS

3-1 A penalty function of one variable sometimes used in parameter optimization is

$$f(\gamma) = \frac{2\epsilon^2}{2m(2m-1)}\left(\frac{\gamma}{\epsilon}\right)^{2m} \qquad \text{for} \quad m = 1, 2, \ldots .$$

Discuss the selection of $\mathbf{\Gamma}$ used in gradient methods by comparing solutions of (3-3) for $\mathbf{\Gamma} = \mathbf{I}$ and $\mathbf{\Gamma} = f_{\gamma\gamma}^{-1}$ with various m and γ_0.

3-2 Compute two iterations of the steepest-descent method for

$$f(\mathbf{c}) = \tfrac{1}{2}\mathbf{c}'\begin{bmatrix} 1 & 1 \\ 1 & 2 \end{bmatrix}\mathbf{c}$$

with $\mathbf{c}_0' = [1 \quad 0]$.

3-3 Derive the Newton–Raphson equation of successive substitution for

$$f(c) = c^3 - 3bc + 2b^{3/2},$$

and determine intervals of c on which the procedure locates a relative maximum and a relative minimum.

3-4 Repeat Problem 3-2 using the conjugate gradient method.

3-5 Compute four iterations of the conjugate gradient method for (3-43) with

$$\mathbf{F} = \begin{bmatrix} 1 & 2 & -1 & 1 \\ 2 & 5 & 0 & 2 \\ -1 & 0 & 6 & 0 \\ 1 & 2 & 0 & 3 \end{bmatrix}, \quad \mathbf{b} = \begin{bmatrix} -65 \\ 24 \\ -11 \\ 6 \end{bmatrix}, \quad \text{and} \quad \mathbf{c}_0 = \begin{bmatrix} 1 \\ 0 \\ 0 \\ 0 \end{bmatrix}.$$

3-6 Show that (3-50) used for the conjugate gradient method can be replaced by

$$\beta_i = -\frac{\mathbf{e}_{i+1}'\mathbf{F}\mathbf{d}_i}{\mathbf{d}_i'\mathbf{F}\mathbf{d}_i}.$$

3-7 When $m = n$ as defined in Theorem 3-3, show that

$$\det \mathbf{F} = \prod_{i=0}^{n-1} \alpha_i^{-1}$$

where α_i is given by (3-47).

3-8 When $m = n$ as defined in Theorem 3-3, show that

$$\mathbf{F}^{-1} = [\phi_{ij}]$$

where

$$\phi_{ij} = \sum_{l=0}^{n-1} \frac{d_{li}d_{lj}}{\mathbf{d}_l'\mathbf{F}\mathbf{d}_l}$$

and d_{lk} is the kth element of $\mathbf{d}_l$.

3-9 Repeat Problem 3-2 using the Fletcher–Powell method with $\mathbf{\Gamma}_0 = \mathbf{I}$.

3-10 When $m = n$ as defined in Theorem 3-4, show that

$$\mathbf{F}^{-1} = \sum_{i=0}^{n-1} \mathbf{A}_i$$

where $\mathbf{A}_i$ is defined by (3-71).

3-11 Write a computer program for the trial procedure summarized in Figure 3-2 assuming $f(\gamma)$, f_γ, and $f_{\gamma\gamma}$ if needed are programmed in a subroutine by the user.

3-12 Write a computer program for the iteration procedure summarized in Figure 3-1 which includes the results of Problem 3-11 and also the steepest-descent, conjugate gradient, Fletcher–Powell, and Newton–Raphson methods of successive approximations as options.

3-13 Compare steepest-descent, conjugate gradient, Fletcher–Powell, and Newton–Raphson methods for

$$f(\mathbf{c}) = 100(c_2 - c_1^2)^2 + (1 - c_1)^2$$

with $\mathbf{c}_0' = [-1.2 \quad 1.0]$ by tabulating f, $f_\mathbf{c}$, number of trials per iteration, and the ratios appearing in (3-38) and (3-42).

3-14 Repeat Problem 3-13 for

$$f(\mathbf{c}) = (c_1 + 10c_2)^2 + 5(c_3 - c_4)^2 + (c_2 - 2c_3)^4 + 10(c_1 - c_4)^4$$

with $\mathbf{c}_0' = [3 \quad -1 \quad 0 \quad 1]$.

3-15 Repeat Problem 3-13 for

$$f(\mathbf{c}) = 100\left[\left(c_3 - \frac{5}{\pi}\tan^{-1}\frac{c_2}{c_1}\right)^2 + \left(\sqrt{c_1^2 + c_2^2} - 1\right)^2\right] + c_3^2$$

with $\mathbf{c}_0' = [-1 \quad 0 \quad 0]$.

3-16 Suppose $f(\mathbf{c})$ is to be minimized subject to the requirement that $|g_i(\mathbf{c})| < \epsilon_i$ for $i = 1, 2, \ldots, m$. A method for formulating such a problem is to introduce a penalty function $h(x)$ and minimize F defined as

$$F(\mathbf{c}) = f(\mathbf{c}) + \sum_{i=1}^{m} h[g_i(\mathbf{c})/\epsilon_i].$$

a) Discuss suitable structural characteristics of penalty function h.
b) Assuming $f(\mathbf{c})$ and $h(x)$ are strictly convex and twice continuously differentiable functions and that g is twice continuously differentiable, what can be said about the convexity of F?
c) Repeat part (b) assuming that

$$\mathbf{g} = \mathbf{G}(\mathbf{c} - \mathbf{a}).$$

d) Presumably, the solution of the above problem approaches the solution of minimizing $f(\mathbf{c})$ subject to $\mathbf{g}(\mathbf{c}) = \mathbf{0}$ when $|\epsilon| \to 0$. What terms in the penalty function formulation are equivalent to Lagrange multipliers? Give a "rule-of-thumb" concerning the sign of Lagrange multipliers for equality constraint problems.

3-17 In conjunction with the min–max method applied to (3-99)
a) Rewrite (3-102) and (3-104) in terms of vectors and matrices given in (3-100) and (3-101).
b) Let c and $\boldsymbol{\lambda}$ be parameterized such that

$$\boldsymbol{\gamma} = \mathbf{c}(t+\alpha) - \mathbf{c}(t) \quad \text{and} \quad \mathbf{l} = \boldsymbol{\lambda}(t+\alpha) - \boldsymbol{\lambda}(t)$$

and then derive differential equations corresponding to the continuous form of the min–max method for equality constraints.

3-18 Verify (3-97).

3-19 Verify (3-98).

3-20
a) Give equations of the Newton–Raphson method of successive substitutions for the Lagrangian formulation of inequality constraints using slack variables defined by (3-90) and (3-91).
b) Give explicit expressions for these equations when $\dim \mathbf{c} = \dim \mathbf{g} = 1$.
c) Discuss the iterative behavior of these equations in terms of the initialization of Lagrange multipliers and slack variables.

3-21 Extend (3-83) to inequality constraints using $\boldsymbol{\Gamma} = \mathbf{I}$.

3-22
a) Use the differential equations derived in Problem 3-21 to minimize

$$f(\mathbf{c}) = \tfrac{1}{2}\mathbf{c}' \begin{bmatrix} 2 & 0 & 0 \\ 0 & 4 & 0 \\ 0 & 0 & 6 \end{bmatrix} \mathbf{c}$$

subject to

$$\mathbf{g} = \begin{bmatrix} -1 & -1 & -1 \\ 1 & 2 & 1 \end{bmatrix} \mathbf{c} + \begin{bmatrix} -1 \\ 2 \end{bmatrix} \leqslant \mathbf{0}$$

with $\mathbf{c}_0 = \mathbf{0}$ and $\boldsymbol{\lambda}_0 = \mathbf{0}$.
b) Sketch g and $\boldsymbol{\lambda}$ as functions of time.

3-23 Using (3-117) and (3-118) instead of (3-104) along with appropriate redefinitions for effective constraints, extend the continuous min–max method to inequality constraints.

3-24 Repeat Problem 3-22 using the differential equations derived in Problem 3-23.

3-25 Repeat Problem 3-24 for

$$\mathbf{g} = \begin{bmatrix} 1 & 1 & 1 \\ 1 & 2 & 1 \end{bmatrix} \mathbf{c} + \begin{bmatrix} 1 \\ 2 \end{bmatrix} \leqslant \mathbf{0}.$$

3-26 Write a computer program for the min–max method with equaltiy constraints. Use the Fletcher–Powell method for the minimization step, and use (3-114) in conjunction with $L[(\mathbf{c} - \mathbf{P}^{-1}\mathbf{Q}'\mathbf{l}), (\boldsymbol{\lambda} + \mathbf{l})]$ to form the maximization step in one dimension.

3-27 Demonstrate the computer program written for Problem 3-26 by solving Problem 2-29.

3-28 Extend the computer program written for Problem 3-26 to apply to both equality and inequality constraints using expressions similar to (3-117) and (3-118).

3-29 Demonstrate the computer program written for Problem 3-28 by minimizing

$$f(\mathbf{c}) = \frac{c_1}{c_2}$$

subject to inequality constraints

$$\frac{c_2^2 + c_1}{c_1 c_2} \leqslant 1 \quad \text{and} \quad c_2 \leqslant 2.$$

3-30 Solve Problem 1-22 using the computer program written in Problem 3-28.

CHAPTER 4

Deterministic design problems

Control system design problems with deterministic signals arise frequently and also arise in many forms. These problems are described by differential equations which are often of rather high order. Moreover, these problems often involve numerous parameters that are to be selected by automated design techniques. Typical dimensionality encountered in these problems is an important factor in the selection of computational methods discussed in Chapter 3.

All design problems discussed in this chapter can be characterized in the following abbreviated form. Closed-loop state equations are linear and are written in the form

$$\dot{\mathbf{x}} = \mathbf{A}\mathbf{x}; \quad \mathbf{x}(0) = \mathbf{x}_0 \tag{4-1}$$

where $\mathbf{A} = \mathbf{A}(\mathbf{c})$ is given and is assumed to be asymptotically stable for all $\mathbf{c}$ of interest. Vector $\mathbf{c}$, of course, includes all parameters that are to be specified by automated design procedures. In addition, a performance integral

$$I = \int_0^\infty \{\tfrac{1}{2}\mathbf{x}'(t)\mathbf{W}\mathbf{x}(t)\}\, \mathrm{d}t \tag{4-2}$$

is assigned to the transient response of (4-1) where $\mathbf{W} = \mathbf{W}(\mathbf{c})$ is assumed to be positive semi-definite for all $\mathbf{c}$ of interest. Use of state transition equation

$$\mathbf{x}(t) = \mathrm{e}^{\mathbf{A}t}\mathbf{x}_0 \tag{4-3}$$

permits rewriting (4-2) in the more convenient form

$$I = \tfrac{1}{2}\mathbf{x}_0'\mathbf{X}\mathbf{x}_0, \tag{4-4}$$

where

$$\mathbf{X} = \int_0^\infty \{\mathrm{e}^{\mathbf{A}'t}\mathbf{W}\,\mathrm{e}^{\mathbf{A}t}\}\, \mathrm{d}t. \tag{4-5}$$

An alternate and more convenient form for **X** is derived from integration by parts occurring in

$$\mathbf{XA} = \int_0^\infty \{e^{\mathbf{A}'t}\mathbf{WA}\, e^{\mathbf{A}t}\}\, dt$$

$$= e^{\mathbf{A}'t}\mathbf{W}\, e^{\mathbf{A}t}\Big|_{t=0}^{t=\infty} - \int_0^\infty \{\mathbf{A}'\, e^{\mathbf{A}'t}\mathbf{W}\, e^{\mathbf{A}t}\}\, dt.$$

In particular, matrix **X** is the unique solution of the well-known relationship

$$\mathbf{XA} + \mathbf{A}'\mathbf{X} + \mathbf{W} = \mathbf{0} \tag{4-6}$$

assuming **A** is asymptotically stable.

All parameter optimization problems arising in this chapter involve not only vector **c** but also matrix **X** and its dependence on **c**. Specifically, a relative minimum of $f = f(\mathbf{c}, \mathbf{X})$ is sought subject to constraints

$$\mathbf{g}_1(\mathbf{c}, \mathbf{X}) = \mathbf{0}_1 \quad \text{and} \quad \mathbf{g}_2(\mathbf{c}, \mathbf{X}) \leqslant \mathbf{0}_2$$

as well as the constraint that **A** is asymptotically stable. Functions f and **g** are assumed to be twice continuously differentiable throughout as are matrices **A** and **W**.

Dependence of f and **g** on **X** and dependence of **X** on **c** pose some new aspects of the computational problem encountered in automated design. These computational considerations are discussed at the outset of this chapter in the absence of a detailed discussion of typical problem formulations. However, the main thrust of this chapter is the formulation of meaningful control system design problems with deterministic signals. Many considerations are pertinent to design problem formulation including the origin and compatibility of design specifications, feasibility of system implementation and control system structure, as well as existence and uniqueness of optimal parameters.

4.1 FORMULATION OF NECESSARY CONDITIONS

In order to conveniently account for the dependence of f and **g** on **X** as well as **c**, the methods illustrated in Section 2.8 are employed here for formulating gradients and hessians [1]. Specifically, equation (4-6) is treated as an additional equality constraint by the introduction of Lagrange multipliers.

Gradients and hessians are expressed conveniently in terms of the Lagrangian

$$L = F + \mathrm{tr}\{\mathbf{\Lambda}\, \mathbf{Z}\} \tag{4-7}$$

where $\mathbf{\Lambda}$ is a matrix of Lagrange multipliers and

$$\mathbf{Z} = \mathbf{XA} + \mathbf{A}'\mathbf{X} + \mathbf{W}. \tag{4-8}$$

Furthermore, function

$$F = f + \hat{\boldsymbol{\lambda}}'\hat{\mathbf{g}} \tag{4-9}$$

is the portion of the Lagrangian that includes both equality and effective inequality constraints contained in $\mathbf{g}$. When the results of this derivation are used for computational purposes, effective inequality constraints are defined by (3-115). Note also that $\boldsymbol{\lambda}'\mathbf{z} = \operatorname{tr}\{\mathbf{\Lambda}'\mathbf{Z}\}$ when vector partitions of $\boldsymbol{\lambda}$ and $\mathbf{z}$ are defined by

$$\mathbf{\Lambda} = [\boldsymbol{\lambda}_1 \quad \boldsymbol{\lambda}_2 \quad \ldots] \quad \text{and} \quad \mathbf{Z} = [\mathbf{z}_1 \quad \mathbf{z}_2 \quad \ldots]$$

respectively.

Conditions for stationarity of Lagrangian L now become

$$L_{\mathbf{\Lambda}} = \mathbf{0}, \quad L_{\mathbf{X}} = \mathbf{0}, \quad L_{\mathbf{c}} = \mathbf{0}, \quad \text{and} \quad L_{\hat{\lambda}} = \mathbf{0} \tag{4-10}$$

where of course

$$L_{\hat{\lambda}} = F_{\hat{\lambda}}. \tag{4-11}$$

These conditions are also necessary for a constrained relative minimum.

All gradient methods are based on a direct calculation of gradients and hence require solution of the following equations. Equations

$$\mathbf{XA} + \mathbf{A}'\mathbf{X} + \mathbf{W} = \mathbf{0} \tag{4-12}$$

and

$$\mathbf{\Lambda A}' + \mathbf{A\Lambda} + F_{\mathbf{X}} = \mathbf{0} \tag{4-13}$$

result from the first two conditions appearing in (4-10). Elements of gradient $L_{\mathbf{c}}$ also become

$$L_{c_i} = F_{c_i} + \operatorname{tr}\{\mathbf{\Lambda}'\mathbf{Z}_{c_i}\} \tag{4-14}$$

where

$$\mathbf{Z}_{c_i} = \mathbf{XA}_{c_i} + \mathbf{A}'_{c_i}\mathbf{X} + \mathbf{W}_{c_i} \tag{4-15}$$

for each i. These equations are solved on each trial in the selection of α for each iteration.

Whenever directly calculated hessians are used for matrices $\mathbf{P}$ and $\mathbf{Q}$ appearing in the min–max method, additional equations must be solved on each iteration. Specifically, the selection corresponding to

$$p_{ij} = \frac{\mathrm{d}}{\mathrm{d}c_j}\left(\frac{\partial L}{\partial c_i}\right) \quad \text{and} \quad q_{ij} = \frac{\mathrm{d}}{\mathrm{d}c_j}\left(\frac{\partial L}{\partial \lambda_i}\right) \tag{4-16}$$

conforms to the methods illustrated in Section 2.8 and requires solution of

$$\mathbf{X}_j\mathbf{A} + \mathbf{A}'\mathbf{X}_j + \mathbf{Z}_{c_j} = \mathbf{0} \tag{4-17}$$

and

$$\mathbf{\Lambda}_j\mathbf{A}' + \mathbf{A}\mathbf{\Lambda}_j + \left(\mathbf{\Lambda}\mathbf{A}'_{c_j} + \mathbf{A}_{c_j}\mathbf{\Lambda} + \frac{\mathrm{d}}{\mathrm{d}c_j} F_{\mathbf{X}}\right) = \mathbf{0} \tag{4-18}$$

for each j. Matrices $\mathbf{X}_j$ and $\mathbf{\Lambda}_j$ are defined by the total derivatives of relationships appearing in (4-12) and (4-13) with respect to c_j and are computed from (4-17) and (4-18) respectively. These matrices then are used to compute

$$p_{ij} = \frac{\mathrm{d}}{\mathrm{d}c_j} F_{c_i} + \mathrm{tr}\left\{\mathbf{\Lambda}'_j\mathbf{Z}_{c_i} + \mathbf{\Lambda}'\left(\frac{\mathrm{d}}{\mathrm{d}c_j}\mathbf{Z}_{c_i}\right)\right\} \tag{4-19}$$

and

$$q_{ij} = \frac{\mathrm{d}}{\mathrm{d}c_j} F_{\lambda_i} \tag{4-20}$$

where

$$\frac{\mathrm{d}}{\mathrm{d}c_j}\mathbf{Z}_{c_i} = \mathbf{Z}_{c_ic_j} + \mathbf{X}_j\mathbf{A}_{c_i} + \mathbf{A}'_{c_i}\mathbf{X}_j \tag{4-21}$$

and

$$\mathbf{Z}_{c_ic_j} = \mathbf{X}\mathbf{A}_{c_ic_j} + \mathbf{A}'_{c_ic_j}\mathbf{X} + \mathbf{W}_{c_ic_j}. \tag{4-22}$$

Terms involving the function F cannot be given explicitly until performance functional f and constraint functions $\mathbf{g}$ are selected for the particular problem at hand.

Equations (4-17) through (4-22) of course are relatively complex, and matrices $\mathbf{A}_{c_ic_j}$ and $\mathbf{W}_{c_ic_j}$ may require large computer memory. In specialized problems, however, some of these equations and matrices may not be needed. The chief advantage of the selection of **P** and **Q** defined by (4-16) is rapid convergence in a suitably small neighborhood of a solution. On the other hand, many design problems of interest are characterized by large dimensions, and hence the Fletcher-Powell method must be used to compute **P** and **Q** indirectly when using the min-max method for problems with constraints.

4.2 METHODS OF SOLVING XA + BX + C = 0

The complexity of computing gradients and hessians given in the previous section is primarily attributed to solving linear matrix equations of the form appearing in (4-6). Therefore, a discussion of efficient computational methods is presented in this section for such equations. Before a discussion of computational methods is presented, however, basic properties of solutions to

$$\mathbf{XA} + \mathbf{BX} + \mathbf{C} = \mathbf{0} \tag{4-23}$$

are discussed here. Coefficient matrices appearing in (4-23) are taken to have dimensions dim $\mathbf{A} = n \times r$, and dim $\mathbf{B} = m \times m$ and hence dim $\mathbf{C}$ = dim $\mathbf{X} = m \times n$. This form of equation is investigated here so that results for any matrix partition of the equations given in (4-3) can be obtained.

The basic result concerning existence and uniqueness of solutions to (4-23) is well-known [2] and is stated here as

Lemma 4-1 *Equation* (*4-23*) *has a unique solution* $\mathbf{X}$ *if and only if* $\lambda_i\{\mathbf{A}\} + \lambda_j\{\mathbf{B}\} \neq 0$ *for all i, j.*

Proof Suppose

$$\lambda = \lambda_i\{\mathbf{A}\} = -\lambda_j\{\mathbf{B}\}$$

for some *i, j.* Then

$$\mathbf{X} = \overline{\mathbf{X}} + \mathbf{mn}'$$

is a solution to (4-23) when $\overline{\mathbf{X}}$ is also a solution where $\mathbf{n}$ and $\mathbf{m}$ are eigenvectors defined by

$$\mathbf{A}'\mathbf{n} = \lambda\mathbf{n} \quad \text{and} \quad \mathbf{Bm} = -\lambda\mathbf{m}.$$

This assertion is verified by direct substitution into (4-23) and hence proves necessity of the eigenvalue condition given in the lemma statement. Sufficiency is proved by a construction that can be also used for computational purposes. Let eigenvalues of $\mathbf{A}$ be denoted as λ_i so that

$$\prod_{i=1}^{n} (\mathbf{A} - \lambda_i\mathbf{I}) = \mathbf{0} \tag{4-24}$$

holds by the Cayley-Hamilton theorem [3]. Then adding and subtracting $\lambda_i\mathbf{X}$ to (4-23) yields

$$\mathbf{0} = \mathbf{X}(\mathbf{A} - \lambda_i\mathbf{I}) + (\lambda_i\mathbf{I} + \mathbf{B})\mathbf{X} + \mathbf{C} \tag{4-25}$$

for any *i.* In addition, equation (4-25) with $i = 1$ is post multiplied by the singular matrix formed by all terms of (4-24) except $(\mathbf{A} - \lambda_1\mathbf{I})$, thereby obtaining

$$\mathbf{0} = (\lambda_1\mathbf{I} + \mathbf{B})\mathbf{X} \prod_{i=2}^{n} (\mathbf{A} - \lambda_i\mathbf{I}) + \mathbf{C} \prod_{i=2}^{n} (\mathbf{A} - \lambda_i\mathbf{I}). \tag{4-26}$$

Factor $\mathbf{X}(\mathbf{A} - \lambda_2\mathbf{I})$ can now be eliminated from (4-26) by use of (4-25) with $i = 2$, thereby obtaining

$$\mathbf{0} = -(\lambda_1\mathbf{I} + \mathbf{B})(\lambda_2\mathbf{I} + \mathbf{B})\mathbf{X} \prod_{i=3}^{n} (\mathbf{A} - \lambda_i\mathbf{I}) - (\lambda_1\mathbf{I} + \mathbf{B})\mathbf{C} \prod_{i=3}^{n} (\mathbf{A} - \lambda_i\mathbf{I})$$
$$+ \mathbf{C} \prod_{i=2}^{n} (\mathbf{A} - \lambda_i\mathbf{I}).$$

This procedure is repeated successively thereby eliminating factors $\mathbf{X}(\mathbf{A} - \lambda_i\mathbf{I})$ for $i = 3, 4, \ldots, n$. The final result is

$$\mathbf{0} = \mathbf{MX} + \mathbf{N} \tag{4-27}$$

where

$$\mathbf{M} = \prod_{i=1}^{n} (\lambda_i\mathbf{I} + \mathbf{B}) \tag{4-28}$$

and

$$\mathbf{N} = \sum_{j=1}^{n} \left\{ (-1)^{(n-j)} \left[\prod_{i=1}^{j-1} (\lambda_i\mathbf{I} + \mathbf{B}) \right] \mathbf{C} \left[\prod_{i=j+1}^{n} (\mathbf{A} - \lambda_i\mathbf{I}) \right] \right\}. \tag{4-29}$$

Equation (4-27) has a unique solution if $\det \mathbf{M} \neq 0$, and $\det \mathbf{M} \neq 0$ if the eigenvalue condition given the lemma statement holds. Finally, by construction, equation (4-23) has a unique solution $\mathbf{X}$ if (4-27) has the unique solution

$$\mathbf{X} = -\mathbf{M}^{-1}\mathbf{N}. \tag{4-30}$$

A somewhat more restrictive condition on the eigenvalues of $\mathbf{A}$ and $\mathbf{B}$ often occurs and leads to the principle result concerning solutions of (4-23); namely,

Theorem 4-1 *Equation (4-23) has the unique solution*

$$\mathbf{X} = \int_0^\infty \{e^{\mathbf{B}t}\mathbf{C}\, e^{\mathbf{A}t}\}\, dt \tag{4-31}$$

if $\mathrm{Re}[\lambda_i\{\mathbf{A}\}] + \mathrm{Re}[\lambda_j\{\mathbf{B}\}] < 0$ *for all* i, j.

Proof The real part condition given in the theorem statement assures a unique solution to (4-23) by Lemma 4-1. Moreover, this real part condition implies that there exist constants $\gamma, \Gamma > 0$ such that

$$\| e^{\mathbf{B}t}\mathbf{C}\, e^{\mathbf{A}t} \| \leqslant \Gamma\, e^{-\gamma t}\ \forall\ t.$$

Norm bounding and integration then yield

$$\left\| \int_0^\infty \{e^{\mathbf{B}t}\mathbf{C}\, e^{\mathbf{A}t}\}\, dt \right\| \leqslant \frac{\Gamma}{\gamma},$$

and hence the integral appearing in (4-31) exists. Finally, evaluation of **XA** with **X** given by (4-31) and integration by parts similar to that used to derive (4-6) yield (4-23).

An immediate consequence of Theorem 4-1 is the special case summarized in

Corollary 4-1 *The solution* **X** *of (4-23) is positive (semi-) definite when* **A** *is asymptotically stable,* $\mathbf{B} = \mathbf{A}'$, *and* **C** *is positive (semi-) definite.*

Proof By Theorem 4-1, solution **X** becomes

$$\mathbf{X} = \int_0^\infty \{e^{\mathbf{A}'t}\mathbf{C}\, e^{\mathbf{A}t}\}\, dt. \tag{4-32}$$

Moreover, a matrix **H** exists such that

$$\mathbf{C} = \mathbf{H}'\mathbf{H}$$

given any positive semi-definite matrix, and **H** is nonsingular when **C** is nonsingular. Now let $\mathbf{x}_0$ be any constant nonzero vector and

$$\mathbf{y}(t) = \mathbf{H}\, e^{\mathbf{A}t}\mathbf{x}_0.$$

Then

$$\mathbf{x}_0'\mathbf{X}\mathbf{x}_0 = \int_0^\infty \mathbf{y}'(t)y(t)\, dt \left(\geq\right) 0$$

holds for all nonzero $\mathbf{x}_0$ and hence **X** is positive (semi-) definite. Solutions given in (4-32) arise frequently in stability theory as well as in optimal control and filter theories.

A number of methods are available for solving (4-23) computationally, and a brief introduction to these methods is presented here. Throughout most of the discussion, the eigenvalue condition of Theorem 4-1 is assumed to hold.

Solutions to (4-23) sometimes are computed by numerical integration as equilibrium solutions to

$$\dot{\mathbf{X}} = \mathbf{XA} + \mathbf{BX} + \mathbf{C}; \quad \mathbf{X}(0) = \mathbf{X}_0. \tag{4-33}$$

Solutions to these linear differential equations can be expressed as

$$\mathbf{X} = e^{\mathbf{B}t}\left[\mathbf{X}_0 + \int_0^t \{e^{-\mathbf{B}\xi}\mathbf{C}\, e^{-\mathbf{A}\xi}\}\, d\xi\right] e^{\mathbf{A}t}$$

or rather as

$$\mathbf{X} = e^{\mathbf{B}t}\mathbf{X}_0\, e^{\mathbf{A}t} + \int_0^t e^{\mathbf{B}\eta}\mathbf{C}\, e^{\mathbf{A}\eta}\, d\eta \tag{4-34}$$

so that

$$\dot{\mathbf{X}} = e^{\mathbf{B}t}(\mathbf{X}_0\mathbf{A} + \mathbf{B}\mathbf{X}_0 + \mathbf{C})\, e^{\mathbf{A}t}.$$

Unfortunately, these relationships reveal that numerical integration generally is very inefficient and requires excessive computer time unless $\mathbf{X}_0$ can be specified *a priori* as a very close approximation to the equilibrium solution of (4-33).

An alternate approach to solving (4-23) is based on direct matrix inversion. In particular, vectors $\mathbf{x}_v$ and $\mathbf{c}_v$ are defined in terms of

$$\mathbf{X} = [\mathbf{x}_1 \quad \mathbf{x}_2 \ldots \mathbf{x}_n] \quad \text{and} \quad \mathbf{C} = [\mathbf{c}_1 \quad \mathbf{c}_2 \ldots \mathbf{c}_n]$$

as

$$\mathbf{x}_v' = [\mathbf{x}_1' \quad \mathbf{x}_2' \ldots \mathbf{x}_n'] \quad \text{and} \quad \mathbf{c}_v' = [\mathbf{c}_1' \quad \mathbf{c}_2' \ldots \mathbf{c}_n'] \tag{4-35}$$

so that (4-23) can be reduced to standard vector-matrix form

$$\boldsymbol{\Delta}\mathbf{x}_v + \mathbf{c}_v = \mathbf{0}. \tag{4-36}$$

Elements of $\boldsymbol{\Delta}$ are easily formed in terms of elements of $\mathbf{A}$ and $\mathbf{B}$ in a digital computer program. Direct matrix inversion then yields

$$\mathbf{x}_v = -\boldsymbol{\Delta}^{-1}\mathbf{c}_v. \tag{4-37}$$

If design problem dimensions are suitably small, this procedure of solving (4-23) is eminently suited to computing gradients and hessians because (4-12) and (4-17) for each j have the same matrix $\boldsymbol{\Delta}$ as do (4-13) and (4-18) for each j. However, matrix $\boldsymbol{\Delta}$ with dim $\boldsymbol{\Delta} = (mn) \times (mn)$ becomes impossibly large to store and invert for many design problems of interest.

Dimensionality difficulties of matrix inversion methods are considerably relieved by use of (4-30) where dim $\mathbf{M} = m \times m$. If $n < m$, this method can be reformulated with an $n \times n$ matrix inversion to solve for $\mathbf{X}'$ by replacing $\mathbf{A}$, $\mathbf{B}$, and $\mathbf{C}$ with $\mathbf{B}'$, $\mathbf{A}'$, and $\mathbf{C}'$ respectively. The primary disadvantage of using (4-30) for computation of gradients and hessians is that $\mathbf{N}$ must be recomputed for each equation. In addition, eigenvalues of $\mathbf{A}$ must be determined, and this step may be a formidable computational task in itself. Finally, all computations involving $\mathbf{M}$ and $\mathbf{N}$ are performed with complex numbers because eigenvalues of $\mathbf{A}$ generally are complex.

These last two computational drawbacks associated with (4-30) are alleviated as follows. Let $p(\lambda)$ be a polynomial in λ and let

$$\mathbf{J} = \begin{bmatrix} -\mathbf{A} & \mathbf{0} \\ \mathbf{C} & \mathbf{B} \end{bmatrix}. \tag{4-38}$$

Also let matrices $\mathbf{L}$, $\mathbf{M}$, and $\mathbf{N}$ be defined by

$$p(\mathbf{J}) = \begin{bmatrix} \mathbf{L} & \mathbf{0} \\ \mathbf{M} & \mathbf{N} \end{bmatrix} \tag{4-39}$$

for the purpose of stating

Theorem 4-2 *Suppose* $\lambda_i\{\mathbf{A}\} + \lambda_j\{\mathbf{B}\} \neq 0$ *for all i, j and* $p(\lambda)$ *is the characteristic polynomial of* **B**. *Then*

$$\mathbf{X} = \mathbf{M}\mathbf{L}^{-1} \tag{4-40}$$

is the solution to (4-23).

Proof Let **X** be the solution to (4-23) and let

$$\mathbf{T} = \begin{bmatrix} \mathbf{0} & \mathbf{I} \\ \mathbf{I} & \mathbf{X} \end{bmatrix}.$$

Then elementary matrix manipulations yield

$$p(\mathbf{J}) = \mathbf{T}p(\mathbf{T}^{-1}\mathbf{J}\mathbf{T})\mathbf{T}^{-1} = \mathbf{T}\begin{bmatrix} \mathbf{0} & \mathbf{0} \\ \mathbf{0} & p(-\mathbf{A}) \end{bmatrix}\mathbf{T}^{-1}$$

so that

$$\mathbf{L} = p(-\mathbf{A}), \quad \mathbf{M} = \mathbf{X}p(-\mathbf{A}), \quad \text{and} \quad \mathbf{N} = \mathbf{0}.$$

The desired result then follows directly because $p(-\mathbf{A})$ is nonsingular.

If the characteristic polynomial of **B** is expressed as

$$p(\lambda) = \sum_{k=0}^{m} \delta_k \lambda^{(m-k)}, \tag{4-41}$$

then coefficients of this polynomial are given by the well-known [4] relationships

$$\begin{Bmatrix} \delta_0 = 1 \\ \mathbf{\Delta}_0 = \mathbf{I} \end{Bmatrix} \quad \text{and} \quad \begin{Bmatrix} \delta_k = -\dfrac{1}{k}\operatorname{tr}(\mathbf{\Delta}_{k-1}\mathbf{B}) \\ \mathbf{\Delta}_k = (\mathbf{\Delta}_{k-1}\mathbf{B}) + \delta_k \mathbf{I} \end{Bmatrix} \quad \text{for} \quad k = 1, 2, \ldots, m. \tag{4-42}$$

This method of computing the coefficients of $p(\lambda)$ can be combined with the computation of $p(\mathbf{J})$ to form an efficient method of computing **L** and **M.** The principle disadvantage of this novel method for computing **X** stems from inaccuracies in the computation of the coefficients of $p(\lambda)$ with large m. If $n < m$, this method can be reformulated with an nth degree polynomial to solve for $\mathbf{X}'$ by replacing **A, B,** and **C** with $\mathbf{B}'$, $\mathbf{A}'$, and $\mathbf{C}'$ respectively.

When design problem dimensions become relatively large, perhaps a truncated infinite series solution [5] constitutes the best method of computing the solution of (4-23) with acceptable numerical accuracy. An infinite series solution of (4-23) is constructed as follows. Let r be any real nonzero scalar that is not equal to $\lambda_i\{\mathbf{A}\}$ or $\lambda_j\{\mathbf{B}\}$ given any i and j. Also let

$$\mathbf{M} = (r\mathbf{I} - \mathbf{B})^{-1}(r\mathbf{I} + \mathbf{B}), \quad \mathbf{N} = (r\mathbf{I} + \mathbf{A})(r\mathbf{I} - \mathbf{A})^{-1} \tag{4-43}$$

and

$$\mathbf{L} = 2r(r\mathbf{I} - \mathbf{B})^{-1}\mathbf{C}(r\mathbf{I} - \mathbf{A})^{-1}. \tag{4-44}$$

Then expansion of terms appearing in

$$\frac{1}{2r}[(r\mathbf{I} + \mathbf{B})\mathbf{X}(r\mathbf{I} + \mathbf{A}) - (r\mathbf{I} - \mathbf{B})\mathbf{X}(r\mathbf{I} - \mathbf{A})] + \mathbf{C} = \mathbf{0}$$

reveals that this equation is equivalent to (4-23), and hence

$$\mathbf{X} = \mathbf{MXN} + \mathbf{L} \tag{4-45}$$

also is equivalent to (4-23). Successive substitutions defined by

$$\mathbf{X}_{i+1} = \mathbf{M}^{(2^i)}\mathbf{X}_i\mathbf{N}^{(2^i)} + \mathbf{X}_i; \quad \mathbf{X}_0 = \mathbf{L} \tag{4-46}$$

then yield

$$\mathbf{X}_i = \sum_{j=1}^{2^i} \mathbf{M}^{(j-1)}\mathbf{L}\mathbf{N}^{(j-1)} \tag{4-47}$$

which is found to be an approximate solution of (4-45) in the sequel.

A convergence criterion for the series given in (4-47) is stated in terms of the *spectral radius* of a matrix which is defined as

$$\rho\{\mathbf{F}\} = \max_i |\lambda_i\{\mathbf{F}\}|.$$

The spectral radius $\rho\{\mathbf{F}\}$ and *spectral norm* $\|\mathbf{F}\|_2$ defined in (1-25) are related by

Proposition 4-1 *There exist constants* $\gamma, \Gamma > 0$ *for any real matrix* $\mathbf{F}$ *such that*

$$\rho^k\{\mathbf{F}\} \leqslant \|\mathbf{F}^k\|_2 \leqslant \Gamma k^\gamma \rho^k\{\mathbf{F}\} \tag{4-48}$$

holds for each $k > 0$. *In addition,* $\gamma = 0$ *and* $\Gamma = 1$ *can be selected whenever* $\mathbf{F}$ *is diagonalizable.*

This property of the spectral radius is used to prove

Theorem 4-3 *Suppose*

i) $\lambda_i\{\mathbf{A}\} + \lambda_j\{\mathbf{B}\} \neq 0$ *for all i, j*

and

ii) $\rho\{\mathbf{M}\}\rho\{\mathbf{N}\} < 1$ *given some real r.*

Then sequence $\{\mathbf{X}_i\}$ *defined by* (*4-46*) *is convergent and* $\mathbf{X}_\infty$ *is the solution of* (*4-23*).

Proof Use of (4-47) and norm bounding yield

$$\|\mathbf{X}_i - \mathbf{M}\mathbf{X}_i\mathbf{N} - \mathbf{L}\| = \|-\mathbf{M}^{(2^i)}\mathbf{L}\mathbf{N}^{(2^i)}\| \leqslant \|\mathbf{L}\|\,\|\mathbf{M}^{(2^i)}\|\,\|\mathbf{N}^{(2^i)}\|.$$

Therefore, there exist constants $\gamma, \Gamma > 0$ such that

$$\|\mathbf{X}_i - \mathbf{M}\mathbf{X}_i\mathbf{N} - \mathbf{L}\|_2 \leqslant \Gamma(2^i)^\gamma\,[\rho\{\mathbf{M}\}\rho\{\mathbf{N}\}]^{(2^i)}.$$

This upper bound tends to zero as i tends to infinity and hence

$$\mathbf{X}_\infty = \mathbf{M}\mathbf{X}_\infty\mathbf{N} + \mathbf{L}.$$

By Lemma 4-1 and the construction of (4-45), matrix $\mathbf{X}_\infty$ thus must exist and is the unique solution of (4-23).

In addition, norm bounding and (4-48) can be used to show that there exists a constant $\Delta > 0$ such that

$$\|\mathbf{X}_i - \mathbf{X}_\infty\|_2 \leqslant \Delta[1 + \gamma(2^i)^\gamma]\,[\rho\{\mathbf{M}\}\rho\{\mathbf{N}\}]^{(2^i)} \tag{4-49}$$

holds for each $i \geqslant 0$ under conditions of Theorem 4-3.

The series given in (4-47) is seen to converge very rapidly when $\rho\{\mathbf{M}\}\rho\{\mathbf{N}\}$ is significantly less than one, and hence an advantageous selection of r is desired. Manipulation of determinants yields

$$\lambda_i\{\mathbf{M}\} = \frac{r + \lambda_i\{\mathbf{B}\}}{r - \lambda_i\{\mathbf{B}\}} \quad \text{and} \quad \lambda_j\{\mathbf{N}\} = \frac{r + \lambda_j\{\mathbf{A}\}}{r - \lambda_j\{\mathbf{A}\}}$$

for each i, j. Hence $\rho\{\mathbf{M}\}\rho\{\mathbf{N}\}$ is seen to depend upon the mapping from the λ-plane into the z-plane which is defined by

$$z = \frac{r + \lambda}{r - \lambda}. \tag{4-50}$$

Table 4-1 Mapping corresponding to (4-50) with r positive real

λ-plane	z-plane
Left-half plane	Interior of unit circle
Imaginary axis	Unit circle
Right-half plane	Exterior of unit circle

This mapping is summarized in Table 4-1 for r positive real. Moreover, minimization of $|z|$ with respect to r for r real yields

$$r = \begin{Bmatrix} |\lambda| & \text{for} & \mathrm{Re}\{\lambda\} < 0 \\ -|\lambda| & \text{for} & \mathrm{Re}\{\lambda\} > 0 \end{Bmatrix},$$

and the minimum value of z becomes

$$|z|_{\min} = \left[\frac{1 - \dfrac{|\mathrm{Re}\{\lambda\}|}{|\lambda|}}{1 + \dfrac{|\mathrm{Re}\{\lambda\}|}{|\lambda|}} \right]^{1/2}$$

so that $|z|_{\min} \to 1$ when $|\mathrm{Re}\{\lambda\}|/|\lambda| \to 0$.

For the computation of gradients and hessians where $\mathbf{B} = \mathbf{A}'$ and $\mathbf{A}$ is asymptotically stable, parameter r is taken to lie between $\rho\{\mathbf{A}\}$ and $1/\rho\{\mathbf{A}^{-1}\}$. Further analysis of $|z|$ suggests the selection

$$r \cong [\rho\{\mathbf{A}\}/\rho\{\mathbf{A}^{-1}\}]^{1/2}$$

assuming the ratio $|\mathrm{Re}\{\lambda_i\}|/|\lambda_i|$ is roughly independent of i where $\lambda_i = \lambda_i\{\mathbf{A}\}$. The primary disadvantage of the series method is a relativey slow rate of convergence when either $\rho\{\mathbf{A}\}$ and $\rho^{-1}\{\mathbf{A}^{-1}\}$ differ by many orders of magnitude or $|\mathrm{Re}\{\lambda_i\}|/|\lambda_i|$ is many orders of magnitude less than one for some i. Moreover, a sequence $\{\mathbf{X}_i\}$ must be computed for each equation of the form of (4-23) required in the computation of gradients and hessians.

4.3 EXCESS POLE SPECIFICATIONS AND TRANSFER-FUNCTION MATRIX SYNTHESIS

Formulation of realistic control problems requires selection of compatible design specifications. Design specifications involve not only the specification of performance integral I and performance functional $f(\mathbf{c}, \mathbf{X})$ but also specification of the closed-loop system structure. Of course, many of these specifications are self-evident for particular design problems. On the other hand, very few guidelines exist for the selection of compatible design specifications given an arbitrary closed-loop structure. The purpose of this section is to examine what can be achieved with an arbitrary closed-loop structure in conjunction with performance integral

$$I = \int_0^\infty \{\tfrac{1}{2}\mathbf{e}'(t)\mathbf{e}(t)\}\, \mathrm{d}t \qquad (4\text{-}51)$$

where **e** is an appropriately selected error vector. Arbitrary closed-loop structure is introduced by considering an open-loop system defined by

$$\dot{\mathbf{x}} = \mathbf{F}\mathbf{x} + \mathbf{G}\mathbf{u}; \quad \mathbf{x}(0) = \mathbf{x}_0 \tag{4-52}$$

and

$$\mathbf{y} = \mathbf{H}\mathbf{x}. \tag{4-53}$$

Closed-loop structure eventually is specified by selecting a linear feedback control equation to minimize I.

Error vector **e** is selected in the very general form

$$\mathbf{e} = (\mathbf{c} - \mathbf{y}_0) - \sum_{k=0}^{K} \mathbf{P}_k \frac{d^k(\mathbf{y} - \mathbf{y}_0)}{dt^k}, \tag{4-54}$$

where **c** denotes a command vector in this section and

$$\mathbf{y}_0 = \mathbf{H}\mathbf{x}_0. \tag{4-55}$$

The matrix of polynomials

$$\mathbf{P}(s) = \sum_{k=0}^{K} \mathbf{P}_k s^k \tag{4-56}$$

formed with square coefficient matrices $\mathbf{P}_k$ is called the *excess pole specification* corresponding to **e**. This matrix of polynomials assumes a special significance whenever $I = 0$.

Conditions which lead to $I = 0$ are deduced by assuming for the present that the $(K-1)$th derivative of **u** exists for all time $t \geqslant 0$. Then successive differentiation of (4-52), (4-53) and substitution into (4-54) yield

$$\mathbf{e} = \mathbf{y}_c - \mathbf{D}\mathbf{x} - \sum_{l=0}^{K-1} \mathbf{E}_l \frac{d^l u}{dt^l} \tag{4-57}$$

where

$$\mathbf{y}_c = \mathbf{c} + (\mathbf{P}_0 - \mathbf{I})\mathbf{y}_0, \tag{4-58}$$

$$\mathbf{D} = \sum_{k=0}^{K} \mathbf{P}_k \mathbf{H}\mathbf{F}^k, \tag{4-59}$$

and

$$\mathbf{E}_l = \left[\sum_{k=0}^{K-1-l} \mathbf{P}_{k+1+l} \mathbf{H}\mathbf{F}^k \right] \mathbf{G} \quad \text{for} \quad l = 0, 1, \ldots, (K-1). \tag{4-60}$$

Equation (4-57) reveals that in general the condition $I = 0$ involves not only **u** and **x** but also $(K-1)$ derivatives of **u**.

In order to achieve $I = 0$ with only gains in the feedback control system, an assumption is introduced; namely

Assumption I Index $l > 0$ implies $\mathbf{E}_l = \mathbf{0}$.
When Assumption I holds, the error vector simplifies to

$$\mathbf{e} = \mathbf{y}_c - \mathbf{D}\mathbf{x} - \mathbf{E}\mathbf{u} \tag{4-61}$$

where

$$\mathbf{E} = \left[\sum_{k=0}^{K-1} \mathbf{P}_{k+1} \mathbf{H}\mathbf{F}^k \right] \mathbf{G}. \tag{4-62}$$

Assumption I also eliminates the need for any differentiability assumptions on $\mathbf{u}$. Unfortunately, a vector $\mathbf{u}$ for which $I = 0$ holds given any $\mathbf{x}$ and $\mathbf{y}_c$ cannot always be found without an additional assumption; namely,

Assumption II The rank of $\mathbf{E}$ equals dim $\mathbf{y}$.
is also invoked so that $I = 0$ for some $\mathbf{u}$.

Assumptions I and II combine to yield the following results concerning controls which give $I = 0$ for all $\mathbf{y}_c$. First, a matrix $\mathbf{V}$ is defined for the case where dim $\mathbf{u} >$ dim y by

Lemma 4-2 *Assumption II implies there exists a matrix* $\mathbf{V}$ *such that* $(\mathbf{EV})$ *is nonsingular.*

Proof Assumption II implies there is a subset of dim y columns of $\mathbf{E}$ which is linearly independent. Thus, matrix $\mathbf{V}$ can always be selected so that $(\mathbf{EV})$ is square and contains only linearly independent columns of $\mathbf{E}$.

Matrix $\mathbf{V}$ is used in conjunction with

$$\mathbf{u} = \mathbf{V}\mathbf{v} \tag{4-63}$$

to define auxiliary control variables denoted by $\mathbf{v}$. Whenever Assumption II holds, condition dim $\mathbf{v}$ = dim $\mathbf{y}$ of course also holds even though dim $\mathbf{u} >$ dim $\mathbf{y}$ may hold. Auxiliary controls $\mathbf{v}$ are used to prove

Theorem 4-4 *Assumptions I and II imply there exists a vector* $\mathbf{u}$ *such that* $I = 0$.

Proof Assumptions I and II imply

$$\mathbf{e} = \mathbf{y}_c - \mathbf{D}\mathbf{x} - \mathbf{E}\mathbf{V}\mathbf{v}$$

with $(\mathbf{EV})$ nonsingular. Therefore, let

$$\mathbf{v} = (\mathbf{EV})^{-1}(\mathbf{y}_c - \mathbf{D}\mathbf{x})$$

so that $\mathbf{e} = \mathbf{0}$ holds for all t and the integral appearing in (4-51) vanishes.

In order to simplify statements of subsequent results, the tacit assumption henceforth is made that the step of introducing auxiliary controls **v** has already been accomplished when (4-52) is first introduced. That is, condition dim **u** = dim **y** henceforth is assumed without loss of generality so that Assumption II now implies that **E** is nonsingular. This tacit assumption then leads to the result $I = 0$ using

$$\mathbf{u} = \mathbf{E}^{-1}(\mathbf{y}_c - \mathbf{D}\mathbf{x}) \tag{4-64}$$

whenever Assumptions I and II are invoked. A flow graph of the closed-loop system based on (4-52), (4-53), and (4-64) is given in Figure 4-1.

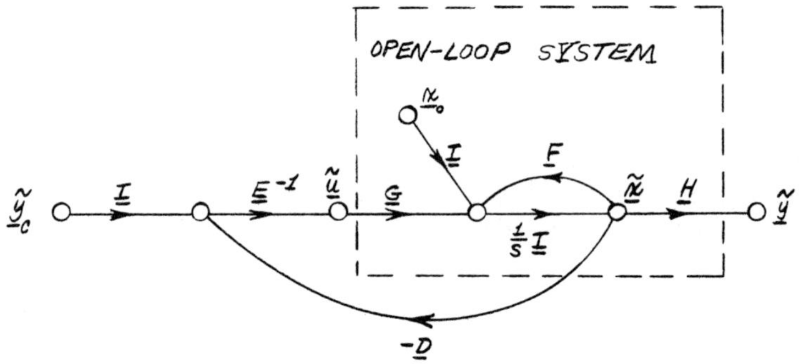

Figure 4-1 Flow graph of open- and closed-loop systems.

Elements of gain matrices $\mathbf{E}^{-1}$ and $\mathbf{E}^{-1}\mathbf{D}$ could of course be regarded as adjustable parameters in a suitably defined automated design problem. Presumably parameter optimization techniques would then yield results equivalent to (4-59) and (4-62). Eventually restrictions on elements of gain matrices $\mathbf{E}^{-1}$ and $\mathbf{E}^{-1}\mathbf{D}$ will be introduced so that the parameter optimization formulation of these problems is in fact needed.

Excess pole specifications [6] originate in practice from the square matrix of desired transfer functions $\mathbf{T}_d(s)$ which appears in

$$\breve{\mathbf{y}} = \mathbf{T}_d(s)\breve{\mathbf{y}}_c \tag{4-65}$$

for $\mathbf{x}_0 = \mathbf{0}$. Vectors $\breve{\mathbf{y}}$ and $\breve{\mathbf{y}}_c$ denote one-sided Laplace transforms of response and response-command vectors $\mathbf{y}$ and $\mathbf{y}_c$ respectively. Feedback control corresponding to $I = 0$ can now be related to the original design goal of realizing some desired matrix of transfer functions $\mathbf{T}_d(s)$. In particular, condition $I = 0$ implies $\breve{\mathbf{e}} = \mathbf{0}$ so that the Laplace transform of both sides of (4-54) then yields

$$\mathbf{0} = \breve{\mathbf{y}}_c - \mathbf{P}(s)\breve{\mathbf{y}}$$

or rather

$$\breve{\mathbf{y}} = \mathbf{P}^{-1}(s)\breve{\mathbf{y}}_c. \tag{4-66}$$

Hence, the matrix of closed-loop transfer functions realized by (4-64) when Assumptions I and II hold is $\mathbf{P}^{-1}(s)$.

A comparison of (4-65) and (4-66) suggests a restriction on $\mathbf{T}_d(s)$ if the synthesis method given previously is to be used. In particular, the desired matrix of transfer functions $\mathbf{T}_d(s)$ is *admissible* if and only if $\mathbf{T}_d^{-1}(s)$ exists as a matrix of polynomials. Whenever $\mathbf{T}_d(s)$ is admissible, the corresponding excess pole specification used to define the error vector $\mathbf{e}$ is

$$\mathbf{P}(s) = \mathbf{T}_d^{-1}(s). \tag{4-67}$$

In addition, matrices $\mathbf{F}$, $\mathbf{G}$, $\mathbf{H}$, and $\mathbf{T}_d(s)$ are *compatible* if and only if $\mathbf{T}_d(s)$ is admissible and Assumptions I and II are satisfied.

Whenever compatible matrices are used with dim $\mathbf{u}$ = dim $\mathbf{y}$, a closed-loop system exhibiting (4-65) can always be realized using feedback control specified by (4-59), (4-62), and (4-64). Additional relationships of importance are stated in terms of the open-loop matrix of transfer functions $\mathbf{T}(s)$ appearing in

$$\breve{\mathbf{y}} = \mathbf{T}(s)\breve{\mathbf{u}} \tag{4-68}$$

for $\mathbf{x}_0 = \mathbf{0}$ and expressed as

$$\mathbf{T}(s) = \mathbf{H}(s\mathbf{I} - \mathbf{F})^{-1}\mathbf{G}. \tag{4-69}$$

Admissibility of $\mathbf{T}_d(s)$ and (4-65) yield

$$\mathbf{P}(s)\breve{\mathbf{y}} = \breve{\mathbf{y}}_c$$

so that (4-68) now yields

$$\mathbf{P}(s)\mathbf{T}(s)\breve{\mathbf{u}} = \breve{\mathbf{y}}_c.$$

Examination of Figure 4-1 also reveals that

$$\breve{\mathbf{u}} = \mathbf{E}^{-1}\breve{\mathbf{y}}_c$$

under the condition $|s| \to \infty$. Relationship

$$\mathbf{E} = \lim_{|s| \to \infty} \mathbf{P}(s)\mathbf{T}(s) \tag{4-70}$$

thus is found to hold under conditions of compatibility when control is given by (4-64).

An elementary example of the synthesis of transfer function matrices using feedback control is provided by the open-loop system.

$$\mathbf{F} = \begin{bmatrix} 0 & 0 & 0 \\ 1 & -3 & -1 \\ 0 & 1 & -1 \end{bmatrix}, \quad \mathbf{G} \begin{bmatrix} 1 \\ 0 \\ 0 \end{bmatrix}, \quad \text{and} \quad \mathbf{H} = [0 \quad 1 \quad 0],$$

which yield the matrix of open-loop transfer functions

$$\mathbf{T}(s) = \left[\frac{(s+1)}{s(s+2)^2}\right],$$

and the matrix of desired closed-loop transfer functions

$$\mathbf{T}_d(s) = \left[\frac{1}{(s+1)^2}\right].$$

This matrix of desired transfer functions is clearly admissible and yields an excess pole specification with

$$\mathbf{P}_0 = [1], \quad \mathbf{P}_1 = [2], \quad \text{and} \quad \mathbf{P}_2 = [1].$$

Moreover, use of (4-60) results in

$$\mathbf{E}_0 = [1] \quad \text{and} \quad \mathbf{E}_1 = [0]$$

so that both Assumptions I and II are satisfied and design specifications are compatible. Finally, use of (4-59) gives

$$\mathbf{D} = [-1 \quad 3 \quad 2],$$

and hence feedback control which yields $I = 0$ for $\mathbf{e}$ defined by (4-54) is

$$\mathbf{u} = \mathbf{y}_c - [-1 \quad 3 \quad 2]\mathbf{x}.$$

This feedback control system is depicted in Figure 4-2.

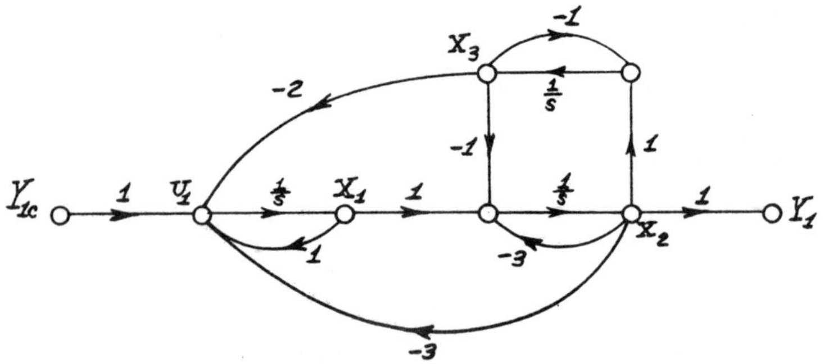

Figure 4-2 Flow graph of example synthesis with $\mathbf{x}_0 = \mathbf{0}$.

4.4 COMPATIBILITY OF DESIGN SPECIFICATIONS AND CLOSED-LOOP STABILITY

A number of results concerning compatibility are derived easily, and many of these results are used computationally in specifying compensation filters. Compensation filters are often required in order to satisfy Assumption I for a given $\mathbf{T}_d(s)$ which is admissible.

Necessary and sufficient conditions for admissibility of $\mathbf{T}_d(s)$ are stated conveniently in terms of

$$\mathbf{T}_d(s) = \frac{1}{D_d(s)} \mathbf{N}_d(s) \tag{4-71}$$

where $D_d(s)$ is a denominator polynomial and $\mathbf{N}_d(s)$ is a matrix of numerator polynomials. Specifically, the diagonally equivalent form of $\mathbf{N}_d(s)$ is used subsequently and is defined [2] by

Proposition 4-2 *Every matrix of polynomials* $\mathbf{N}(s)$ *can be reduced to diagonally equivalent form* $\boldsymbol{\Delta}(s)$ *such that*

$$\mathbf{N}(s) = \boldsymbol{\Phi}(s)\boldsymbol{\Delta}(s)\boldsymbol{\Psi}(s) \tag{4-72}$$

where

i) $\boldsymbol{\Phi}(s)$ *and* $\boldsymbol{\Psi}(s)$ *are matrices of polynomials with constant nonzero determinants*

and

ii) $\boldsymbol{\Delta}(s)$ *is a diagonal matrix of polynomials which is uniquely defined by*

$$a)\ \delta_{ii}(s) = \begin{Bmatrix} \delta_i(s) \text{ for } i = 1, 2, \ldots, r \\ 0 \text{ for } i = r + 1, \ldots \end{Bmatrix},$$

$$b)\ \delta_{i+1}(s)/\delta_i(s) \text{ is polynomial for each } i = 1, 2, \ldots, (r-1),$$

and

$$c)\ \delta_i(s) \text{ has unit leading coefficient for each } i = 1, 2, \ldots, r.$$

In conjunction with the diagonally equivalent form of $\mathbf{N}(s)$, integer r is called the *rank* of $\mathbf{N}(s)$ and $\delta_i(s)$ is called the ith *invariant polynomial* of $\mathbf{N}(s)$. The diagonally equivalent form of $\mathbf{N}(s)$ can be calculated with a finite number of elementary row and column operations. For example, matrix

$$\mathbf{N}(s) = \begin{bmatrix} (s^2 + s + 1) & -(s + 1) \\ -(s + 1) & (s + 1)^2 \end{bmatrix}$$

has the diagonally equivalent form corresponding to

$$\boldsymbol{\Phi}(s) = \begin{bmatrix} (s^2 + s + 1) & s \\ -(s + 1) & -1 \end{bmatrix}, \quad \boldsymbol{\Delta}(s) = \begin{bmatrix} 1 & 0 \\ 0 & s(s + 1)^3 \end{bmatrix}$$

and

$$\boldsymbol{\Psi}(s) = \begin{bmatrix} 1 & (s + 1)(s^2 + s - 1) \\ 0 & -1 \end{bmatrix}.$$

Unfortunately, elementary row and column operations may be tedious when done by hand or numerically inaccurate when done by digital computer.

The main result concerning admissibility of $\mathbf{T}_d(s)$ is stated as

Theorem 4-5 *Matrix* $\mathbf{T}_d(s)$ *is admissible if and only if*

i) $\mathbf{N}_d(s)$ *has full rank*

and

ii) $D_d(s)/N_d(s)$ *is a polynomial where* $N_d(s)$ *is the invariant polynomial of* $\mathbf{N}_d(s)$ *of highest degree.*

Proof Let $\boldsymbol{\Delta}_d(s)$ be the diagonally equivalent form of $\mathbf{N}_d(s)$ so that

$$\mathbf{N}_d(s) = \boldsymbol{\Phi}_d(s)\boldsymbol{\Delta}_d(s)\boldsymbol{\Psi}_d(s).$$

Because $\mathbf{N}_d(s)$ is square and has full rank, matrix $\boldsymbol{\Delta}_d(s)$ has an inverse so that

$$\mathbf{T}_d^{-1}(s) = D_d(s)\boldsymbol{\Psi}_d^{-1}(s)\boldsymbol{\Delta}_d^{-1}(s)\boldsymbol{\Phi}_d^{-1}(s).$$

By construction, nonsingular matrices $\boldsymbol{\Phi}_d^{-1}(s)$ and $\boldsymbol{\Psi}_d^{-1}(s)$ have polynomial elements only, and hence $\mathbf{T}_d^{-1}(s)$ is a matrix of polynomials if and only if all of the unique elements of $D_d(s)\boldsymbol{\Delta}_d^{-1}(s)$ are polynomials so that this matrix can be reduced to the diagonally equivalent form of $\mathbf{T}_d^{-1}(s)$.

Condition i) of Theorem 4-5 is equivalent to the condition that det $\mathbf{N}_d(s) \neq 0$ for some value of s.

Some examples of admissible $\mathbf{T}_d(s)$ are cited here because they arise frequently in control. All statements concerning these examples are given under the assumption that condition i) of Theorem 4-5 is satisfied. Non-interacting design results from selecting $\mathbf{T}_d(s)$ to be diagonal, and admissibility then results when each major-diagonal element has a constant numerator. For instance, an admissible non-interacting design is

$$\mathbf{T}_d(s) = \begin{bmatrix} \dfrac{1}{(s + 1)} & 0 \\ 0 & \dfrac{2}{(s + 2)(s + 3)} \end{bmatrix}.$$

Semi-interacting design results from selecting $\mathbf{T}_d(s)$ to be triangular. Admissibility then results when each major-diagonal element has a constant numerator and every denominator factor of each ij-element with $i \neq j$ is also a minimal common denominator factor of either the ith row or the jth column. For instance, an admissible semi-interacting design is

$$\mathbf{T}_d(s) = \begin{bmatrix} \dfrac{1}{(s+1)} & \dfrac{(s+3)(s+4)}{(s+1)^2(s+2)} \\ 0 & \dfrac{2}{(s+1)(s+2)} \end{bmatrix}.$$

An example of admissible interacting designs occurs when all elements of $\mathbf{T}_d(s)$ have constant numerators and every denominator factor of each ij-element is also a common denominator factor of either the ith row or jth column. For instance, an admissible interacting design is

$$\mathbf{T}_d(s) = \begin{bmatrix} \dfrac{1}{(s+1)(s+3)} & \dfrac{1}{(s+2)(s+3)} \\ \dfrac{1}{(s+1)} & \dfrac{2}{s+2} \end{bmatrix}.$$

Once admissibility of $\mathbf{T}_d(s)$ has been verified, the next step in establishing compatibility normally is the verification that Assumption I is satisfied. If Assumption I is not satisfied, then an appropriate compensation filter must be introduced. For these purposes, constant matrices $\mathbf{Z}_l$ are defined by

$$\lim_{|s| \to \infty} \left[\mathbf{P}(s)\mathbf{T}(s) - \sum_{l=0}^{K-1} \mathbf{Z}_l s^l \right] = \mathbf{0}, \tag{4-73}$$

and a diagonal matrix of transfer functions $\mathbf{V}(s)$ is defined with major-diagonal elements

$$v_{ii}(s) = \left(\sum_{k=0}^{K_i} \alpha_{ik} s^k \right)^{-1}, \quad \text{where} \quad \alpha_{i,K_i} \neq 0. \tag{4-74}$$

The first result concerning Assumption I given here is useful computationally and is

Lemma 4-3 *Suppose* $\mathbf{T}_d(s)$ *is admissible and let* K_j *be the maximum value of* l *for which the* j*th column of* $\mathbf{Z}_l$ *contains a nonzero element. Then Assumption I is satisfied if and only if each* $K_j = 0$.

Proof Let the minimal polynomial of **F** be

$$M(s) = \sum_{k=0}^{n} m_k s^k \quad \text{with} \quad m_n = 1.$$

Then use of relationship [3]

$$(s\mathbf{I} - \mathbf{F})^{-1} = \frac{1}{M(s)} \sum_{j=0}^{n-1} s^j \left[\sum_{i=j+1}^{n} m_i \mathbf{F}^{(i-j-1)} \right]$$

and (4-56) in conjunction with (4-73) yields

$$\mathbf{Z}_l = \sum_{i=l}^{K-1} m_{n+l-i} \mathbf{E}_i \quad \text{for} \quad K \leqslant n \tag{4-75}$$

after some tedious calculations [7]. The desired result follows directly from an inspection of (4-75).

Selection of appropriate compensation filters if needed then is based on the result.

Theorem 4-6 *Suppose* $\mathbf{T}_d(s)$ *is admissible and let* K_j *be the maximum value of* l *for which the jth column of* $\mathbf{Z}_l$ *contains a nonzero element. Then the open-loop system corresponding to auxiliary controls* $\mathbf{v}$ *defined by*

$$\breve{\mathbf{u}} = \mathbf{V}(s)\breve{\mathbf{v}}$$

satisfies Assumption I.

Proof Laplace transforms yield

$$\breve{\mathbf{y}} = \mathbf{T}(s)\mathbf{V}(s)\breve{\mathbf{v}}$$

with $\mathbf{x}_0 = \mathbf{0}$ so that the open-loop matrix of transfer functions becomes $\mathbf{T}(s)\mathbf{V}(s)$ for the composite system. Recomputation of $\mathbf{Z}_l$ using (4-73) demonstrates that the maximum value of l for which the jth column of $\mathbf{Z}_l$ contains a nonzero element has been reduced by K_j with the introduction of these compensation filters. Thus, index $l > 0$ implies $\mathbf{Z}_l = \mathbf{0}$ for the composite system and hence Assumption I is satisifed by Lemma 4-3.

Moreover, compensation filters corresponding to $\mathbf{V}(s)$ are realized easily in state-equation form using for example companion matrices. Open-loop equations which satisfy Assumption I for any admissible $\mathbf{T}_d(s)$ thus can always be expressed in the form of (4-52) and (4-53).

Once admissibility of $\mathbf{T}_d(s)$ and conditions $\mathbf{E}_l = \mathbf{0}$ for each $l > 0$ have been verified, the final step in establishing compatibility is the verification that Assumption II is satisfied. The principle result concerning Assumption II is stated in terms of $\mathbf{Z}_0$ defined by (4-73) and is

Theorem 4-7 *Suppose* $\mathbf{T}_d(s)$ *is admissible and Assumption I is satisfied with* $\dim \mathbf{u} = \dim \mathbf{y}$. *Then matrix* $\mathbf{Z}_0$ *is defined, and Assumption II is satisfied if and only if* $\mathbf{Z}_0$ *is nonsingular.*

Proof Assumption I and admissibility of $\mathbf{T}_d(s)$ imply

$$\mathbf{Z}_0 = \mathbf{E}_0 = \mathbf{E}$$

by (4-75) and the definition of **E** given in (4-62).

A second result which follows from Theorem 4-7 is

Corollary 4-2 *Suppose* $\mathbf{T}_d(s)$ *is admissible and Assumption I is satisifed with* $\dim \mathbf{u} = \dim \mathbf{y}$. *Then Assumption II is satisfied only if* $\mathbf{T}(s)$ *is invertible for some s.*

Proof Assumption I and admissibility of $\mathbf{T}_d(s)$ imply **E** is given by (4-70). Moreover, given any constant vector $\mathbf{f}_0$, post multiplication and the indicated limit can be interchanged yielding

$$\mathbf{E}\mathbf{f}_0 = \lim_{|s|\to\infty} [\mathbf{P}(s)\mathbf{T}(s)\mathbf{f}_0] = \lim_{|s|\to\infty} [\mathbf{P}(s)\mathbf{T}(s)\mathbf{f}(s)]$$

where

$$\mathbf{f}(s) = \sum_{k=0}^{n} \mathbf{f}_k s^{-k}.$$

Use of the diagonally equivalent form of $\mathbf{N}(s)$ now yields

$$\mathbf{E}\mathbf{f}_0 = \lim_{|s|\to\infty} \frac{1}{s^n D(s)} \mathbf{P}(s)\mathbf{\Phi}(s)\mathbf{\Delta}(s)\mathbf{d}_0]$$

where

$$\mathbf{d}_0 = s^n \mathbf{\Psi}(s)\mathbf{f}(s)$$

is an arbitrary constant corresponding to some n and some choice of $\mathbf{f}_k$. Previously stated properties of $\mathbf{\Psi}(s)$ lead to

$$\mathbf{f}_0 = \lim_{|s|\to\infty} \left[\frac{1}{s^n} \mathbf{\Psi}^{-1}(s)\mathbf{d}_0\right]$$

so that $\mathbf{d}_0 \neq \mathbf{0}$ implies $\mathbf{f}_0 \neq \mathbf{0}$. Finally, if $\mathbf{T}(s)$ is singular for all s, then

$$\mathbf{\Delta}(s)\mathbf{d}_0 = \mathbf{0} \; \forall \; s$$

given some $\mathbf{d}_0 \neq \mathbf{0}$ so that

$$\mathbf{E}\mathbf{f}_0 = \mathbf{0}$$

for some $\mathbf{f}_0 \neq \mathbf{0}$.

Matrix $\mathbf{T}(s)$ must be invertible for some s if compatibility is ever to be achieved. Moreover, the matrix of open-loop transfer functions is invertible for some s only if matrices $\mathbf{G}$ and $\mathbf{H}$ have rank equal to dim $\mathbf{u}$ = dim $\mathbf{y}$.

Suppose $\mathbf{T}_d(s)$ is admissible, Assumption I is satisfied, and $\mathbf{T}(s)$ is invertible for some s. Also suppose that Assumption II is not satisfied. Then matrix $\mathbf{T}_d(s)$ can often be altered by increasing degrees of the elements of $\mathbf{P}(s)$ so that eventually Assumption II is satisfied. In order to identify such alterations, the following definitions are introduced under the assumption that $\mathbf{T}(s)$ is invertible for some s. Let L_i be the maximum integers for which the matrix $\mathbf{T}_\infty$ exists where

$$\mathbf{T}_\infty = \lim_{|s|\to\infty} \mathbf{\Lambda}(s)\mathbf{T}(s) \tag{4-76}$$

and $\mathbf{\Lambda}(s)$ is the diagonal matrix having major diagonal elements

$$\lambda_{ii}(s) = s^{L_i}. \tag{4-77}$$

Also let $\mathbf{P}_\infty$ be defined by

$$\mathbf{P}_\infty = \lim_{|s|\to\infty} \mathbf{P}(s)\mathbf{\Lambda}^{-1}(s) \tag{4-78}$$

whenever such a matrix exists. The significance of these definitions stems from the result

Theorem 4-8 *Suppose* $\mathbf{T}_d(s)$ *is admissible, Assumption I is satisfied, and* $\mathbf{T}(s)$ *is invertible for some s. Then Assumption II is satisfied only if* $\mathbf{T}_\infty$ *is nonsingular, and*

$$\mathbf{E} = \mathbf{P}_\infty\mathbf{T}_\infty \tag{4-79}$$

holds whenever $\mathbf{T}_\infty$ *is nonsingular.*

Proof Assumption I and admissibility of $\mathbf{T}_d(s)$ imply $\mathbf{E}$ exists and is given by (4-70). By construction, matrix $\mathbf{T}_\infty$ exists and

$$\begin{aligned}\mathbf{E} &= \lim_{|s|\to\infty} [\mathbf{P}(s)\mathbf{\Lambda}^{-1}(s)\mathbf{\Lambda}(s)\mathbf{T}(s)]\\ &= \lim_{|s|\to\infty} [\mathbf{P}(s)\mathbf{\Lambda}^{-1}(s)\mathbf{T}_\infty].\end{aligned} \tag{4-80}$$

Therefore, matrix $\mathbf{E}$ is singular whenever $\mathbf{T}_\infty$ is singular. Now let k be the smallest nonnegative integer for which

$$\mathbf{\Pi}_k = \lim_{|s|\to\infty} [s^{-k}\mathbf{P}(s)\mathbf{\Lambda}^{-1}(s)]$$

exists. Then equation (4-80) becomes

$$\mathbf{E} = \lim_{|s|\to\infty} [s^k\mathbf{\Pi}_k\mathbf{T}_\infty].$$

However, matrix $\mathbf{E}$ exists so that

$$\mathbf{\Pi}_k \mathbf{T}_\infty = \mathbf{0}$$

if $k > 0$, and hence $\mathbf{\Pi}_k = \mathbf{0}$ for $k > 0$ when $\mathbf{T}_\infty$ is nonsingular. Thus, relationship

$$\mathbf{E} = \mathbf{\Pi}_0 \mathbf{T}_\infty$$

must hold if $\mathbf{T}_\infty$ is nonsingular, and $\mathbf{P}_\infty = \mathbf{\Pi}_0$ by definition.

Of course, equation (4-79) implies that $\mathbf{P}_\infty$ must exist and be nonsingular if compatibility is ever to be achieved. The relationship given in (4-78) indicates a singular $\mathbf{P}_\infty$ can often be corrected by increasing degrees of the elements of $\mathbf{P}(s)$. Although not discussed here in any detail, decreasing degrees of the elements of $\mathbf{P}(s)$ is of course a method for satisfying Assumption I without the addition of compensation filters.

An example usage of the frequency domain relationships given in this section is provided by the matrix of open-loop transfer functions

$$\mathbf{T}(s) = \begin{bmatrix} \dfrac{1}{(s+1)} & -\dfrac{(s-1)}{s(s+1)} \\[2ex] \dfrac{(s+2)}{s^2(s+1)} & \dfrac{1}{s^2} \end{bmatrix}$$

and the matrix of desired closed-loop transfer functions

$$\mathbf{T}_d(s) = \begin{bmatrix} \dfrac{1}{(s+1)} & 0 \\[2ex] 0 & \dfrac{1}{(s+1)^2} \end{bmatrix}.$$

This matrix of desired transfer functions is clearly admissible and yields an excess pole specification

$$\mathbf{P}(s) = \begin{bmatrix} (s+1) & 0 \\ 0 & (s+1)^2 \end{bmatrix}.$$

Use of (4-73) now yields

$$\mathbf{Z}_1 = \begin{bmatrix} 0 & 0 \\ 0 & 0 \end{bmatrix} \quad \text{and} \quad \mathbf{Z}_0 = \begin{bmatrix} 1 & -1 \\ 1 & 1 \end{bmatrix}$$

so that both Assumptions I and II hold and hence compatibility has been verified. As a matter of interest, the matrix of numerator polynomials corresponding to

$$\mathbf{T}(s) = \frac{1}{D(s)}\mathbf{N}(s) \tag{4-81}$$

is

$$\mathbf{N}(s) = \begin{bmatrix} s^2 & -s(s-1) \\ (s+2) & (s+1) \end{bmatrix}$$

and yields

$$\det \mathbf{N}(s) = 2s(s^2 + s - 1).$$

Therefore, matrix $\mathbf{T}(s)$ is invertible for almost all s. Moreover, matrix $\mathbf{\Lambda}(s)$, which is defined by (4-76) and (4-77), is found to be

$$\mathbf{\Lambda}(s) = \begin{bmatrix} s & 0 \\ 0 & s^2 \end{bmatrix}.$$

Elementary calculations then yield

$$\mathbf{P}_\infty = \begin{bmatrix} 1 & 0 \\ 0 & 1 \end{bmatrix} \quad \text{and} \quad \mathbf{T}_\infty = \begin{bmatrix} 1 & -1 \\ 1 & 1 \end{bmatrix}$$

which of course correspond to $\mathbf{Z}_0$.

Unlike pole placement procedures which are based on Reference [8], the procedure given here for synthesis of transfer function matrices does not involve a complete pole-placement specification. Instead, closed-loop poles not included in the minimal common denominator polynomial of $\mathbf{T}_d(s)$ are placed automatically by $\mathbf{F}$, $\mathbf{G}$, and $\mathbf{H}$. These additional poles of course may be very troublesome and may correspond to an unstable subsystem which is either uncontrollable, unobservable or both. For instance, the previous example results in $(s+1)^2(s^2+s-1$ as the minimal polynomial of the closed-loop system based on any minimal realization [9] of $\mathbf{T}(s)$, and hence closed-loop instability results.

Conditions for closed-loop stability are stated conveniently in terms of $\mathbf{T}(s)$ given by (4-69) in the form of (4-81). A sufficient but not necessary condition for asymptotic stability of the closed-loop system is

Theorem 4-9 *Suppose* $\mathbf{F}$, $\mathbf{G}$, $\mathbf{H}$, *and* $\mathbf{T}_d(s)$ *are compatible and* $\mathbf{u}$ *is given by* (4-64 *Let* $D_d(s)$ *be the minimal common denominator polynomial of* $\mathbf{T}_d(s)$ and $D(s)$ *be the minimal polynomial of* $\mathbf{F}$. *Then the closed-loop system is asymptotically stable if all zeros of* $D_d(s)$ *and* det $\mathbf{N}(s)$ *have negative real parts.*

Proof Design under compatibility conditions results in

$$\breve{\mathbf{u}} = \mathbf{T}^{-1}(s)\mathbf{T}_d(s)\breve{\mathbf{y}}_c$$

with $\mathbf{x}_0 = \mathbf{0}$ and hence

$$\breve{\mathbf{x}} = (s\mathbf{I} - \mathbf{F})^{-1}\mathbf{G}\mathbf{T}^{-1}(s)\mathbf{T}_d(s)\breve{\mathbf{y}}_c. \tag{4-82}$$

By definition of $D(s)$ as the minimal polynomial of $\mathbf{F}$, every eigenvalue of the closed-loop system is seen to correspond to a factor of either det $\mathbf{N}(s)$ or $D_d(s)$.

A related necessary but not sufficient condition for asymptotic stability of the closed-loop system is

Theorem 4-10 *Suppose* $\mathbf{F}$, $\mathbf{G}$, $\mathbf{H}$, *and* $\mathbf{T}_d(s)$ *are compatible and* $\mathbf{u}$ *is given by (4-64). Let* $D_d(s)$ *be the minimal common denominator polynomial of* $\mathbf{T}_d(s)$. *Then the closed-loop system is asymptotically stable only if all zeros of* $D_d(s)$ *and the minimal common denominator polynomial of* $(s\mathbf{I} - \mathbf{F})^{-1}\mathbf{G}\mathbf{T}^{-1}(s)$ *have negative real parts.*

Proof Because design under compatibility conditions results in (4-82) and because

$$\mathbf{H}(s\mathbf{I} - \mathbf{F})^{-1}\mathbf{G}\mathbf{T}^{-1}(s) = \mathbf{I},$$

no factor of the minimal common denominator polynomial of $(s\mathbf{I} - \mathbf{F})^{-1}\mathbf{G}\mathbf{T}^{-1}(s)$ can also be a common factor of the elements of $\mathbf{N}_d(s)$. Thus, all factors of this denominator polynomial must correspond to eigenvalues of the closed-loop system.

Unfortunately, necessary and sufficient conditions for asymptotic stability of the closed-loop system involve structure of the product $(s\mathbf{I} - \mathbf{F})^{-1}\mathbf{G}\mathbf{T}^{-1}(s)$ and hence cannot be established in the general case. In the previous example, for instance, $(s^2 + s - 1)$ is a factor of both det $\mathbf{N}(s)$ and the minimal polynomial of the closed-loop system whereas s is a factor of det $\mathbf{N}(s)$ but is not a factor of the minimal polynomial of the closed-loop system.

When the closed-loop system is unstable for a given $\mathbf{F}$, $\mathbf{G}$, and $\mathbf{H}$, then these matrices must be modified by selecting alternate response and control variables if possible or by state augmentation. A state-augmentation procedure which may be useful for these purposes is described in the next section.

4.5 STATE AUGMENTATION AND POLE SUPPRESSION SPECIFICATIONS

State augmentation has a number of applications in addition to those already mentioned. Of particular importance is the addition of compensation filters for the purposes of permitting incomplete state-variable feedback and of reducing the residues of poles which are known to depend upon system operating conditions. The method of state augmentation presented here has the property of preserving compatibility.

Suppose **F**, **G**, and **H** are given and the corresponding open-loop system is adjoined as follows. Let the ith row of **H** be denoted as $\mathbf{h}_i$ so that $\mathbf{H}' = [\mathbf{h}_1 \ \mathbf{h}_2 \ \ldots]$ and hence $y_i = \mathbf{h}_i'\mathbf{x}$ for each i. Also let additional open-loop state equations corresponding to each response variable have the form

$$\dot{\mathbf{x}}_i = \mathbf{F}_i\mathbf{x}_i + \mathbf{G}_i\mathbf{x}; \quad \mathbf{x}_i(0) = \mathbf{x}_{i0} \tag{4-83}$$

where dim $\mathbf{x}_i = n_i$,

$$\mathbf{F}_i = \begin{bmatrix} \mathbf{0} & \mathbf{I} \\ -p_{i0} & -\mathbf{p}_i' \end{bmatrix} \quad \text{with} \quad \mathbf{p}_i' = [p_{i1} \quad p_{i2} \quad \ldots \quad p_{i,\, n_{i-1}}], \tag{4-84}$$

and

$$\mathbf{G}_i' = [\mathbf{0} \quad \mathbf{0} \quad \ldots \quad \mathbf{q}_i] \quad \text{with} \quad \mathbf{q}_i' = [q_{i1} \quad q_{i2} \quad \ldots]. \tag{4-85}$$

Finally, let auxiliary response variables z_i be defined as

$$z_i = y_i + \mathbf{r}_i'\mathbf{x}_a \quad \text{for each } i \tag{4-86}$$

where $\mathbf{x}_a' = [\mathbf{x}_1' \quad \mathbf{x}_2' \quad \ldots]$ and

$$\mathbf{r}_i' = [r_{i1} \quad r_{i2} \quad \ldots]. \tag{4-87}$$

Then composite open-loop equations are written in the form

$$\begin{bmatrix} \dot{\mathbf{x}} \\ \dot{\mathbf{x}}_a \end{bmatrix} = \begin{bmatrix} \mathbf{F} & \mathbf{0} \\ \mathbf{G}_a & \mathbf{F}_a \end{bmatrix} \begin{bmatrix} \mathbf{x} \\ \mathbf{x}_a \end{bmatrix} + \begin{bmatrix} \mathbf{G} \\ \mathbf{0} \end{bmatrix} \mathbf{u}; \quad \begin{bmatrix} \mathbf{x}(0) \\ \mathbf{x}_a(0) \end{bmatrix} = \begin{bmatrix} \mathbf{x}_0 \\ \mathbf{x}_{a0} \end{bmatrix} \tag{4-88}$$

and

$$\mathbf{z} = [\mathbf{H} \quad \mathbf{H}_a] \begin{bmatrix} \mathbf{x} \\ \mathbf{x}_a \end{bmatrix} \tag{4-89}$$

so that

$$\mathbf{F}_a = \begin{bmatrix} \mathbf{F}_1 & \mathbf{0} & \mathbf{0} & \ldots \\ \mathbf{0} & \mathbf{F}_2 & \mathbf{0} & \ldots \\ \ldots & \ldots & \ldots & \ldots \end{bmatrix}, \quad \mathbf{G}_a = \begin{bmatrix} \mathbf{G}_1 \\ \mathbf{G}_2 \\ \vdots \end{bmatrix}, \quad \text{and} \quad \mathbf{H}_a' = [\mathbf{r}_1 \quad \mathbf{r}_2 \quad \ldots]. \tag{4-90}$$

Coefficient matrices of (4-88) and (4-89) henceforth are referred to as the *augmented* **F**, **G**, and **H** matrices.

Auxiliary response variables now are found to be

$$\breve{\mathbf{z}} = [\mathbf{\Phi}(s) + \mathbf{H}]\breve{\mathbf{x}} \tag{4-91}$$

with $\mathbf{x}_0 = \mathbf{0}$ and $\mathbf{x}_{a0} = \mathbf{0}$. Moreover, every element of $\mathbf{\Phi}(s)$ is of the form

$$\left(\sum_{k=0}^{n-1} \alpha_k s^k\right)\left(\sum_{k=0}^{n} \beta_k s^k\right)^{-1}$$

with $\beta_n = 1$ for some n, and the property $\lim_{|s|\to\infty} \mathbf{\Phi}(s) = \mathbf{0}$ leads to the following result for a given $\mathbf{T}_d(s)$

Lemma 4-4 *Compatibility of* **F**, **G**, **H**, *and* $\mathbf{T}_d(s)$ *implies compatibility of the augmented* **F**, **G**, **H** *and* $\mathbf{T}_d(s)$. *Moreover, compatibility and use of (4-64) for the augmented system imply*

$$\lim_{|s|\to\infty} \mathbf{T}_c(s) = \lim_{|s|\to\infty} \mathbf{T}_d(s)$$

when $\mathbf{T}_c(s)$ *is defined by*

$$\breve{\mathbf{y}} = \mathbf{T}_c(s)\breve{\mathbf{z}}_c$$

for $\mathbf{x}_0 = \mathbf{0}$ *and* $\mathbf{x}_{a0} = \mathbf{0}$.

Proof The first result follows directly from using (4-73) to verify that $\mathbf{Z}_l$ remains unchanged for each l because $\lim_{|s|\to\infty} \mathbf{\Phi}(s) = \mathbf{0}$. Similarly, the second result follows from

$$\breve{\mathbf{z}} = \mathbf{T}_d(s)\breve{\mathbf{z}}_c \tag{4-92}$$

and

$$\breve{\mathbf{z}} = \breve{\mathbf{y}} + \mathbf{\Phi}(s)\breve{\mathbf{x}}.$$

A broad class of postfilters thus can be introduced without altering design specifications.

Even though (4-92) has been accomplished for the augmented closed-loop system, the desired relationship given in (4-65) cannot be reconstructed from this design in a usable form. However, this problem is rectified when augmentation is specialized as follows. System augmentation is *exact* when there exists a matrix $\mathbf{\Psi}(s)$ such that

$$\mathbf{\Phi}(s) = \mathbf{\Psi}(s)\mathbf{H}. \tag{4-93}$$

Exact system augmentation of course results in

$$\breve{\mathbf{z}} = \mathbf{S}(s)\breve{\mathbf{y}} \tag{4-94}$$

where

$$\mathbf{S}(s) = \mathbf{\Psi}(s) + \mathbf{I}. \tag{4-95}$$

In addition, exact augmentation results in

Theorem 4-11 *Suppose* **F**, **G**, **H**, *and* $\mathbf{T}_d(s)$ *are compatible. Then use of* (4-64) *for a system with exact augmentation yields* (4-65) *where*

$$\tilde{\mathbf{z}}_c = \mathbf{R}(s)\tilde{\mathbf{y}}_c \tag{4-96}$$

and

$$\mathbf{R}(s) = \mathbf{P}(s)\mathbf{S}(s)\mathbf{T}_d(s). \tag{4-97}$$

Proof The desired result follows from (4-94) and

$$\tilde{\mathbf{z}}_c = \mathbf{T}_d^{-1}(s)\mathbf{S}(s)\tilde{\mathbf{y}} = \mathbf{P}(s)\mathbf{S}(s)\mathbf{T}_d(s)\tilde{\mathbf{y}}_c$$

assuming design is accomplished under compatibility conditions.

In other words, whenever auxiliary response variables $\tilde{\mathbf{z}}$ can be expressed directly in terms of a postfilter operating on response variables $\tilde{\mathbf{y}}$, then the matrix $\mathbf{T}_d(s)$ is realized exactly for $\tilde{\mathbf{y}}$ with the addition of a prefilter defined by $\mathbf{R}(s)$. A flow graph of the closed-loop system with exact augmentation is shown in Figure 4-3 where the matrix **D** is taken to be $[\mathbf{D} \;\; \mathbf{D}_a]$ for the augmented system. Prefilters defined by $\mathbf{R}(s)$ can always be realized in minimal state-equation form [9], and they simplify in accordance with $\mathbf{R}(s) = \mathbf{S}(s)$ whenever $\mathbf{P}(s)$ and $\mathbf{S}(s)$ commute.

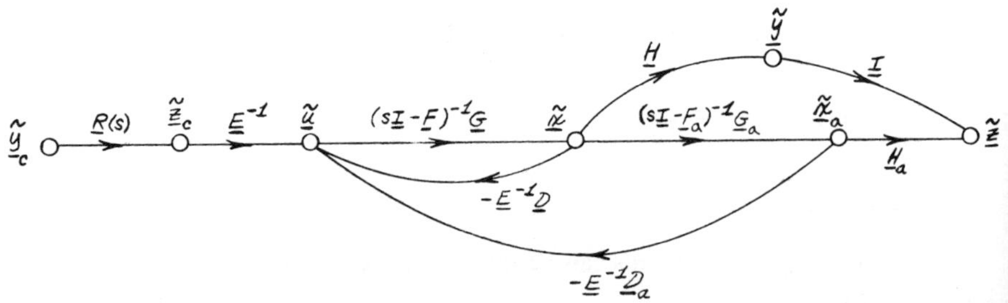

Figure 4-3 Flow graph of closed-loop system with exact augmentation and zero initial state.

A useful example of exact augmentation is constructed as follows. Let

$$\mathbf{q}_i = \mathbf{h}_i \quad \text{for each } i$$

and

$$\mathbf{H}_a' = \begin{bmatrix} \mathbf{r}_{11} & \mathbf{0} & \mathbf{0} & \cdots \\ \mathbf{0} & \mathbf{r}_{22} & \mathbf{0} & \cdots \\ \cdots & \cdots & \cdots & \cdots \end{bmatrix} \quad \text{with dim } \mathbf{r}_{ii} = n_i.$$

Then matrix $\mathbf{\Phi}(s)$ is not only of the form given in (4-93) but also $\mathbf{\Psi}(s)$ is diagonal with major-diagonal elements

$$\psi_{ii} = \left(\sum_{k=0}^{n_i - 1} r_{i,k+m_i} s^k \right) \left(\sum_{k=0}^{n_i} p_{ik} s^k \right)^{-1}$$

where

$$m_i = 1 + \sum_{j=0}^{i-1} n_j$$

and $p_{i,n_i} = 1$.

An important application of system augmentation concerns incomplete state-variable feedback and is accomplished as follows. The matrix $\mathbf{D}$ given in (4-59) for the unaugmented system now becomes the matrix $[\mathbf{D} \quad \mathbf{D}_a]$ where

$$\mathbf{D} = \sum_{k=0}^{K} \mathbf{P}_k \left[\mathbf{H}\mathbf{F}^k + \mathbf{H}_a \left(\sum_{l=0}^{k-1} \mathbf{F}_a^l \mathbf{G}_a \mathbf{F}^{k-1-l} \right) \right] \tag{4-98}$$

and

$$\mathbf{D}_a = \sum_{k=0}^{K} \mathbf{P}_k \mathbf{H}_a \mathbf{F}_a^k \tag{4-99}$$

for the augmented system corresponding to (4-88) and (4-89). Then, for instance, matrix $\mathbf{H}_a$ can sometimes be selected so that $\mathbf{d}_j = \mathbf{0}$ where $\mathbf{D} = [\mathbf{d}_1 \quad \mathbf{d}_2 \quad \ldots]$. This selection would result in feedback control that does not require measurement of x_j, but this selection of $\mathbf{H}_a$ requires the solution of a set of linear equations. An example of this procedure is given in the next section.

Another application of system augmentation concerns sensitivity of closed-loop transfer functions to variable poles of the open-loop system. Variability of open-loop poles as well as high frequency gains $\mathbf{T}_\infty$ arises frequently when the control system is subjected to a wide range of operating conditions. Although not discussed in detail here, residues of closed-loop poles corresponding to these variable open-loop poles can often be reduced to zero for nominal design conditions by the introduction of exact state-augmentation. Specifically, the matrix of open-loop transfer functions becomes $\mathbf{S}(s)\mathbf{T}(s)$ with exact augmentation. Therefore, the added closed-loop poles which have zero residues correspond to the determinant of the matrix of numerator polynomials of $\mathbf{S}(s)$. The matrix of numerator polynomials of $\mathbf{S}(s)$ hence is called the *pole suppression specification* in this context. Application of pole suppression specifications to the design of insensitive feedback control systems is largely ad hoc and is illustrated in a subsequent example.

4.6 EXAMPLE OF TRANSFER-FUNCTION MATRIX SYNTHESIS

Design of a lateral command-augmentation system is a classic problem of flight control [10, 11]. This problem is considered here in a somewhat over-simplified form for the purposes of illustration. The basic equations of lateral motion and actuator dynamics are given in accordance with state and control variable assignments summarized in Table 4-2. Coefficient matrices corresponding to these

Table 4-2 State and control variable assignment for equations of lateral motion

Variable	Common symbol	Description
x_1	β	Side slip rate
x_2	β	Side slip angle
x_3	p	Roll rate
x_4	r	Yaw rate
x_5	δ_r	Rudder deflection
x_6	δ_a	Aileron deflection
u_1	δ_{rc}	Rudder deflection command
u_2	δ_{ac}	Aileron deflection command

equations for an aircraft of the X-15 type at 147,000 feet altitude and Mach No. 5.5 are taken [10] to be

$$\mathbf{F} = \begin{bmatrix} -0.0119 & -2.9586 & 0.00905 & 0.0103 & 1.4945 & 0.2496 \\ 1.0 & 0 & 0 & 0 & 0 & 0 \\ 0 & 0.514 & -0.0625 & 0.0376 & -2.325 & -5.116 \\ 0 & 2.9586 & -0.0037 & -0.0103 & -1.432 & -0.2496 \\ 0 & 0 & 0 & 0 & -25.0 & 0 \\ 0 & 0 & 0 & 0 & 0 & -6.667 \end{bmatrix}$$

and

$$\mathbf{G} = \begin{bmatrix} -0.0625 & 0 \\ 0 & 0 \\ 0 & 0 \\ 0 & 0 \\ 25.0 & 0 \\ 0 & 6.667 \end{bmatrix}$$

The specific problem considered here is the design of a roll rate-yaw rate command augmentation system so that coefficient matrix

$$\mathbf{H} = \begin{bmatrix} 0 & 0 & 1.0 & 0 & 0 & 0 \\ 0 & 0 & 0 & 1.0 & 0 & 0 \end{bmatrix}$$

is selected.

Table 4-3 Open-loop numerator and denominator polynomials

$N_{11}(s) = -58.1(s + 6.67)s[s^2 + 2(0.0141)(1.63)s + (1.63)^2]$
$N_{21}(s) = 35.8(s + 6.67)(s + 0.31)[s^2 - 2(0.434)(0.272)s + (0.272)^2]$
$N_{12}(s) = -34.1(s + 25.0)s[s^2 + 2(0.00702)(1.71)s + (1.71)^2]$
$N_{22}(s) = -1.66(s + 25.0)(s + 0.685)[s^2 - 2(0.501)(0.685)s + (0.685)^2]$
$D(s) = (s + 25.0)(s + 6.67)(s + 0.0642)(s - 0.00328)[s^2 + 2(0.00691)(1.72)s + (1.72)^2]$

Open-loop transfer functions corresponding to these response and control variables are summarized in Table 4-3. In particular, polynomials appearing in (4-81) are given. Examination of these polynomials reveals

$$\mathbf{\Lambda}(s) = \begin{bmatrix} s^2 & 0 \\ 0 & s^2 \end{bmatrix} \quad \text{and} \quad \mathbf{T}_\infty = \begin{bmatrix} -58.1 & -34.1 \\ 35.8 & -1.66 \end{bmatrix}$$

so that all compatible excess-pole specifications have elements of second or lower degrees. The matrix of desired closed-loop transfer functions is taken to be

$$\mathbf{T}_d(s) = \begin{bmatrix} \dfrac{-34.0}{(s + 7.0)(s + 1.4)} & 0 \\ 0 & \dfrac{-35.8}{(s + 25.0)(s + 1.4)} \end{bmatrix}$$

so that

$$\mathbf{P}_\infty = \begin{bmatrix} \dfrac{-1}{34.0} & 0 \\ 0 & \dfrac{-1}{35.8} \end{bmatrix}.$$

Compatibility of these design specifications is easily verified.

Expressions given in (4-59) and (4-62) are evaluated easily on the digital computer which yields

$$\mathbf{E}^{-1} = \begin{bmatrix} -0.0503 & 1.09 \\ 1.08 & -1.85 \end{bmatrix}$$

and

$$\mathbf{D} = \begin{bmatrix} -0.0151 & -0.129 & -0.273 & -0.00921 & -1.14 & 0.252 \\ -0.0826 & -2.18 & 0.00272 & -0.970 & 0.0553 & 0.137 \end{bmatrix}$$

to three significant figures. In addition, the feedback gain matrix is

$$\mathbf{E}^{-1}\mathbf{D} = \begin{bmatrix} -0.089 & -2.36 & 0.0167 & -1.05 & 0.117 & 0.136 \\ 0.137 & 3.90 & -0.300 & 1.79 & -1.33 & 0.0190 \end{bmatrix},$$

and the closed-loop numerator and denominator polynomials corresponding to responses $\tilde{\mathbf{y}}$ and inputs $\tilde{\mathbf{y}}_c$ are summarized in Table 4-4.

Table 4-4 Closed-loop numerator and denominator polynomials for $\tilde{\mathbf{y}}$ and $\tilde{\mathbf{y}}_c$

$N_{11}(s) = -34.0(s + 25.0)(s + 1.40)(s + 0.0175)s$
$N_{21}(s) = (7.30 \times 10^{-9})(s + 11.0)(s - 3.74)[s^2 + 2(0.0921)(0.463)s + (0.463)^2]$
$N_{12}(s) = -(5.92 \times 10^{-7})(s + 2.68)s[s^2 + 2(0.00993)(7.33)s + (7.33)^2]$
$N_{22}(s) = -35.8(s + 7.00)(s + 1.40)(s + 0.0175)s$
$D(s) = (s + 25.0)(s + 7.00)(s + 1.40)^2(s + 0.0175)s$

This design illustrates a number of the aspects of closed-loop system stability given previously. In particular, Table 4-3 yields

$$\det \mathbf{N}(s) = N_1(s)N_2(s)N_3(s),$$

where

$$N_1(s) = (s + 25.0), \quad N_2(s) = (s + 0.0175)s,$$

and

$$N_3(s) = (s + 6.67)[s^2 + 2(0.00908)(1.31)s + (1.31)^2].$$

Polynomial $N_1(s)$ is a factor of $D_d(s)$ and hence also a factor of the closed-loop characteristic polynomial. Polynomial $N_2(s)$ also is a factor of the closed-loop characteristic polynomial but is not a factor $D_d(s)$. Thus, polynomial $N_2(s)$ is not a factor of the minimal closed-loop $D(s)$ under exact numerical conditions. On the other hand, polynomial $N_3(s)$ is not a factor of the closed-loop characteristic polynomial and hence is not a factor of the minimal closed-loop $D(s)$ even with numerical inaccuracy.

The closed-loop system is of course not asymptotically stable, and the pole at $s = 0$ corresponds to the desired spiral mode of lateral motion. In addition, this pole moves into both the right- and left-half s-plane depending upon flight condition. As long as this pole is within a radius of say 0.005 of the origin, the pilot-handling qualities of the augmented lateral dynamics are considered acceptable. Sensitivity of the design determines the range of flight conditions for which this design is acceptable. Sensitivity is not discussed here in detail.

Perhaps the most severe difficulty of the previous design is the required measurement of side slip rate x_1. An examination of (4-98) reveals that condition $\mathbf{d}_1 = \mathbf{0}$ is achieved by the selection

$$\mathbf{F}_a = [-5.0], \quad \mathbf{G}_a = [0 \quad 1.0 \quad 0 \quad 0 \quad 0], \quad \text{and} \quad \mathbf{H}_a = \begin{bmatrix} -0.514 \\ -2.9586 \end{bmatrix}.$$

This selection corresponds to a single first order compensation filter associated with both response variables, and this form of compensation can be expressed in terms of

$$\mathbf{\Phi}(s) = \begin{bmatrix} \dfrac{-0.514}{s+5.0} & 0 \\ 0 & \dfrac{-2.9586}{s+5.0} \end{bmatrix} \begin{bmatrix} 0 & 1.0 & 0 & 0 & 0 & 0 \\ 0 & 1.0 & 0 & 0 & 0 & 0 \end{bmatrix}.$$

The augmented system is of course compatible with the original matrix of desired transfer functions so that matrices

$$\mathbf{E}^{-1} = \begin{bmatrix} -0.0503 & 1.09 \\ 1.08 & -1.85 \end{bmatrix},$$

$$\mathbf{D} = \begin{bmatrix} 0 & -0.0779 & -0.273 & -0.00921 & -1.14 & 0.252 & -0.109 \\ 0 & -0.412 & 0.00272 & -0.970 & 0.0553 & 0.137 & -5.95 \end{bmatrix}$$

and

$$\mathbf{E}^{-1}\mathbf{D} = \begin{bmatrix} 0 & -0.444 & 0.0167 & -1.05 & 0.117 & 0.136 & -6.46 \\ 0 & 0.679 & -0.300 & 1.79 & -1.33 & 0.0190 & 10.9 \end{bmatrix}$$

can be computed directly for the composite system. In addition, closed-loop numerator and denominator polynomials corresponding to responses $\tilde{\mathbf{z}}$ and inputs $\tilde{\mathbf{z}}_c$ are summarized in Table 4-5. The closed-loop system is of course unstable, but pilot-handling qualities are still considered acceptable. Finally, closed-loop numerator and denominator polynomials corresponding to responses $\tilde{\mathbf{y}}$ and inputs $\tilde{\mathbf{z}}_c$ are summarized in Table 4-6. These polynomials are seen to have

Table 4-5 Closed-loop numerator and denominator polynomials for $\tilde{\mathbf{z}}$ and $\tilde{\mathbf{z}}_c$ with side-slip rate suppressed

$$N_{11}(s) = -34.0(s + 25.0)(s + 4.30)(s + 1.40)(s + 0.709)(s - 0.000898)$$
$$N_{21}(s) = (3.59 \times 10^{-7})(s + 45.1)(s + 4.31)(s - 0.577)[s^2 + 2(0.863)(0.492)s + (0.492)^2]$$
$$N_{12}(s) = (3.34 \times 10^{-5})(s + 26.4)(s + 4.95)(s + 0.000189)[s^2 - 2(0.243)(1.51)s + (1.51)^2]$$
$$N_{22}(s) = -35.8(s + 7.00)(s + 4.30)(s + 1.40)(s + 0.709)(s - 0.000898)$$

$$D(s) = (s + 25.0)(s + 7.00)(s + 4.30)(s + 1.40)^2(s + 0.709)(s - 0.000898)$$

Table 4-6 Closed-loop numerator and denominator polynomials for $\tilde{\mathbf{y}}$ and $\tilde{\mathbf{z}}_c$ with side-slip rate suppressed

$$N_{11}(s) = -34.0(s + 25.0)(s + 4.30)(s + 1.40)(s + 0.708)(s + 0.00000337)$$
$$N_{21}(s) = (3.59 \times 10^{-7})(s + 26000.0)(s + 25.0)(s + 7.60)(s + 1.40)(s - 7.61)$$
$$N_{12}(s) = (3.34 \times 10^{-5})(s + 7.00)(s + 1.40)s[s^2 - 2(0.689)(743.0)s + (743.0)^2]$$
$$N_{22}(s) = -35.8(s + 7.00)(s + 5.00)(s + 1.40)(s + 0.0338)(s - 0.0162)$$

$$D(s) = (s + 25.0)(s + 7.00)(s + 4.30)(s + 1.40)^2(s + 0.709)(s - 0.000898)$$

the same high-frequency characteristics as the polynomials appearing in Table 4-5. The augmented design hence is non-interacting from $\tilde{\mathbf{z}}_c$ to $\tilde{\mathbf{y}}$ under exact numerical conditions even without a prefilter.

Prefilters used to obtain (4-65) for the given $\mathbf{T}_d(s)$ are not considered to be acceptable for this augmented design because a pole at 0.0162 would be required. However, the prefilter corresponding to

$$\mathbf{R}(s) = \begin{bmatrix} \dfrac{s - 0.000898}{s + 0.00000337} & 0 \\ 0 & 1.00 \end{bmatrix}$$

yields a non-interacting design from $\tilde{\mathbf{y}}_c$ to $\tilde{\mathbf{y}}$, and (4-65) is obtained for the given $\mathbf{T}_d(s)$ both as $|s| \to 0$ and $|s| \to \infty$. An acceptable design thus is obtained even though augmentation is not exact.

4.7 SELECTION OF PERFORMANCE FUNCTIONALS AND OPTIMAL GAIN CONTROL

Most control system design problems of course do not result in $I = 0$ corresponding to the performance integral defined in (4-51). Therefore, condition $\mathbf{X} \neq \mathbf{0}$ occurs for most problems and hence a performance functional f must also be selected. Naturally there are design problems for which performance functional f is specified directly from curve fitting or other analyses of the design problem in question. More often than not, however, this functional is unspecified and instead is selected in accordance with general response properties desired of the closed-loop system.

Many typical choices of performance functional f are included in the class defined by

$$f = \operatorname{tr}\{\mathbf{B}'\mathbf{X}^l\mathbf{B}\}. \tag{4-100}$$

For example, selection of

$$l = 1 \quad \text{and} \quad \mathbf{B} = \mathbf{i}_i$$

where $\mathbf{i}_i$ denotes the ith standard basis vector, yields

$$f = \mathbf{i}_i'\mathbf{X}\mathbf{i}_i = x_{ii} = I_i$$

where I_i denotes the value of I corresponding to $\mathbf{x}_0 = \mathbf{i}_i$. In other words, this specialization of (4-100) can be used to minimize the initial-state equivalent of step response and the like. A similar selection of f corresponds to

$$l = 1 \quad \text{and} \quad \mathbf{B} = \mathbf{I}$$

so that now

$$f = \operatorname{tr}\{\mathbf{X}\} = \sum_{i=1}^{n} I_i = \sum_{i=1}^{n} \lambda_i\{\mathbf{X}\}$$

where $n = \dim \mathbf{x}$. Another selection of f which can be used for worst-case design corresponds to

$$l > 1 \quad \text{and} \quad \mathbf{B} = \mathbf{I}.$$

In particular, the performance functional becomes

$$f = \operatorname{tr}\{\mathbf{X}^l\} = \sum_{i=1}^{n} \lambda_i^l\{\mathbf{X}\} \to k\lambda_{\max}^l\{\mathbf{X}\} \quad \text{as} \quad l \to \infty$$

where k is some positive constant of proportionality when $\mathbf{X}$ is positive semi-definite. This performance functional is selected in order to approximate system response characteristics that would be induced by

$$f = \lambda_{\max}\{\mathbf{X}\} = \sup_{|\mathbf{x}_0|=1} \{\mathbf{x}_0'\mathbf{X}\mathbf{x}_0\}. \tag{4-101}$$

The performance functional given in (4-101) corresponds to the least upper bound of performance integral I computed for all initial states belonging to $\{\mathbf{x}_0 : \mathbf{x}_0 \in \mathscr{R}^n$ and $|\mathbf{x}_0| = 1\}$ and hence might be useful for worst-case design.

In order to examine implications of performance functionals belonging to the class defined by (4-100), the special case of optimal gain control is introduced here. Specifically, the open-loop system is again defined by (4-52) and (4-53). In addition, linear control equations employing response-variable feedback are taken to be

$$\mathbf{u} = -\mathbf{C}\mathbf{y}, \tag{4-102}$$

where the adjustable parameter vector $\mathbf{c}$ has been assigned to the feedback gain matrix $\mathbf{C}$ by vector partitions as

$$\mathbf{C} = [\mathbf{c}_1 \quad \mathbf{c}_2 \quad \ldots]. \tag{4-103}$$

Finally, the performance integral is defined in (4-51) where now the error vector is taken to be

$$\mathbf{e} = \mathbf{D}\mathbf{x} + \mathbf{E}\mathbf{u}. \tag{4-104}$$

Coefficient matrices $\mathbf{D}$ and $\mathbf{E}$ are considered here to be general in origin and they may of course be specified from procedures used for transfer-function matrix synthesis.

Elimination of controls from (4-52) using (4-102) and then elimination of responses using (4-53) yield closed-loop state equations in the form of (4-1) where

$$\mathbf{A} = \mathbf{F} - \mathbf{G}\mathbf{C}\mathbf{H}. \tag{4-105}$$

Similarly, the error vector defined in (4-104) becomes

$$\mathbf{e} = (\mathbf{D} - \mathbf{E}\mathbf{C}\mathbf{H})\mathbf{x}$$

so that the performance integral given in (4-51) can be reduced to the form of (4-2) where

$$\mathbf{W} = (\mathbf{D} - \mathbf{E}\mathbf{C}\mathbf{H})'(\mathbf{D} - \mathbf{E}\mathbf{C}\mathbf{H}). \tag{4-106}$$

Necessary conditions corresponding to (4-10) can now be written explicitly for optimal gain problems without constraints.

Gradient L_c is rewritten conveniently in matrix form $L_{\mathbf{C}}$ for optimal gain problems. Gradient $L_{\mathbf{C}}$ then becomes

$$L_{\mathbf{C}} = 2[(\mathbf{E}'\mathbf{E})\mathbf{C}(\mathbf{H}\mathbf{\Lambda}\mathbf{H}') - (\mathbf{E}'\mathbf{D} + \mathbf{G}'\mathbf{X})(\mathbf{\Lambda}\mathbf{H}')]$$

using Lagrangian L defined by (4-7), (4-8), and (4-9). If $(\mathbf{E}'\mathbf{E})$ and $(\mathbf{H}\mathbf{\Lambda}\mathbf{H}')$ are nonsingular, then stationarity condition $L_{\mathbf{C}} = \mathbf{0}$ implies that

$$\mathbf{C} = (\mathbf{E}'\mathbf{E})^{-1}(\mathbf{E}'\mathbf{D} + \mathbf{G}'\mathbf{X})(\mathbf{\Lambda}\mathbf{H}')(\mathbf{H}\mathbf{\Lambda}\mathbf{H}')^{-1}. \qquad (4\text{-}107)$$

This expression for optimal gains also simplifies to a number of frequently occurring special cases when additional assumptions are imposed.

For example, optimal gains become

$$\mathbf{C} = (\mathbf{E}'\mathbf{E})^{-1}(\mathbf{E}'\mathbf{D} + \mathbf{G}'\mathbf{X})\mathbf{H}^{-1}$$

when $\mathbf{H}$ is nonsingular. This case corresponds to linear optimal control which is derived from the calculus of variations, and matrices $\mathbf{X}_j$ and $\mathbf{\Lambda}_j$ as well as $\mathbf{\Lambda}$ no longer need be computed when the Newton–Raphson method is used for computational purposes. In addition, optimal gains become

$$\mathbf{C} = \mathbf{E}^{-1}\mathbf{D}$$

when conditions $\mathbf{H} = \mathbf{I}$, $\mathbf{X} = \mathbf{0}$, and $\mathbf{E}$ nonsingular are imposed. This case corresponds to conditions for compatibility which arise in the synthesis of transfer function matrices.

Properties of matrices $(\mathbf{E}'\mathbf{E})$ and $(\mathbf{H}\mathbf{\Lambda}\mathbf{H}')$ are of interest due to assumptions which lead to (4-107). Matrix $(\mathbf{E}'\mathbf{E})$ is of course nonsingular if and only if rank $\mathbf{E} = \dim \mathbf{e}$, and this condition generally presents no problem because $\mathbf{E}$ is independent of other variables. On the other hand, matrix $(\mathbf{H}\mathbf{\Lambda}\mathbf{H}')$ is nonsingular only if rank $\mathbf{H} = \dim \mathbf{y}$ and is nonsingular if, in addition, $\mathbf{\Lambda}$ is nonsingular. A necessary and sufficient condition for $(\mathbf{H}\mathbf{\Lambda}\mathbf{H}')$ to be nonsingular is somewhat more complicated and is related to conditions for complete controllability.

By way of example, suppose performance functional f is given by (4-100) with $l = 1$. Then equation (4-13) becomes

$$\mathbf{\Lambda}\mathbf{A}' + \mathbf{A}\mathbf{\Lambda} + \mathbf{B}\mathbf{B}' = \mathbf{0},$$

and hence the Lagrange multiplier matrix is given by

$$\mathbf{\Lambda} = \int_0^\infty \{e^{\mathbf{A}t}\mathbf{B}\mathbf{B}'\, e^{\mathbf{A}'t}\}\, dt$$

when **A** is asymptotically stable. The matrix in question can then be written in the form

$$\mathbf{H\Lambda H}' = \int_0^\infty (\mathbf{H}\,e^{\mathbf{A}t}\mathbf{B})(\mathbf{H}\,e^{\mathbf{A}t}\mathbf{B})'\,dt. \tag{4-108}$$

Matrix

$$\mathbf{R} = [\mathbf{HB} \quad \mathbf{HAB} \quad \ldots \quad \mathbf{HA}^{(n-1)}\mathbf{B}] \tag{4-109}$$

and vector

$$\mathbf{r} = \int_0^\infty \mathbf{H}\,e^{\mathbf{A}t}\mathbf{B}\mathbf{v}(t)\,dt \tag{4-110}$$

are also pertinent to conditions for the invertibility of $(\mathbf{H\Lambda H}')$.

The concept of complete responsiveness of a linear time invariant system is introduced in a fashion analogous to output controllability [3] as

Definition 4-1 Let $m = \dim \mathbf{y}$ and **A** be asymptotically stable. Then the ordered triple $\{\mathbf{A}, \mathbf{B}, \mathbf{H}\}$ is *completely responsive* if and only if, given any $\boldsymbol{\rho} \in \mathscr{R}^m$, there exists a piecewise continuous $\mathbf{v}(t)$ on $[0, \infty)$ for which $\mathbf{r} = \boldsymbol{\rho}$.

The principle result concerning the invertibility of $(\mathbf{H\Lambda H}')$ as given in (4-108) then can be stated as

Theorem 4-12 *Suppose* $m = \dim \mathbf{y}$, $n = \dim \mathbf{x}$, *and* **A** *is asymptotically stable. The conditions*

i) *rank* $\mathbf{R} = m$,
ii) $(\mathbf{H\Lambda H}')$ *nonsingular,*

and

iii) $\{\mathbf{A}, \mathbf{B}, \mathbf{H}\}$ *completely responsive*

are equivalent.

Proof In order to prove that i) implies ii), assume rank $\mathbf{R} = m$ and $(\mathbf{H\Lambda H}')$ is singular. Then there exists a nonzero vector $\boldsymbol{\rho} \in \mathscr{R}^m$ such that

$$0 = \boldsymbol{\rho}'(\mathbf{H\Lambda H}')\boldsymbol{\rho} = \int_0^\infty (\boldsymbol{\rho}'\mathbf{H}\,e^{\mathbf{A}t}\mathbf{B})(\boldsymbol{\rho}'\mathbf{H}\,e^{\mathbf{A}t}\mathbf{B})'\,dt$$

and hence

$$\boldsymbol{\rho}'\mathbf{H}\,e^{\mathbf{A}t}\mathbf{B} = \mathbf{0}' \text{ on } [0, \infty).$$

In addition, repeated differentiation and evaluation at $t = 0$ yields

$$\boldsymbol{\rho}'\mathbf{HA}^{(i-1)}\mathbf{B} = \mathbf{0}' \quad \text{for} \quad i = 1, 2, \ldots, n$$

so that rank $\mathbf{R} < m$ holds which is a contradiction. In order to prove that ii) implies iii), let $\mathbf{v}$ be the piecewise continuous function

$$\mathbf{v}(t) = \mathbf{B}' \, e^{\mathbf{A}'t}\mathbf{H}'\mathbf{w}.$$

Then substitution into (4-110) yields

$$\mathbf{r} = (\mathbf{H}\mathbf{\Lambda}\mathbf{H}')\mathbf{w} = \boldsymbol{\rho}$$

given any $\boldsymbol{\rho} \in \mathscr{R}^m$ when constant vector

$$\mathbf{w} = (\mathbf{H}\mathbf{\Lambda}\mathbf{H}')^{-1}\boldsymbol{\rho}$$

is selected. In order to prove that iii) implies i), assume $\{\mathbf{A}, \mathbf{B}, \mathbf{H}\}$ is completely responsive and rank $\mathbf{R} < m$. Then there exists a nonzero $\boldsymbol{\rho} \in \mathscr{R}^m$ such that

$$\boldsymbol{\rho}'\mathbf{H}\mathbf{A}^{(i-1)}\mathbf{B} = \mathbf{0}' \quad \text{for} \quad i = 1, 2, \ldots, n.$$

Moreover, the inverse Laplace transform of the relationship used to derive (4-75) yields

$$e^{\mathbf{A}t} = \sum_{i=1}^{n} \mathbf{A}^{(i-1)} f_i(t)$$

for suitably defined functions $f_i(t)$. Substitution into (4-110) now yields

$$0 = \boldsymbol{\rho}'\mathbf{r} = \sum_{i=1}^{n} \boldsymbol{\rho}'\mathbf{H}\mathbf{A}^{(i-1)}\mathbf{B} \int_0^\infty f_i(t)\mathbf{v}(t)\,dt$$

for all piecewise continuous $\mathbf{v}(t)$. Thus, all response vectors are constrained to lie in the subspace of $\mathscr{R}^m$ that is orthogonal to $\boldsymbol{\rho}$, and hence the assumption of complete responsiveness is contradicted.

These criteria for the invertibility of $(\mathbf{H}\mathbf{\Lambda}\mathbf{H}')$ unforunately cannot be applied in general before optimal gains are computed because the criteria involve $\mathbf{A}$ which depends on $\mathbf{X}$ and $\mathbf{\Lambda}$.

4.8 EXAMPLE OF OPTIMAL GAIN CONTROL

An example of optimal control is provided by a simplified aircraft pitch-channel design problem. Open-loop actuator, aircraft, and compensator dynamics are defined by

$$\mathbf{F} = \begin{bmatrix} 0 & 0 & 0 & 0 & 0 & 0 \\ 1 & 0 & 0 & 0 & 0 & 0 \\ 0 & a_1 & -a_3 & 0 & 0 & -a_2 \\ 0 & 0 & 1 & 0 & 0 & 0 \\ 0 & 0 & 0 & 1 & 0 & 0 \\ 0 & 0 & 1 & 0 & 0 & -a_4 \end{bmatrix} \quad \text{and} \quad \mathbf{G} = \begin{bmatrix} 1 \\ 0 \\ 0 \\ 0 \\ 0 \\ 0 \end{bmatrix}.$$

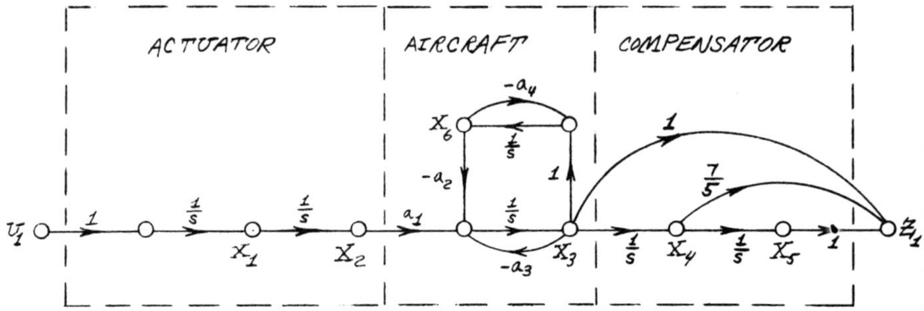

Figure 4-4 Flow graph of open-loop dynamics with zero initial state for pitch channel design

Open-loop dynamics also are depicted in Figure 4-4, and state and control variable assignments are summarized in Table 4-7. In addition, elements of **F** are summarized in Table 4-8 for three flight conditions.

Table 4-7 State and control variables for pitch channel design

Variable	Description
x_1	Elevator rate
x_2	Elevator deflection
x_3	Pitch angle
x_4	Compensator state
x_5	Compensator state
x_6	Angle of attack
u_1	Elevator command

Table 4-8 Flight conditions for pitch channel example

Flight condition	a_1	a_2	a_3	a_4
FC1	5.00	3.00	1.00	1.00
FC2	1.60	1.20	0.22	0.10
FC3	0.50	0.50	0.05	0.01

Open-loop actuator dynamics have the transfer function

$$\frac{X_2}{U_1} = \frac{1}{s^2}.$$

This transfer function is equivalent to those defined by (4-74) and is required in order to achieve admissibility with a third order excess pole specification. In addition, open-loop aircraft dynamics have the transfer function

$$\frac{X_3}{X_2} = \frac{a_1(s + a_4)}{s^2 + (a_3 + a_4)s + (a_2 + a_3a_4)}$$

which exhibits a wide range of characteristics for the three flight conditions given in Table 4-8. Finally, open-loop compensator dynamics have the transfer function

$$\frac{Z_1}{X_3} = \frac{s^2 + 2(0.7)(1.0)s + (1.0)^2}{s^2}.$$

This transfer function is the postfilter $\mathbf{S}(s)$ that has been introduced for the purpose of achieving insensitivity with respect to changes in flight condition. Pole suppression specification $(s^2 + 1.4s + 1.0)$ corresponds roughly to the variable poles of aircraft transfer function X_2/X_3 for FC2 which is taken to be the nominal flight condition.

The performance integral selected here for design purposes is defined by excess pole specification

$$\mathbf{P}(s) = \left[\left(\frac{s}{10.0} + 1\right)\right]\left[\left(\frac{s}{5.0}\right)^2 + 2(0.3)\left(\frac{s}{5.0}\right) + 1\right].$$

This selection is introduced for the purpose of transfer function synthesis and results in coefficient matrices appearing in (4-104)

$$\mathbf{D} = [0.09075 \quad 0.4472 \quad 1.231 \quad 1.620 \quad 1.000 \quad -0.3287]$$

and

$$\mathbf{E} = [0.006400]$$

for FC2.

For the purposes of comparison, five different cases of optimal gain control for FC2 are introduced here. Case 1 is defined by $\mathbf{H} = \mathbf{I}$, $l = 1$, and $\mathbf{B} = \mathbf{I}$ so that the corresponding system has complete state-variable feedback and results in exact synthesis of the desired transfer-function matrix corresponding to $\mathbf{P}(s)$. The remaining four cases correspond to systems having incomplete state-variable feedback defined by

$$\mathbf{H} = \begin{bmatrix} 1 & 0 & 0 & 0 & 0 & 0 \\ 0 & 1 & 0 & 0 & 0 & 0 \\ 0 & 0 & 1 & 0 & 0 & 0 \\ 0 & 0 & 0 & 1 & 0 & 0 \\ 0 & 0 & 0 & 0 & 1 & 0 \end{bmatrix}$$

under the assumption that angle of attack x_6 cannot be measured. Case 2 is further defined by l and $\mathbf{B}$ selected so that $f = x_{33}$, and hence the initial-state equivalent of step response in pitch angle is to be minimized. Cases 3, 4, and 5 are further defined by $\mathbf{B} = \mathbf{I}$ and $l = 1$, $l = 8$, and $l = 16$ respectively so that various approximations of $f = \lambda_{\max}\{\mathbf{X}\}$ are to be minimized.

The performance of these five systems are to be analyzed in part by the location of zeros of the characteristic polynomial of the closed-loop system. This polynomial is written in the form

$$\det(s\mathbf{I} - \mathbf{A}) = p(s)z(s),$$

where

$$p(s) = (s + \alpha_p)(s^2 + 2\zeta_p\omega_p s + \omega_p^2)$$

and

$$z(s) = (s + \alpha_z)(s^2 + 2\zeta_z\omega_z s + \omega_z^2)$$

for convenience. Zeros of $p(s)$ are to be associated with those of the design specification $\mathbf{P}(s)$, and zeros of $z(s)$ are to be associated with the numerator zeros of the transfer function from U_1 to Z_1.

Feedback gains corresponding to Cases 1 through 5 are summarized in Table 4-9. Cases 2 through 5 do not include feedback from x_6, and hence feedback gain c_{16} is not defined for these cases. Table 4-10 summarizes the performance of Cases 1 through 5 for FC2 in terms of x_{33}, $\mathrm{tr}\{\mathbf{X}\}$, and $\lambda_{\max}\{\mathbf{X}\}$ as well as in terms of parameters of the polynomials $p(s)$ and $z(s)$. Case 1 is noted to yield $\mathbf{X} = \mathbf{0}$ and exact pole placement in accordance with the zeros of $\mathbf{P}(s)$ and the numerator zeros of the transfer function from U_1 to Z_1. Case 2 is noted to yield

Table 4-9 Feedback gains for cases 1 through 5

Case	c_{11}	c_{12}	c_{13}	c_{14}	c_{15}	c_{16}
1	14.18	69.88	192.3	253.1	156.3	−51.35
2	14.17	69.96	192.2	202.7	160.7	–
3	22.31	79.04	315.4	343.6	317.3	–
4	25.31	83.48	361.8	417.6	362.4	–
5	25.52	83.97	365.3	424.8	365.0	–

Table 4-10 Performance of cases 1 through 5 for FC2

Case	x_{33}	tr $\{\mathbf{X}\}$	$\lambda_{max}\{\mathbf{X}\}$	α_p	ζ_p	ω_p	α_z	ζ_z	ω_z
1	0.000	0.000	0.000	10.0	0.300	5.00	0.100	0.700	1.00
2	0.00000287	1.055	1.055	9.97	0.301	5.00	0.104	0.706	0.996
3	0.0394	0.466	0.321	19.5	0.179	4.81	0.102	0.610	1.05
4	0.0662	0.496	0.259	22.7	0.154	4.81	0.102	0.643	1.04
5	0.0683	0.501	0.255	22.9	0.153	4.81	0.102	0.643	1.04

a very small value of x_{33} as well as accurate pole placement. However, the system corresponding to this case results in a rather large value of $\lambda_{max}\{\mathbf{X}\}$. Differences between these cases indicate that worst-case design is nearly achieved by Case 5. Unfortunately, worst-case design does not place the real pole corresponding to α_p with any degree of accuracy.

Approximations to worst-case design also require actuator bandwidths that are considered to be excessive. Actuator bandwidth is $\omega_a = c_{12}^{1/2}$, and acceptable designs must satisfy $\omega_a \leqslant 8.5$. Therefore, Cases 3, 4, and 5 violate the inequality constraint

$$g_1 = c_{12} - 72.25 \leqslant 0.$$

Feedback gains for Cases 3, 4, and 5 subject to this bandwidth constraint are given in Table 4-11. The performance of the systems corresponding to these three cases with FC2 is summarized in Table 4-12. A very small increase in $\lambda_{max}\{\mathbf{X}\}$ is noted, and pole placement is not altered significantly.

A typical example of convergence properties of the method of successive substitutions defined by (3-102), (3-117), and (3-118) with $\alpha = 1$, $\mathbf{P} = L_{\mathbf{cc}}$, and $\mathbf{Q} = L_{\hat{\lambda}\mathbf{c}}$ is summarized in Table 4-13 for this inequality constraint. These data

Table 4-11 Feedback gains for cases 3, 4, and 5 subject to a bandwidth constraint

Case	c_{11}	c_{12}	c_{13}	c_{14}	c_{15}
3	20.21	72.25	285.7	309.7	287.9
4	21.63	72.25	309.8	357.0	311.5
5	21.67	72.25	310.9	361.3	311.8

Table 4-12 Performance of cases 3, 4, and 5 subject to a bandwidth constraint for FC2

Case	x_{33}	tr $\{\mathbf{X}\}$	$\lambda_{\max}\{\mathbf{X}\}$	α_p	ξ_p	ω_p	α_z	ξ_z	ω_z
3	0.00237	0.470	0.328	17.5	0.171	4.84	0.102	0.607	1.05
4	0.00440	0.504	0.264	19.1	0.141	4.84	0.102	0.641	1.04
5	0.00467	0.511	0.259	19.2	0.139	4.84	0.102	0.647	1.04

Table 4-13 Convergence of min-max method for case 4 with a bandwidth constraint

Iteration	$f \times 10^5$	c_{11}	c_{12}	c_{13}	c_{14}	c_{15}	$\lambda_1 \times 10^7$
0	2.7984	25.31	83.48	361.8	417.6	362.4	0.000
1	3.2529	21.64	72.25	311.6	360.4	313.1	5.452
2	3.1920	21.61	72.25	310.0	357.3	311.6	7.853
3	3.1916	21.63	72.25	309.8	357.0	311.5	8.198

Table 4-14 Performance of cases 1 through 5 for FC1

Case	α_p	ξ_p	ω_p	α_z	ξ_z	ω_z
1	14.0	0.00135	8.36	0.855	0.666	0.967
2	14.0	0.00089	8.37	1.25	0.563	0.811
3	22.0	0.0133	8.57	1.13	0.536	0.934
4	24.9	0.0158	8.62	1.12	0.569	0.935
5	25.0	0.0160	8.63	1.12	0.574	0.933

Table 4-15 Performance of cases 1 through 5 for FC3

Case	α_p	ξ_p	ω_p	α_z	ξ_z	ω_z
1	0.890	0.904	7.02	0.01	0.239	1.33
2	0.908	0.902	7.04	0.01	0.230	1.33
3	18.5	0.812	1.78	0.01	0.304	1.64
4	21.8	0.783	1.53	0.01	0.296	1.88
5	22.1	0.790	1.51	0.01	0.291	1.90

correspond to Case 4 when the point given in Table 4-9 for this case is used as the starting point.

Sensitivity properties of Cases 1 through 5 without bandwidth constraints are illustrated by parameters of the closed-loop characteristic polynomial. These data are given in Tables 4-14 and 4-15 which result from evaluating the design based on FC2 with aircraft dynamics corresponding to FC1 and FC3 respectively. All cases are seen to maintain system stability for both FC1 and FC3. However, damping ratio ζ_p is particularly unacceptable for Cases 1 and 2 with FC1 as is α_p for Cases 1 and 2 with FC3. Undesirable effects illustrated by Tables 4-14 and 4-15 are primarily due to large changes in high-frequency gain a_1 and not to the variable poles of open-loop transfer function X_3/X_2. Large changes in high-frequency gain are counteracted in practice by controller gain changing.

The final data presented for this example is based on gain changing normally employed to obtain increased system insensitivity. In particular, a single gain changer is implemented in such a manner that feedback gains c_{13}, c_{14}, c_{15} and perhaps c_{16} are multiplied by the ratio of a_1 corresponding to the existing flight condition to a_1 corresponding to the nominal design flight condition. The results of the implementation of this gain changer are summarized in Tables 4-16 and 4-17 for FC1 and FC3 respectively. These data indicate that all cases are nearly invariant over the entire range of flight conditions except with respect to α_z. However, this variation in α_z is required so that $\alpha_z \cong a_4$ holds and hence

Table 4-16 Performance obtained with gain changing for FC1

Case	α_p	ζ_p	ω_p	α_z	ζ_z	ω_z
1	10.2	0.317	5.54	1.56	0.632	0.716
2	10.2	0.314	5.55	1.85	0.503	0.666
3	14.3	0.237	5.25	1.45	0.504	0.813
4	22.7	0.219	5.21	1.39	0.536	0.819
5	23.0	0.219	5.21	1.40	0.541	0.818

Table 4-17 Performance obtained with gain changing for FC3

Case	α_p	ζ_p	ω_p	α_z	ζ_z	ω_z
1	9.97	0.307	4.87	0.00998	0.618	1.03
2	9.94	0.308	4.87	0.0100	0.611	1.04
3	19.5	0.173	4.69	0.0100	0.564	1.09
4	22.7	0.145	4.69	0.0100	0.603	1.08
5	22.9	0.143	4.69	0.0100	0.609	1.08

insensitivity is achieved by cancellation of the variable zero occurring in the open-loop transfer function X_3/X_2.

4.9 LINEAR OPTIMAL CONTROL

Perhaps the most significant case of optimal gain control occurs with $(\mathbf{E}'\mathbf{E})$ nonsingular and $\{\mathbf{A}, \mathbf{B}, \mathbf{H}\}$ completely responsive for some nonsingular $\mathbf{H}$. As mentioned previously, this case is characterized by optimal gains that do not depend upon Lagrange multiplier matrix $\mathbf{\Lambda}$ and hence do not depend upon $\mathbf{B}$. In other words, optimal gains now in effect do not depend upon which initial states are deemed relevant to the design problem in question. This case is also characterized by design problems for which dim $\mathbf{e} \geqslant$ dim $\mathbf{u}$ often occurs so that $\mathbf{E}$ may not be square and hence may not be invertible. Moreover, condition $\mathbf{X} \neq \mathbf{0}$ almost always occurs when dim $\mathbf{e} >$ dim $\mathbf{u}$. Because $\mathbf{H}$ has been assumed to be nonsingular, these problems can be reformulated with response vector $\mathbf{y}$ treated as the system state vector. Therefore, no generality in problem formulation is lost by assuming that $\mathbf{H} = \mathbf{I}$. This special case of optimal gain control also results from the variational formulation of comparable automated design problems with respect to control function $\mathbf{u}(t)$ and henceforth is referred to as *linear optimal control.*

Equations which correspond to linear optimal control can be derived directly from those of optimal gain control. Specifically, equations (4-105), (4-106), and (4-107) become

$$\mathbf{A} = \mathbf{F} - \mathbf{GC}, \tag{4-111}$$

$$\mathbf{W} = (\mathbf{D} - \mathbf{EC})'(\mathbf{D} - \mathbf{EC}),$$

and

$$\mathbf{C} = (\mathbf{E}'\mathbf{E})^{-1}(\mathbf{E}'\mathbf{D} + \mathbf{G}'\mathbf{X}) \tag{4-112}$$

respectively. Substitution of these relationships into (4-12) then yields

$$\mathbf{0} = (\mathbf{D}'\mathbf{D}) + \mathbf{XF} + \mathbf{F}'\mathbf{X} - \mathbf{C}'(\mathbf{E}'\mathbf{E})\mathbf{C} \tag{4-113}$$

which is a matrix algebraic equation that is quadratic in $\mathbf{X}$.

The problem of existence and uniqueness of solutions to these quadratic equations is related to both algebraic and successive approximation methods of solution. This problem hence is of particular significance to the design of optimal controls and is examined in terms of related quadratic equations. Specifically, equation (4-113) can be reduced to the form

$$\mathbf{0} = \mathbf{P} + \mathbf{XQ} + \mathbf{Q}'\mathbf{X} - \mathbf{XRX} \tag{4-114}$$

or alternately to the form

$$\mathbf{0} = \mathbf{P} + \mathbf{XA} + \mathbf{A}'\mathbf{X} + \mathbf{XRX} \tag{4-115}$$

using

$$\mathbf{A} = \mathbf{Q} - \mathbf{RX}. \tag{4-116}$$

For the purposes of investigating questions of existence and uniqueness of solution **X**, however, the other matrices appearing in (4-114), (4-115), and (4-116) are taken to be arbitrary real matrices except that **P** and **R** are nominally assumed to be positive semi-definite. Similar investigations of these equations are to be found in References [12, 13].

Properties of quadratic equations (4-114) are stated in terms of the real parts of eigenvalues which are denoted as $\lambda_R = \text{Re}\{\lambda\}$ for convenience. Also, the *complex null space* of a real matrix **F** is needed frequently in the sequel and is defined as

$$\mathcal{N}(\mathbf{F}) = \{\mathbf{f}\colon \mathbf{f} \in \mathscr{C}^n \quad \text{and} \quad \mathbf{Ff} = \mathbf{0}\}.$$

The first preliminary result that is required for the eventual theorem of interest is

Lemma 4-5 *Suppose*

i) **P**, **R**, *and* **X** *exist as positive semi-definite matrices*

and

ii) $\mathbf{m}\{\mathbf{Q}\} \in \mathcal{N}(\mathbf{P})$ *implies that the corresponding* $\lambda_R\{\mathbf{Q}\} < 0$.

Then **A** *is asymptotically stable.*

Proof Let $\mathbf{n} = \mathbf{n}\{\mathbf{A}\}$ and correspondingly $\lambda = \lambda\{\mathbf{A}\}$. Then (4-115) yields

$$0 = \mathbf{n}^\dagger \mathbf{Pn} + 2\lambda_R \mathbf{n}^\dagger \mathbf{Xn} + \mathbf{n}^\dagger \mathbf{XRXn}.$$

First assume $\mathbf{n}^\dagger\mathbf{Xn} = 0$ so that $\mathbf{Xn} = \mathbf{0}$ because **X** is positive semi-definite. Then $\mathbf{n}^\dagger\mathbf{Pn} = 0$ must follow so that $\mathbf{Pn} = \mathbf{0}$ because **P** is also positive semi-definite and hence $\mathbf{n} \in \mathcal{N}(\mathbf{P})$. Moreover, use of (4-116) yields

$$\mathbf{An} = \mathbf{Qn} = \lambda\mathbf{n}, \tag{4-117}$$

and hence $\mathbf{n} = \mathbf{n}\{\mathbf{Q}\}$ so that $\lambda_R < 0$ must hold by assumption. Now assume $\mathbf{n}^\dagger\mathbf{Xn} \neq 0$ but both $\mathbf{n}^\dagger\mathbf{Pn} = 0$ and $\mathbf{n}^\dagger\mathbf{XRXn} = 0$ hold. Then $\mathbf{Pn} = \mathbf{0}$ must follow so that $\mathbf{n} \in \mathcal{N}(\mathbf{P})$. Moreover, $\mathbf{RXn} = \mathbf{0}$ so that (4-117) again holds and hence $\lambda_R < 0$. All other **n**, if there are any, simultaneously satisfy

$$\mathbf{n}^\dagger\mathbf{Xn} \neq 0 \quad \text{and} \quad \mathbf{n}^\dagger\mathbf{Pn} + \mathbf{n}^\dagger\mathbf{XRXn} \neq 0$$

so that

$$\lambda_R = -\frac{1}{2}\frac{\mathbf{n}^\dagger\mathbf{Pn} + \mathbf{n}^\dagger\mathbf{XRXn}}{\mathbf{n}^\dagger\mathbf{Xn}} < 0.$$

An interesting consequence of this lemma can be stated as follows.

Suppose conditions i) and ii) of Lemma 4-5 hold. Then $\lambda_R\{\mathbf{Q}'\} < 0$ holds whenever $\mathbf{n}\{\mathbf{Q}'\} \in N(\mathbf{R})$ holds.

This assertion is proved by assuming $\mathbf{n}\{\mathbf{Q}'\} \in N(\mathbf{R})$. Then

$$\mathbf{A}'\mathbf{n} = \mathbf{Q}'\mathbf{n} = \lambda\mathbf{n}$$

must follow so that $\lambda_R\{\mathbf{A}'\} = \lambda_R\{\mathbf{A}\} < 0$ is established by Lemma 4-5.

The second preliminary result concerning matrix quadratic equations is based on

Proposition 4-3 *Let $\mathcal{N}$ be some subset of $\mathcal{C}^n$ for which $\mathbf{n} \in \mathcal{N}$ implies that $\mathbf{Fn} \in \mathcal{N}$. Then there exists an eigenvector $\mathbf{m}\{\mathbf{F}\} \in \mathcal{N}$.*

This property [14] of subsets which are closed under $\mathbf{F}$ is allied with the fact that $\mathbf{F}$ has at least one eigenvector belonging to $\mathcal{C}^n$, and this property is used in the proof of

Lemma 4-6 Suppose

i) $\mathbf{P}$ *and* $\mathbf{X}$ *exist as positive semi-definite matrices*

and

ii) no $\mathbf{m}\{\mathbf{Q}\}$ *belongs to* $N(\mathbf{P})$.

Then $\mathbf{X}$ *is nonsingular.*

Proof Assume i) and ii) of the lemma hold but $\mathbf{X}$ is singular, and let $\mathbf{n}$ be any nonzero vector which belongs to $N(\mathbf{X})$. Then, by use of (4-114), $\mathbf{Xn} = \mathbf{0}$ implies that $\mathbf{n}^\dagger\mathbf{Pn} = 0$ and hence $\mathbf{Pn} = \mathbf{0}$ holds for every such $\mathbf{n}$. In addition, use of (4-114) yields $\mathbf{XQn} = \mathbf{0}$ so that $\mathbf{Qn} \in N(\mathbf{X})$ also holds for every such $\mathbf{n}$. Therefore, null space $N(\mathbf{X})$ is closed under $\mathbf{Q}$, and there exists an eigenvector $m\{\mathbf{Q}\} \in N(\mathbf{X})$ for which $\mathbf{Pm} = \mathbf{0}$ holds by (4-114). Condition ii) thus is contradicted.

The third preliminary result concerning matrix quadratic equations is based on

Proposition 4-4 *Every infinite sequence $\{\mathbf{X}_i\}$ of positive semi-definite matrices with $(\mathbf{X}_{i+1} - \mathbf{X}_i)$ negative semi-definite for each i converges to some positive semi-definite limit.*

This property [15] of positive semi-definite matrices is used in the proof of

Lemma 4-7 *Suppose*

i) $\mathbf{P}$ *and* $\mathbf{R}$ *are positive semi-definite,*

ii) $\mathbf{m}\{\mathbf{Q}\} \in N(\mathbf{P})$ *implies that the corresponding* $\lambda_R\{\mathbf{Q}\} < 0$,

and

iii) there exists a positive semi-definite matrix $\mathbf{Y}$ *for which* $(\mathbf{Q} - \mathbf{RY})$ *is asymptotically stable.*

Then there exists one and only one positive semi-definite solution to (4-114).

Proof Existence of a positive semi-definite solution is proved by introducing matrices $\mathbf{X}_i$, $\mathbf{T}_i$, $\mathbf{A}_i$ and Σ_i which are given by

$$\mathbf{0} = \mathbf{T}_i + \mathbf{P} + \mathbf{X}_i\mathbf{Q} + \mathbf{Q}'\mathbf{X}_i - \mathbf{X}_i\mathbf{R}\mathbf{X}_i \tag{4-118}$$

$$\mathbf{A}_i = \mathbf{Q} - \mathbf{R}\mathbf{X}_i \tag{4-119}$$

$$\mathbf{\Sigma}_i\mathbf{A}_i + \mathbf{A}_i'\mathbf{\Sigma}_i = \mathbf{T}_i \tag{4-120}$$

and

$$\mathbf{X}_{i+1} = \mathbf{X}_i + \mathbf{\Sigma}_i \tag{4-121}$$

for each i. Assume there exists a positive semi-definite $\mathbf{X}_i$ for which $\mathbf{T}_i$ is positive semi-definite. Then $N(\mathbf{T}_i + \mathbf{P}) \subset N(\mathbf{P})$ holds so that $\mathbf{A}_i$ is asymptotically stable by Lemma 4-5. Moreover, matrix $\mathbf{\Sigma}_i$ is negative semi-definite by Corollary 4-1, and the elimination of $\mathbf{T}_i$ from (4-118) and (4-120) yields

$$\mathbf{0} = \mathbf{X}_{i+1}\mathbf{A}_i + \mathbf{A}_i'\mathbf{X}_{i+1} + (\mathbf{P} + \mathbf{X}_i\mathbf{R}\mathbf{X}_i)$$

so that $\mathbf{X}_{i+1}$ is also positive semi-definite by Corollary 4-1. Finally,

$$\mathbf{T}_{i+1} = \mathbf{\Sigma}_i\mathbf{R}\mathbf{\Sigma}_i \tag{4-122}$$

holds so that $\mathbf{T}_{i+1}$ is positive semi-definite. In other words,

$$\mathbf{X} = \lim_{i \to \infty} \mathbf{X}_i$$

exists by Proposition 4-4 under the previous assumption, and (4-114) holds because $\lim_{i \to \infty} \mathbf{\Sigma}_i = \mathbf{0}$ implies that $\lim_{i \to \infty} \mathbf{T}_i = \mathbf{0}$. Now, given any positive semi-definite $\mathbf{Y}$ for which $(\mathbf{Q} - \mathbf{R}\mathbf{Y})$ is asymptotically stable, let $\mathbf{W}$ be any positive semi-definite matrix such that

$$[\mathbf{W} - (\mathbf{P} + \mathbf{Y}\mathbf{Q} + \mathbf{Q}'\mathbf{Y} - \mathbf{Y}\mathbf{R}\mathbf{Y})]$$

is also positive semi-definite. In addition, let $\mathbf{Z}$ be the unique positive semi-definite solution to

$$\mathbf{0} = \mathbf{Z}(\mathbf{Q} - \mathbf{R}\mathbf{Y}) + (\mathbf{Q} - \mathbf{R}\mathbf{Y})'\mathbf{Z} + \mathbf{W}$$

in accordance with Corollary 4-1. Finally, let

$$\mathbf{X}_0 = \mathbf{Y} + \mathbf{Z} \tag{4-123}$$

so that (4-118) yields

$$\mathbf{T}_0 = [\mathbf{W} - (\mathbf{P} + \mathbf{Y}\mathbf{Q} + \mathbf{Q}'\mathbf{Y} - \mathbf{Y}\mathbf{R}\mathbf{Y})] + \mathbf{Z}\mathbf{R}\mathbf{Z}.$$

Thus, both $\mathbf{X}_0$ and $\mathbf{T}_0$ have been constructed to be positive semi-definite matrices, and existence hence is established. In order to establish uniqueness, let $\mathbf{X}_1$ and $\mathbf{X}_2$ be two distinct positive semi-definite solutions to (4-114). Then $\mathbf{A}_1$ and $\mathbf{A}_2$ are asymptotically stable by Lemma 4-5. Now let

$$\mathbf{Z} = \mathbf{X}_2 - \mathbf{X}_1$$

so that (4-115) can be used to eliminate **P**, thereby obtaining

$$\mathbf{0} = \mathbf{Z}\mathbf{A}_1 + \mathbf{A}_1'\mathbf{Z} - \mathbf{W} = \mathbf{Z}\mathbf{A}_2 + \mathbf{A}_2'\mathbf{Z} + \mathbf{W}$$

where

$$\mathbf{W} = \mathbf{ZRZ}.$$

Because **W** is positive semi-definite, matrix **Z** is both positive and negative semi-definite by Corollary 4-1 and hence $\mathbf{Z} = \mathbf{0}$.

The wording of Lemma 4-7 of course does not state that the positive semi-definite solution is the only solution to (4-114).

The final preliminary result concerning matrix quadratic equations is stated as

Lemma 4-8 *Suppose*

i) **R** is *positive semi-definite*

and

ii) $\mathbf{m}\{\mathbf{Q}'\} \in N(\mathbf{R})$ *implies that the corresponding* $\lambda_R\{\mathbf{Q}'\} < 0$.

Then there exists a positive semi-definite matrix **Y** *for which* $(\mathbf{Q} - \mathbf{RY})$ *is asymptotically stable.*

Proof If **Q** is asymptotically stable, let $\mathbf{Y} = \mathbf{0}$. Otherwise, define a nonsingular matrix **T** for the similarity transformation

$$\bar{\mathbf{Q}}' = \mathbf{T}^{-1}\mathbf{Q}'\mathbf{T} = \begin{bmatrix} \bar{\mathbf{Q}}_{11} & \mathbf{0}_{12} \\ \mathbf{0}_{21} & \bar{\mathbf{Q}}_{22} \end{bmatrix}'$$

where all $\lambda_R\{\bar{\mathbf{Q}}_{11}\} \geqslant 0$ and all $\lambda_R\{\bar{\mathbf{Q}}_{22}\} < 0$. Also let

$$\bar{\mathbf{R}} = \mathbf{T}'\mathbf{RT}$$

so that $\bar{\mathbf{R}}_{11}$ is positive semi-definite. Now consider any $\mathbf{m}\{\mathbf{Q}'\}$ for which $\lambda_R\{\mathbf{Q}'\} \geqslant 0$ so that $\mathbf{m} \notin N\{\mathbf{R}\}$ holds by assumption and hence $\mathbf{m}^\dagger\mathbf{Rm} > 0$. By construction,

$$\mathbf{Q}'\mathbf{m} = \lambda\mathbf{m}$$

holds and hence

$$\mathbf{T}^{-1}\mathbf{Q}'\mathbf{Tn} = \bar{\mathbf{Q}}'\mathbf{n} = \lambda\mathbf{n}$$

holds where $\mathbf{n} = \mathbf{T}^{-1}\mathbf{m}$. Because $\lambda_R \geqslant 0$ by assumption, $\mathbf{n}$ must be an eigenvector corresponding to $\overline{\mathbf{Q}}'_{11}$ so that $\mathbf{n}' = [\mathbf{n}'_1 \quad \mathbf{0}'_2]$. Moreover,

$$\mathbf{n}_1^\dagger \overline{\mathbf{R}}_{11}\mathbf{n}_1 = \mathbf{n}^\dagger \overline{\mathbf{R}}\mathbf{n} = \mathbf{m}^\dagger \mathbf{R}\mathbf{m} > 0$$

holds and hence $\overline{\mathbf{R}}_{11}\mathbf{n}_1 \neq \mathbf{0}_1$ because $\overline{\mathbf{R}}_{11}$ is positive semi-definite. Furthermore, $\mathbf{n}_1\{-\overline{\mathbf{Q}}'\} \notin N(\overline{\mathbf{R}}_{11})$ for all such vectors. A matrix $\overline{\mathbf{Z}}_{11}$ can always be found such that $(-\overline{\mathbf{Q}}'_{11} - \mathbf{I}_{11}\overline{\mathbf{Z}}_{11})$ is asymptotically stable. Thus, by Lemma 4-7, matrix $\mathbf{Z}_{11}$ exists as a unique positive semi-definite matrix such that

$$\mathbf{0}_{11} = \overline{\mathbf{R}}_{11} + \overline{\mathbf{Z}}_{11}(-\overline{\mathbf{Q}}'_{11}) + (-\overline{\mathbf{Q}}'_{11})'\overline{\mathbf{Z}}_{11} - \overline{\mathbf{Z}}_{11}\mathbf{I}_{11}\overline{\mathbf{Z}}_{11}.$$

Moreover, matrix $\overline{\mathbf{Z}}_{11}$ is nonsingular by Lemma 4-6 so that

$$\overline{\mathbf{Y}}_{11} = \overline{\mathbf{Z}}_{11}^{-1}$$

is also positive definite and satisfies

$$\mathbf{0}_{11} = \mathbf{I}_{11} + \overline{\mathbf{Y}}_{11}\overline{\mathbf{Q}}_{11} + \overline{\mathbf{Q}}'_{11}\overline{\mathbf{Y}}_{11} - \overline{\mathbf{Y}}_{11}\overline{\mathbf{R}}_{11}\overline{\mathbf{Y}}_{11}.$$

Matrix $(\overline{\mathbf{Q}}_{11} - \overline{\mathbf{R}}_{11}\overline{\mathbf{Y}}_{11})$ therefore is asymptotically stable by Lemma 4-5. Finally, let

$$\mathbf{Y} = \mathbf{T}\begin{bmatrix} \overline{\mathbf{Y}}_{11} & \mathbf{0}_{12} \\ \mathbf{0}_{21} & \mathbf{0}_{22} \end{bmatrix}\mathbf{T}'$$

so that

$$\mathbf{T}'(\mathbf{Q} - \mathbf{R}\mathbf{Y})(\mathbf{T}')^{-1} = \begin{bmatrix} (\overline{\mathbf{Q}}_{11} - \overline{\mathbf{R}}_{11}\overline{\mathbf{Y}}_{11}) & \mathbf{0}_{12} \\ -\overline{\mathbf{R}}_{21}\overline{\mathbf{Y}}_{11} & \overline{\mathbf{Q}}_{22} \end{bmatrix}$$

is clearly asymptotically stable and hence $(\mathbf{Q} - \mathbf{R}\mathbf{Y})$ is also asymptotically stable.

These four preliminary results pertinent to solutions of (4-114) are summarized in terms of

Definition 4-2 The ordered triple $\{\mathbf{P}, \mathbf{Q}, \mathbf{R}\}$ is called regular if and only if

i) $\mathbf{P}$ and $\mathbf{R}$ are positive semi-definite,

ii) $\mathbf{m}\{\mathbf{Q}\} \in N(\mathbf{P})$ implies that the corresponding $\lambda_R\{\mathbf{Q}\} < 0$,

and

iii) $\mathbf{m}\{\mathbf{Q}'\} \in N(\mathbf{R})$ implies that the corresponding $\lambda_R\{\mathbf{Q}'\} < 0$.

In particular, the principal result is stated as

Theorem 4-13 *Suppose* $\{\mathbf{P}, \mathbf{Q}, \mathbf{R}\}$ *is regular. Then there exists one and only one positive semi-definite solution to (4-114).*

Proof By Lemma 4-8, there exists a positive semi-definite matrix $\mathbf{X}$ for which $\mathbf{A}$ is asymptotically stable. Then the desired result follows directly from Lemma 4-7.

This sufficiency condition for the existence and uniqueness of a positive semi-definite solution to (4-114) applies to most linear optimal control problems of interest and is stated conveniently in terms of only coefficient matrices **P, Q,** and **R.**

Two computational methods for solving (4-114) are directly related to Theorem 4-13. First of all, equations (4-118) through (4-121) constitute a method of successive substitution, and this method exhibits quadratic convergence when initialized in accordance with (4-123) under the preconditions of Lemma 4-7. This method also exhibits quadratic convergence under somewhat different preconditions such as those given in

Theorem 4-14 *Suppose*

i) **R** *is positive semi-definite*

and

ii) a solution $\mathbf{X}_e$ *of (4-114) exists for which there also exist constants* $\Gamma, \gamma,$ $\epsilon > 0$ *such that* $\|\mathbf{X} - \mathbf{X}_e\| < \epsilon$ *implies*

$$\|e^{\mathbf{A}t}\| \leqslant \Gamma\, e^{-\gamma t}$$

and

$$\delta = \frac{\Gamma^4}{4\gamma^2} \|\mathbf{R}\|\, \|\mathbf{T}\| < 1.$$

Then use of (4-118) through (4-121) with $\mathbf{X}_0$ *given results in a convergent sequence with limit* $\mathbf{X}_e$ *when*

$$\delta_0 < 1 \quad \textit{and} \quad \frac{\Gamma^2}{2\gamma} \frac{\|\mathbf{T}_0\|}{1-\delta_0} < \epsilon. \tag{4-124}$$

Proof First assume $\mathbf{X}_\infty$ exists given some $\mathbf{X}_i$ so that

$$\mathbf{X}_\infty = \mathbf{X}_i + \sum_{j=i}^{\infty} \mathbf{\Sigma}_j.$$

If $\mathbf{X}_\infty$ is also a solution to (4-114) and $\|\mathbf{X}_\infty - \mathbf{X}_e\| < \epsilon$, then $\mathbf{X}_\infty = \mathbf{X}_e$ can be shown to follow by a uniqueness proof similar to that used in the proof of Lemma 4-7. Now assume that $\|\mathbf{X}_i - \mathbf{X}_e\| < \epsilon$ holds for all $i \geqslant 0$ given some $\mathbf{X}_0$. Then use of (4-31) yields

$$\|\mathbf{\Sigma}_i\| = \left\| - \int_0^\infty \{e^{\mathbf{A}_i' t}\mathbf{T}_i\, e^{\mathbf{A}_i t}\}\, dt \right\| \leqslant \frac{\Gamma^2}{2\gamma} \|\mathbf{T}_i\|$$

so that use of (4-122) now yields

$$\|\mathbf{T}_{i+1}\| \leqslant \frac{\Gamma^4}{4\gamma^2} \|\mathbf{R}\| \, \|\mathbf{T}_i\|^2$$

or rather

$$\|\mathbf{T}_{i+1}\| \leqslant \delta_i \|\mathbf{T}_i\|.$$

Thus, $\delta_{i+1} \leqslant \delta_i < 1$ holds for all $i \geqslant 0$ and hence

$$\|\mathbf{T}_i\| \leqslant \delta_0^i \|\mathbf{T}_0\| \;\forall\; i \geqslant 0.$$

Moreover,

$$\left\| \sum_{j=i}^{\infty} \mathbf{\Sigma}_j \right\| \leqslant \frac{\Gamma^2}{2\gamma} \sum_{j=i}^{\infty} \|\mathbf{T}_j\| \leqslant \frac{\Gamma^2}{2\gamma} \|\mathbf{T}_0\| \delta_0^i \sum_{j=i}^{\infty} \delta_0^{(j-i)} \leqslant \frac{\Gamma^2}{2\gamma} \frac{\|\mathbf{T}_0\|}{1-\delta_0}$$

holds for all $i \geqslant 0$ under this assumption. In other words, every such sequence is convergent to $\mathbf{X}_e$ and

$$\|\mathbf{X}_i - \mathbf{X}_e\| \leqslant \frac{\Gamma^2}{2\gamma} \frac{\|\mathbf{T}_0\|}{1-\delta_0}$$

holds for all $i \geqslant 0$. Therefore, the conditions given in (4-124) imply convergence with $\mathbf{X}_\infty = \mathbf{X}_e$.

Of course, the successive substitution method defined by (4-118) through (4-121) may converge to a solution of (4-114) even when Theorem 4-14 and when the conditions and initialization procedure of Lemma 4-7 do not apply.

The primary difficulty of the successive substitution method defined by (4-118) through (4-121) is the requirement to initialize such that $\mathbf{A}_0$ is asymptotically stable. An alternate computational method which circumvents this difficulty is based on an extension of Theorem 4-2 and the result given in Reference [16]. For convenience, let

$$\mathbf{Y}(\mathbf{X}) = \mathbf{P} + \mathbf{X}\mathbf{Q} + \mathbf{Q}'\mathbf{X} - \mathbf{X}\mathbf{R}\mathbf{X}, \tag{4-125}$$

$$\mathbf{J} = \begin{bmatrix} -\mathbf{Q} & \mathbf{R} \\ \mathbf{P} & \mathbf{Q}' \end{bmatrix}, \tag{4-126}$$

and

$$p(\mathbf{J}) = \begin{bmatrix} \mathbf{K} & \mathbf{L} \\ \mathbf{M} & \mathbf{N} \end{bmatrix} \tag{4-127}$$

where $p(\lambda)$ is a polynomial in λ. The result then can be stated as

Theorem 4-15 *Suppose* $\{\mathbf{P}, \mathbf{Q}, \mathbf{R}\}$ *is regular. Then* $p(\lambda)$ *is the* nth *degree strictly*

Hurwitz polynomial defined by

$$\det(\lambda \mathbf{I} - \mathbf{J}) = (-1)^n p(\lambda) p(-\lambda), \tag{4-128}$$

and **X** *is the* $n \times n$ *positive semi-definite solution to* $\mathbf{Y}(\mathbf{X}) = \mathbf{0}$ *given by*

$$\mathbf{X} = (\mathbf{MK}' + \mathbf{NL}')(\mathbf{KK}' + \mathbf{LL}')^{-1}. \tag{4-129}$$

Proof Let **X** be any solution to $\mathbf{Y}(\mathbf{X}) = \mathbf{0}$ for which **A** given by (4-116) is asymptotically stable. Such a matrix exists by Theorem 4-13 and Lemma 4-5. Properties of determinants then are used to show

$$\det(\lambda \mathbf{I} - \mathbf{J}) = \det \begin{bmatrix} (\lambda \mathbf{I} + \mathbf{A}) & -\mathbf{R} \\ \mathbf{0} & (\lambda \mathbf{I} - \mathbf{A}') \end{bmatrix} = \det(\lambda \mathbf{I} + \mathbf{A}) \times \det(\lambda \mathbf{I} - \mathbf{A}')$$

and hence

$$p(\lambda) = \det(\lambda \mathbf{I} - \mathbf{A})$$

is the strictly Hurwitz polynomial appearing in (4-128). Moreover, there is only one such matrix **X**, as shown by the uniqueness proof used in the proof of Lemma 4-7, and hence **X** is also positive semi-definite by Lemma 4-7. Now let

$$\mathbf{T} = \begin{bmatrix} \mathbf{0} & \mathbf{I} \\ \mathbf{I} & \mathbf{X} \end{bmatrix}$$

so that

$$p(\mathbf{J}) = \mathbf{T} p\left(\begin{bmatrix} \mathbf{A}' & \mathbf{0} \\ \mathbf{R} & -\mathbf{A} \end{bmatrix}\right) \mathbf{T}^{-1} = \mathbf{T} \begin{bmatrix} \mathbf{0} & \mathbf{0} \\ \mathbf{L} & p(-\mathbf{A}) \end{bmatrix} \mathbf{T}^{-1},$$

where **L** is an appropriately defined matrix. Elements of $p(\mathbf{J})$ which are defined in (4-127) therefore satisfy

$$\mathbf{X}[\mathbf{K} \quad \mathbf{L}] = [\mathbf{M} \quad \mathbf{N}] \tag{4-130}$$

and

$$(\mathbf{K} + \mathbf{LX}) = p(-\mathbf{A}).$$

Moreover, $p(-\mathbf{A})$ must be nonsingular because **A** is asymptotically stable. Matrix $(\mathbf{K} + \mathbf{LX})$ must also be nonsingular and hence $[\mathbf{K} \quad \mathbf{L}]$ must be of rank n. Therefore, matrix $(\mathbf{KK}' + \mathbf{LL}')$ is nonsingular so that postmultiplication of (4-130) by $[\mathbf{K} \quad \mathbf{L}]'$ and matrix inversion yield (4-129).

The *spectral factorization* method of solving (4-114) under regularity conditions consists of computing the coefficients of and factoring $\det(\lambda \mathbf{I} - \mathbf{J})$ to determine $p(\lambda)$, evaluating $p(\mathbf{J})$, and then solving (4-129) in accordance with (4-127). A number of standard computational procedures can, of course, be employed in conjunction with the spectral factorization method.

A number of linear optimal control problems cannot be reduced to the form of (4-114) with $\{\mathbf{P}, \mathbf{Q}, \mathbf{R}\}$ regular. This situation is typified by design problems that involve independent signals such as nonzero desired responses. All signals which arise in this context are generated as an initial-state response of linear constant-coefficient differential equations. Problems involving independent signals of this class are formulated as a system that includes an uncontrolled subsystem. Moreover, the uncontrollable subsystem often is not asymptotically stable corresponding to step, ramp, and other types of signals.

Matrix quadratic equations corresponding to problems with an uncontrolled subsystem can always be reduced to the following form. If state-vector partitions $\mathbf{x}_1$ and $\mathbf{x}_2$ correspond to the controlled and uncontrolled subsystems respectively, then coefficient matrices appearing in (4-114) are partitioned accordingly as

$$\mathbf{P} = \begin{bmatrix} \mathbf{P}_{11} & \mathbf{P}_{12} \\ \mathbf{P}_{21} & \mathbf{P}_{22} \end{bmatrix}, \quad \mathbf{Q} = \begin{bmatrix} \mathbf{Q}_{11} & \mathbf{Q}_{12} \\ \mathbf{0}_{21} & \mathbf{Q}_{22} \end{bmatrix}$$

and (4-131)

$$\mathbf{R} = \begin{bmatrix} \mathbf{R}_{11} & \mathbf{0}_{12} \\ \mathbf{0}_{21} & \mathbf{0}_{22} \end{bmatrix}.$$

Correspondingly, the partitioned form of (4-116) becomes

$$\mathbf{A} = \begin{bmatrix} (\mathbf{Q}_{11} - \mathbf{R}_{11}\mathbf{X}_{11}) & (\mathbf{Q}_{12} - \mathbf{R}_{11}\mathbf{X}_{12}) \\ \mathbf{0}_{21} & \mathbf{Q}_{22} \end{bmatrix}. \tag{4-132}$$

In addition, the corresponding partitioned form of (4-114) becomes

$$\mathbf{0}_{11} = \mathbf{P}_{11} + \mathbf{X}_{11}\mathbf{Q}_{11} + \mathbf{Q}'_{11}\mathbf{X}_{11} - \mathbf{X}_{11}\mathbf{R}_{11}\mathbf{X}_{11}, \tag{4-133}$$

$$\mathbf{0}_{12} = (\mathbf{P}_{12} + \mathbf{X}_{11}\mathbf{Q}_{12}) + \mathbf{X}_{12}\mathbf{Q}_{22} + \mathbf{A}'_{11}\mathbf{X}_{12}, \tag{4-134}$$

and

$$\mathbf{0}_{22} = (\mathbf{P}_{22} + \mathbf{X}'_{12}\mathbf{Q}_{12} + \mathbf{Q}'_{12}\mathbf{X}_{12} - \mathbf{X}'_{12}\mathbf{R}_{11}\mathbf{X}_{12}) + \mathbf{X}_{22}\mathbf{Q}_{22} + \mathbf{Q}'_{22}\mathbf{X}_{22} \tag{4-135}$$

assuming **P** and **R** are symmetric so that the fourth partition can be omitted.

The partitioned form of (4-114) reveals a number of simplifications caused by the uncontrolled subplant. For purposes of illustration, linear optimal control corresponding to these equations is depicted in Figure 4-5 with an example of controlled and uncontrolled subplants. Matrix $\mathbf{X}_{11}$, corresponding to feedback control of the controlled subplant, is independent of $\mathbf{X}_{12}$ and $\mathbf{X}_{22}$, and hence

(4-133) can be solved directly. Moreover, Theorems 4-13 and 4-15 apply to the solution of this matrix quadratic equation when $\{\mathbf{P}_{11}, \mathbf{Q}_{11}, \mathbf{R}_{11}\}$ is regular. Once (4-133) is solved for $\mathbf{X}_{11}$, then all coefficient matrices of (4-134) are specified. This equation also does not depend on $\mathbf{X}_{22}$. Thus, this equation is of the linear type given in (4-23), and the results of Section 4.2 apply to (4-134). Moreover, feed-forward control of the controlled subplant is determined from solution $\mathbf{X}_{12}$ as depicted in Figure 4-5. Once (4-133) and (4-134) are solved for $\mathbf{X}_{11}$ and $\mathbf{X}_{12}$

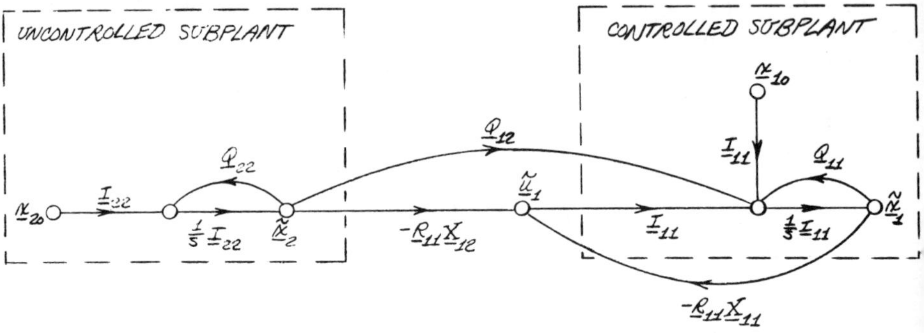

Figure 4-5 Flow graph of linear optimal control example with independent signal generation.

respectively, then all coefficient matrices of (4-135) are specified. This equation is also of the linear type given in (4-23). However, this equation need not be solved because linear optimal control does not depend upon $\mathbf{X}_{22}$. In fact, equation (4-135) does not have a unique solution for many problems of interest because $\mathbf{Q}_{22}$ has one or more zero eigenvalues. In this situation, a solution $\mathbf{X}_{22}$ must be determined from other considerations.

4.10 SUMMARY

The formulation of control system design problems with deterministic signals is presented here in considerable detail. Systematic methods of problem formulation are developed from the point of view of the synthesis of transfer function matrices. The methods presented here for the synthesis of transfer function matrices can of course be applied to a large class of design problems arising in feedback control which do not require parameter optimization methods. However, these synthesis concepts and methods also apply directly in the selection of performance integral and hence in the formulation of a large class of design problems that do require parameter optimization techniques.

The method of transfer function matrix synthesis presented here is accompanied by a precise theory of the compatibility of design specifications. Both the tests for compatibility and the actual computation of feedback gains, once compatibility has been achieved, are accomplished easily using a digital computer. In order to achieve compatibility, open-loop dynamics may have to be altered by the addition of compensation filters. Moreover, the open-loop system can be further augmented by a class of postfilters having low-pass characteristics without altering compatibility. These post-filters are used for a number of design applications including incomplete state-variable feedback and sensitivity reduction. Moreover, many of the compensation filters belonging to this class correspond to usable prefilters that yield exact design.

A suitable performance functional f in general must be selected when exact synthesis of a transfer function matrix is not possible. A number of factors in the selection of f are exposed in the context of optimal gain-control problems. These factors include uniqueness conditions which are summarized as the notion of complete responsiveness. In addition, objectives such as worst-case design can also be accounted for in the selection of f. Relatively realistic and sophisticated design examples are included.

Linear optimal control is a special case of optimal gain control and is used to expose additional aspects of typical design problems. In particular, properties of the matrix quadratic equations appearing in (4-114) are investigated and are summarized as the notion of regularity. Simplification of these quadratic equations occurs and is identified for systems which include an uncontrolled subsystem. Uncontrolled subsystems arise in design problems that involve a class of deterministic signals.

Computational aspects of design problems formulated in this chapter are also discussed. Computation of gradients and hessians which appear in necessary conditions for relative minima is formulated in an efficient form. A matrix of Lagrange multipliers is introduced for the purposes of this formulation. Properties of linear matrix equations appearing in (4-23) are derived as well as efficient computational methods of solution. These methods include both finite series solutions and approximations based on a truncated infinite series. Finally, computational methods of solving quadratic matrix equations given in (4-114) are derived. An efficient method of successive substitutions is included as well as a method which requires factoring a polynomial.

BIBLIOGRAPHY

1. Merriam, C. W., III and D. Jordan: "A computational method for parameter optimization problems arising in control", *International Journal of Control*, **14**, 385–397.

2. Gantmacher, F. R.: *The Theory of Matrices*, Vol. 1, Chelsea Publishing Company, New York, 1960.
3. Ogata, K.: *State Space Analysis of Control Systems*, Prentice-Hall, Inc., Englewood Cliffs, 1967.
4. Faddeeva, V. N.: *Computational Methods of Linear Algebra*, Dover Publications, Inc., New York, 1959.
5. Smith, R. A.: "Matrix equation **XA** + **BX** = **C**", *SIAM Journal on Applied Mathematics*, **16**, No. 1, 1968.
6. Jordan, D. and C. W. Merriam III: "Synthesis of Transfer Function Matrices in Control", *Proc. of 1972 IFAC Congress,* Paris, 1972.
7. Jordan, D.: "A new method for control system design", Ph.D. Thesis, Cornell University, Ithaca, 1970.
8. Wonham, W. M.: "On pole assignment in multi-input controllable linear systems", *IEEE Transactions on Automatic Control*, **AC-12**, No. 6, 1967.
9. Kalman, R. E.: "Irreducible realizations and the degree of a rational matrix", *SIAM Journal on Applied Mathematics*, **13**, No. 2, 1965.
10. Rynaski, E. G. *et al.*: "Design of linear flight control systems using optimal control theory", Report No. IH-1648-F-1, Cornell Aeronautical Laboratory, Inc., Buffalo, 1963.
11. Rediess, H. A.: "A new model performance index for the engineering design of control systems", Report No. TE-16, Experimental Astronomy Laboratory, M.I.T., Cambridge, 1968.
12. Potter, J. E.: "A matrix equation arising in statistical filter theory", Report No. RE-9, Experimental Astronomy Laboratory, M.I.T., Cambridge, 1965.
13. Wonham, W. M.: "On a matrix Ricatti equation of stochastic control", *SIAM Journal on Control*, **6**, No. 4, 1968.
14. Gel'Fand, I.M.: *Lectures on Linear Algebra*, Interscience Publishers, Inc., New York, 1963.
15. Kantorovich, L. V. and G. P. Akilov: *Functional Analysis in Normed Spaces*, The Macmillan Company, New York, 1964.
16. Roth, W. E.: "On the matrix equation $\mathbf{X}^2 + \mathbf{AX} + \mathbf{XB} + \mathbf{C} = \mathbf{0}$", *Proceedings of the American Mathematical Society*, **1**, 1950.

PROBLEMS

4-1 The integral defined by (1-89) and (1-90) is to be minimized. Identify the corresponding matrices **A** and **W** appearing in (4-6) as well as the corresponding performance functional $f(\mathbf{c}, \mathbf{X})$ when

$$\mathbf{F} = \mathbf{\Gamma} - \gamma \mathbf{c}'.$$

4-2 Repeat Problem 4-1 for the following design problem. Open-loop state equations are taken to be

$$\dot{\mathbf{x}} = \mathbf{Fx} + \mathbf{Gu}; \quad \mathbf{x}(0) = \mathbf{x}_0$$

and control is given by

$$\mathbf{u} = -\mathbf{Cx}.$$

The performance integral is taken to be

$$I = \int_0^\infty \{\tfrac{1}{2}\mathbf{x}'\mathbf{\Phi x} + \mathbf{u}'\mathbf{\Psi x} + \tfrac{1}{2}\mathbf{u}'\mathbf{\Omega u}\}\, dt$$

and the performance functional is given by

$$f = \sum_{j=1}^{m} I \Big|_{\mathbf{x}_0 = \mathbf{b}_j}.$$

4-3 Give expressions for all entries in (4-13) through (4-22) which depend directly upon the design problem defined in Problem 4-1. Also give an expression for $L_\mathbf{c}$.

4-4 Give expressions for all entries in (4-13) through (4-22) which depend directly upon the design problem defined in Problem 4-2. Also give an expression for $L_\mathbf{c}$.

4-5 Consider the performance functional

$$\mathbf{F}(\mathbf{c}) = f(\mathbf{c}, \mathbf{X}),$$

where $\mathbf{X}$ is defined by (4-6) with $\mathbf{A} = \mathbf{A}(\mathbf{c})$ and $\mathbf{W} = \mathbf{W}(\mathbf{c})$.

a) Give expressions for $F_\mathbf{c}$ and $F_\mathbf{cc}$ without use of Lagrange multipliers.

b) Compare the results of part a) with those of Section 4.1 from the computational point of view of solving equations in the form of (4-23).

4-6 Matrix $\mathbf{X}$ as given in (4-31) under conditions of Theorem 4-1 can be evaluated analytically as follows. Let

$$e^{\mathbf{A}t} = \sum_{k=1}^{K} \left[\sum_{l=0}^{L_k - 1} \frac{1}{l!} t^l \mathbf{A}_{kl} \right] e^{-s_k t}$$

where K is the number of distinct eigenvalues of $\mathbf{A}$ and L_k is the size of the maximal Jordan block corresponding to the kth distinct eigenvalue of $\mathbf{A}$. Then (4-31) can be written as

$$\mathbf{X} = \sum_{k=1}^{K} \left[\sum_{l=0}^{L_k - 1} \mathbf{B}_{kl}\mathbf{C}\mathbf{A}_{kl} \right].$$

Specify $\mathbf{B}_{kl}$ in terms of s_k, $\mathbf{B}$, and l.

4-7 Determine $\mathbf{X}$ as defined by (4-23) when

$$\mathbf{A} = \begin{bmatrix} -2 & 1 \\ 0 & -1 \end{bmatrix}, \quad \mathbf{B} = \mathbf{A}', \quad \text{and} \quad \mathbf{C} = \begin{bmatrix} 4 & 2 \\ 2 & 2 \end{bmatrix}.$$

4-8 Repeat Problem 4-7 for

$$\mathbf{A} = [1], \quad \mathbf{B} = \begin{bmatrix} -4 & -4 \\ 1 & 0 \end{bmatrix}, \quad \text{and} \quad \mathbf{C} = \begin{bmatrix} -2 \\ 1 \end{bmatrix}.$$

4-9 Write a computer program for determining the elements of $\mathbf{\Delta}$ and for determining $\mathbf{X}$ by the direct matrix inversion method defined by (3-36).

4-10 Write a computer program for determining $\mathbf{X}$ by the method defined by (4-40).

4-11 Demonstrate that Proposition 4-1 is true by first noting

a) $\rho\{\mathbf{F}\} = \rho\{\mathbf{J}\} = \rho\{\mathbf{\Lambda}\}$ where **J** is the Jordan canonical form of **F** and $\mathbf{\Lambda}$ is a diagonal matrix of the eigenvalues of **F**,

b) $\rho^k\{\mathbf{F}\} = \rho\{\mathbf{F}^k\}$,

c) $\|\mathbf{F}\|_2 = \|\mathbf{J}\|_2$,

d) $\rho\{\mathbf{F}\} \leqslant \|\mathbf{F}\|_2$ and the equality holds when **F** can be diagonalized,

and

e) $\|\mathbf{F}^k\|_2 \leqslant \sum_{l=0}^{m} \dfrac{k!}{l!(k-l)!} \|\mathbf{\Lambda}^{(k-l)}\|_2$ where m is the maximal Jordan block size of **F**.

4-12 Write a computer program for determining **X** by the method defined by (4-46).

4-13 Specialize the computer program written for Problem 4-12 to the case where $\mathbf{B} = \mathbf{A}'$ and **C** is symmetric. Also write a computer program for evaluating (4-14). Then incorporate these programs into the computer program written for Problem 3-28.

4-14 Find a feedback control system that yields

$$\mathbf{T}_d(s) = \begin{bmatrix} \dfrac{1}{s^2} \end{bmatrix}$$

for the open-loop system depicted in Figure 4-2.

4-15 Find a feedback control system that yields

$$\mathbf{T}_d(s) = \begin{bmatrix} \dfrac{1}{(s+1)} & 0 \\ 0 & \dfrac{1}{(s+1)^2} \end{bmatrix}$$

for

$$\mathbf{F} = \begin{bmatrix} 0 & 1 & 0 & 0 \\ 0 & 0 & 0 & 0 \\ 0 & 0 & 0 & 0 \\ 0 & 0 & 0 & -1 \end{bmatrix}, \quad \mathbf{G} = \begin{bmatrix} -1 & 0 \\ 2 & 1 \\ 1 & 0 \\ 1 & 0 \end{bmatrix}, \quad \text{and} \quad \mathbf{H} = \begin{bmatrix} 0 & 1 & -2 & 0 \\ 1 & 0 & 0 & 1 \end{bmatrix}.$$

Find the characteristic polynomial of the closed-loop system and test for asymptotic stability.

4-16 Repeat Problem 4-15 for

$$\mathbf{F} = \begin{bmatrix} 0 & 1 & 0 & 0 & 0 \\ 0 & 0 & 0 & 0 & 0 \\ 0 & 0 & 0 & 0 & 0 \\ 0 & 0 & 0 & -1 & 0 \\ 0 & 0 & 0 & 0 & -1 \end{bmatrix}, \quad \mathbf{G} = \begin{bmatrix} -1 & 0 \\ 2 & 1 \\ 1 & 0 \\ 1 & -1 \\ 0 & 2 \end{bmatrix},$$

and

$$\mathbf{H} = \begin{bmatrix} 0 & 1 & -2 & 1 & -1 \\ 1 & 0 & 0 & 0 & 0 \end{bmatrix}.$$

4-17 Write a computer program for checking Assumptions I and II and for computing feedback system coefficients when $\mathbf{T}_d(s)$ can be realized.

4-18 Find the diagonally equivalent form of $(s\mathbf{I} - \mathbf{F})$ for the matrix given in Problem 1-2, and explain the significance of invariant polynomials found for this matrix.

4-19 Repeat Problem 4-18 for the matrix given in Problem 1-3.

4-20 Repeat Problem 4-14 for

$$\mathbf{T}_d(s) = \left[\frac{1}{(s+1)^3}\right]$$

by adding an appropriate compensation filter if necessary.

4-21 Repeat Problem 4-15 for

$$\mathbf{T}_d(s) = \frac{1}{(s+1)^2}\begin{bmatrix} 1 & 0 \\ 0 & 1 \end{bmatrix}$$

by adding an appropriate compensation filter if necessary.

4-22 Determine whether Assumptions I and II can ever be satisfied for the system corresponding to

$$\mathbf{T}(s) = \begin{bmatrix} \dfrac{1}{(s+2)} & \dfrac{1}{(s+1)} \\[2ex] \dfrac{(s+3)}{(s+1)(s+2)} & \dfrac{(s+3)}{(s+1)^2} \end{bmatrix}$$

and explain the answer.

4-23 Repeat Problem 4-22 for

$$\mathbf{T}(s) = \begin{bmatrix} \dfrac{1}{s} & \dfrac{-1}{s(s+1)} \\[2ex] \dfrac{1}{s^2} & \dfrac{1}{s^2(s+1)} \end{bmatrix}.$$

4-24 Determine whether excess pole specification

$$\mathbf{P}(s) = \begin{bmatrix} s^2 & s(s+1) \\ s(s+1) & (s+1)^2 \end{bmatrix}$$

can ever be realized and explain the answer.

4-25 Repeat Problem 4-24 for

$$\mathbf{P}(s) = \begin{bmatrix} s^2(s+1) & s(s+1) \\ (s+1)^2 & (s+1) \end{bmatrix}.$$

4-26 Repeat Problem 4-14 using a postfilter

$$\mathbf{S}(s) = \left[\frac{s+\alpha}{s}\right]$$

and selecting α so that the feedback gain multiplying x_3 is zero if possible. Explain the results and draw a complete flow graph of the system including the corresponding exact prefilter if possible.

4-27 Repeat Problem 4-26 for

$$\mathbf{T}_d(s) = \left[\frac{1}{(s+1)^2}\right].$$

4-28 Repeat Problem 4-26 for

$$\mathbf{F} = \begin{bmatrix} -3 & -1 \\ 1 & -1 \end{bmatrix}, \quad \mathbf{G} = \begin{bmatrix} 1 \\ 0 \end{bmatrix}, \quad \mathbf{H} = [1 \quad 0], \quad \text{and} \quad \mathbf{T}_d(s) = \left[\frac{1}{(s+1)}\right].$$

4-29 Repeat the design of a lateral command-augmentation system given in Section 4.6 for

$$\mathbf{F} = \begin{bmatrix} -0.0036 & -0.9325 & 0.0066 & 0.0028 & 0.367 & 0.0311 \\ 1.0 & 0 & 0 & 0 & 0 & 0 \\ 0 & 0.0637 & -0.0123 & -0.0004 & 0.353 & -0.6478 \\ 0 & 0.9325 & -0.0011 & -0.0028 & -0.3495 & -0.0311 \\ 0 & 0 & 0 & 0 & -25.0 & 0 \\ 0 & 0 & 0 & 0 & 0 & -6.667 \end{bmatrix}$$

and

$$\mathbf{G} = \begin{bmatrix} -0.0175 & 0 \\ 0 & 0 \\ 0 & 0 \\ 0 & 0 \\ 25.0 & 0 \\ 0 & 6.667 \end{bmatrix}$$

which correspond to an aircraft of the X-15 type at 211,000 feet altitude and Mach No. 5.6.

4-30 Repeat Problem 4-29 for

$$\mathbf{F} = \begin{bmatrix} -0.0018 & -0.4927 & 0.0061 & 0.0015 & 0.1857 & 0.0161 \\ 1.0 & 0 & 0 & 0 & 0 & 0 \\ 0 & 0.0208 & -0.0062 & -0.0002 & 0.15 & -0.3355 \\ 0 & 0.4927 & -0.00058 & -0.0015 & -0.1782 & -0.0161 \\ 0 & 0 & 0 & 0 & -25.0 & 0 \\ 0 & 0 & 0 & 0 & 0 & -6.667 \end{bmatrix}$$

and

$$\mathbf{G} = \begin{bmatrix} -0.0075 & 0 \\ 0 & 0 \\ 0 & 0 \\ 0 & 0 \\ 25.0 & 0 \\ 0 & 6.667 \end{bmatrix}$$

which correspond to an aircraft of the X-15 type at 226,000 feet altitude and Mach. No. 5.3.

4-31 An alternate design concept for lateral command-augmentation systems is based on controlling side-slip angle and roll rate instead of roll rate and yaw rate so that response variables now are selected in accordance with

$$\mathbf{H} = \begin{bmatrix} 0 & 0 & 1 & 0 & 0 & 0 \\ 0 & 1 & 0 & 0 & 0 & 0 \end{bmatrix}.$$

Repeat the design given in Section 4.6 if possible using this matrix and

$$\mathbf{T}_d(s) = \begin{bmatrix} \dfrac{-34.0}{(s+7.0)(s+1.4)} & \dfrac{-25.0}{(s+25.0)(s+1.4)} \\[2ex] \dfrac{-3.4}{(s+7.0)(s^2+2.4s+4.0)} & \dfrac{25.0}{(s+25.0)(s^2+2.4s+4.0)} \end{bmatrix}.$$

4-32 Repeat Problem 4-31 for an aircraft of the X-15 type at 211,000 feet altitude and Mach No. 5.6 as defined in Problem 4-29.

4-33 Repeat Problem 4-31 for an aircraft of the X-15 type at 226,000 feet altitude and Mach No. 5.3 as defined in Problem 4-30.

4-34 Repeat the design problem given in Section 4.6 using

$$\mathbf{H} = \begin{bmatrix} 0 & 0 & 1 & 0 & 0 & 0 \\ 0 & 1 & \alpha & \beta & 0 & 0 \end{bmatrix}$$

and

$$\mathbf{T}_d(s) = \begin{bmatrix} \dfrac{-34.0}{(s+7.0)(s+1.4)} & 0 \\[2ex] 0 & \dfrac{25.0}{(s+25.0)(s^2+2.4s+4.0)} \end{bmatrix},$$

where α and β are selected so that compatibility is established.

4-35 Repeat Problem 4-34 for an aircraft of the X-15 type at 211,000 feet altitude and Mach No. 5.6 as defined in Problem 4-29.

4-36 Repeat Problem 4-34 for an aircraft of the X-15 type at 226,000 feet altitude and Mach No. 5.3 as defined in Problem 4-30.

4-37 Determine whether $\{\mathbf{A}, \mathbf{B}, \mathbf{H}\}$ is completely responsive where

$$\mathbf{A} = \begin{bmatrix} 0 & 0 & 0 \\ 1 & -3 & -1 \\ 0 & 1 & -1 \end{bmatrix}, \quad \mathbf{B} = \begin{bmatrix} 1 \\ 0 \\ 0 \end{bmatrix}, \quad \text{and} \quad \mathbf{H} = [0 \quad 0 \quad 1].$$

4-38 Repeat Problem 4-37 for

$$\mathbf{A} = \begin{bmatrix} 0 & -\frac{1}{2} & -\frac{1}{2} \\ 1 & -\frac{3}{2} & -\frac{1}{2} \\ 1 & -\frac{1}{2} & -\frac{3}{2} \end{bmatrix}, \quad \mathbf{B} = \begin{bmatrix} 0 \\ 1 \\ -1 \end{bmatrix}, \quad \text{and} \quad \mathbf{H} = [1 \quad 1 \quad 1].$$

4-39 The ordered pair $\{\mathbf{A}, \mathbf{B}\}$ with $\mathbf{A}$ asymptotically stable is completely controllable if and only if $\{\mathbf{A}, \mathbf{B}, \mathbf{H}\}$ is completely responsive for some nonsingular $\mathbf{H}$. Give two necessary and sufficient conditions for complete controllability with $\mathbf{A}$ asymptotically stable.

4-40 Suppose $\{\mathbf{A}, \mathbf{B}, \mathbf{H}\}$ is completely responsive.

a) Prove that matrix

$$\int_0^T \{\mathbf{H}\, e^{(T-t)\mathbf{A}}\mathbf{B}\mathbf{B}'\, e^{(T-t)\mathbf{A}'}\mathbf{H}'\}\, dt$$

is nonsingular for all $T > 0$.

b) Prove that there exists a piecewise continuous control $\mathbf{u}$ such that

$$\dot{\mathbf{x}} = \mathbf{A}\mathbf{x} + \mathbf{B}\mathbf{u} \quad \text{and} \quad \mathbf{r} = \mathbf{H}\mathbf{x}$$

can be controlled from any $\mathbf{r}(0) \in \mathscr{R}^m$ to any other $\mathbf{r}(T) \in \mathscr{R}^m$ with T finite.

4-41 Repeat the design of the aircraft pitch channel given in Section 4.8 using FC1 as the nominal flight condition.

4-42 Repeat Problem 4-41 using FC3 as the nominal flight condition.

4-43 Repeat the design of the aircraft pitch channel given in Section 4.8 using

$$\mathbf{F} = \begin{bmatrix} 0 & 0 & 0 & 0 \\ 1 & 0 & 0 & 0 \\ 0 & a_1 & -a_3 & -a_2 \\ 0 & 0 & 1 & -a_4 \end{bmatrix}, \quad \mathbf{G} = \begin{bmatrix} 1 \\ 0 \\ 0 \\ 0 \end{bmatrix}, \quad \text{and} \quad z_1 = x_3.$$

Compare the sensitivity of this system with that given in Section 4.8.

4-44 Repeat Problem 4-43 using FC1 as the nominal flight condition.

4-45 Repeat Problem 4-43 using FC3 as the nominal flight condition.

4-46 Find **P, Q,** and **R** appearing in (4-114) for optimal gain control corresponding to (4-11? and (4-113).

4-47 Specialize the results of the Problem 4-46 to the case where **E** is nonsingular. Under this condition, what can be said about the relationship between regularity and compatibility?

4-48 Find **P, Q,** and **R** appearing in (4-114) for linear optimal control corresponding to Problem 4-2.

4-49 The ordered triple $\{\mathbf{P}, \mathbf{Q}, \mathbf{R}\}$ is strictly regular when the triple is regular and no $\mathbf{m}\{\mathbf{Q}\}$ belongs to $N(\mathbf{P})$. Restate Theorem 4-13 for the condition of strict regularity.

4-50 In the notation of Problem 4-2, determine whether linear optimal control corresponding to

$$\mathbf{F} = \begin{bmatrix} 0 & 0 & \cdots & 0 & 0 \\ 1 & 0 & \cdots & 0 & 0 \\ 0 & 1 & \cdots & 0 & 0 \\ \cdots & \cdots & \cdots & \cdots & \cdots \\ 0 & 0 & \cdots & 1 & 0 \end{bmatrix}, \quad \mathbf{G} = \begin{bmatrix} 1 \\ 0 \\ 0 \\ \vdots \\ 0 \end{bmatrix}, \quad \mathbf{\Phi} = \begin{bmatrix} 0 & \cdots & 0 & 0 \\ 0 & \cdots & 0 & 0 \\ \cdots & \cdots & \cdots & \cdots \\ 0 & \cdots & 0 & 0 \\ 0 & \cdots & 0 & 1 \end{bmatrix},$$

$$\mathbf{\Psi} = \mathbf{0}, \quad \text{and} \quad \mathbf{\Omega} = \mathbf{I}$$

is (strictly) regular.

4-51 Determine the values of α for which linear optimal control corresponding to

$$\mathbf{P} = (n + \alpha)\mathbf{1}, \quad \mathbf{Q} = \frac{\alpha}{2}\mathbf{I}, \quad \text{and} \quad \mathbf{R} = \mathbf{I}$$

is (strictly) regular where $\mathbf{1}$ denotes an $n \times n$ matrix with only unit elements.

4-52 In the notation of Problem 4-2, let

$$\mathbf{F} = -n\mathbf{I} + \mathbf{1}, \quad \mathbf{G} = \mathbf{I}, \quad \mathbf{\Phi} = n\mathbf{1}, \quad \mathbf{\Psi} = \mathbf{0}, \quad \text{and } \mathbf{\Omega} = \mathbf{I}$$

where $\mathbf{1}$ denotes an $n \times n$ matrix with only unit elements. Also initialize the iterative method used in the proof of Lemma 4-7 with

$$\mathbf{X}_0 = \alpha_0 \mathbf{1}.$$

a) Verify that all preconditions of Lemma 4-7 hold, and specify all α_0 such that the method is known *a priori* to be convergent to a positive semi-definite solution.

4-53 Write a computer program for determining the solution to (4-114) by the spectral factorization method.

4-54 Suppose

$$\mathbf{F} = \begin{bmatrix} 0 & 0 & 0 \\ 1 & 0 & 0 \\ 0 & 0 & 0 \end{bmatrix}, \quad \mathbf{G} = \begin{bmatrix} 1 \\ 0 \\ 0 \end{bmatrix}, \quad \mathbf{\Phi} = \begin{bmatrix} 2 & 0 & 0 \\ 0 & 1 & -1 \\ 0 & -1 & 1 \end{bmatrix}, \quad \mathbf{\Psi} = \mathbf{0}, \quad \text{and} \quad \mathbf{\Omega} = \mathbf{I}$$

in the notation of Problem 4-2. Determine the linear optimal controller if possible and discuss the significance of this design problem.

4-55 Repeat Problem 4-54 for

$$\mathbf{F} = \begin{bmatrix} 0 & 0 & 0 & 0 & 0 & 0 \\ 1 & 0 & 0 & 0 & 0 & 0 \\ 0 & 0 & 0 & 1 & 0 & 0 \\ 0 & 0 & 0 & 0 & 1 & 0 \\ 0 & 0 & 0 & 0 & 0 & 1 \\ 0 & 0 & 0 & 0 & 0 & 0 \end{bmatrix}, \quad \mathbf{G} = \begin{bmatrix} 1 \\ 0 \\ 0 \\ 0 \\ 0 \\ 0 \end{bmatrix},$$

$$\mathbf{\Phi} = \begin{bmatrix} 5 & 0 & 0 & 0 & 0 & 0 \\ 0 & 4 & -4 & 0 & 0 & 0 \\ 0 & -4 & 4 & 0 & 0 & 0 \\ 0 & 0 & 0 & 0 & 0 & 0 \\ 0 & 0 & 0 & 0 & 0 & 0 \\ 0 & 0 & 0 & 0 & 0 & 0 \end{bmatrix}, \quad \mathbf{\Psi} = \mathbf{0}, \quad \text{and} \quad \mathbf{\Omega} = \mathbf{I}.$$

CHAPTER 5

Stochastic design problems

Control system design problems involving stochastic signals also arise frequently in practice. These problems are subject to the same difficulties encountered with deterministic signals. However, additional difficulties, which are primarily computational for problems of high dimension, occur with some stochastic systems of interest here. The origin of these additional difficulties is identified in the following brief introduction.

All design problems discussed in this chapter involve linear closed-loop state equations that can be reduced to the stochastic form

$$d\mathbf{x} = (\mathbf{A}\mathbf{x})\, dt + d\mathbf{u} + \sum_{k=1}^{m} (\mathbf{B}_k\mathbf{x})\, dw_k; \quad \mathbf{x}(0) = \mathbf{x}_0. \tag{5-1}$$

Matrices $\mathbf{A} = \mathbf{A}(c)$ and $\mathbf{B}_k = \mathbf{B}_k(\mathbf{c})$ are assumed to correspond to the stochastic equivalent of asymptotic stability for all $\mathbf{c}$ of interest. Vector $\mathbf{c}$ of course includes all parameters which are to be specified by optimization procedures. Initial state $\mathbf{x}_0$ is taken to be an independent Gaussian vector with

$$E\{\mathbf{x}_0\} = \mathbf{0} \quad \text{and} \quad E\{\mathbf{x}_0\mathbf{x}_0'\} = \mathbf{E}_0. \tag{5-2}$$

In addition, stochastic forcing functions $\mathbf{u}(t)$ and $\mathbf{w}(t)$ are taken to be independent Brownian motions with

$$E\{\mathbf{u}(t)\} = \mathbf{0}, \quad E\{\mathbf{w}(t)\} = \mathbf{0} \tag{5-3}$$

and

$$E\{\mathbf{u}(t)\mathbf{u}'(\tau)\} = \mathbf{C} \min\{t, \tau\}, \quad E\{\mathbf{w}(t)\mathbf{w}'(\tau)\} = \mathbf{I} \min\{t, \tau\} \tag{5-4}$$

where $\mathbf{C} = \mathbf{C}(\mathbf{c})$.

The solution of these state equations can of course be written in terms of a stochastic integral [1] as

$$\mathbf{x}(t) = e^{\mathbf{A}t}\mathbf{x}_0 + \int_0^t e^{(t-\xi)\mathbf{A}} \{d\mathbf{u}(\xi) + \sum_{k=1}^{m} [\mathbf{B}_k\mathbf{x}(\xi)]\, dw_k(\xi)\}. \tag{5-5}$$

Moreover, stochastic calculus [2] can be used to show that

$$E\{\mathbf{x}(t)\} = \mathbf{0} \tag{5-6}$$

and

$$\mathbf{E}(t) = e^{\mathbf{A}t}\mathbf{E}_0\, e^{\mathbf{A}'t} + \int_0^t [e^{(t-\xi)\mathbf{A}}\{\mathbf{C} + D[\mathbf{B}, \mathbf{E}(\xi)\}\, e^{(t-\xi)\mathbf{A}'}]\, d\xi \tag{5-7}$$

where

$$\mathbf{E}(t) = E\{\mathbf{x}(t)\mathbf{x}'(t)\} \tag{5-8}$$

and

$$D[\mathbf{B}, \mathbf{E}(t)] = \sum_{k=1}^{m} \mathbf{B}_k\mathbf{E}(t)\mathbf{B}_k'. \tag{5-9}$$

The covariance matrix of state vector $\mathbf{x}(t)$ is also given by

$$\dot{\mathbf{E}} = \mathbf{E}\mathbf{A}' + \mathbf{A}\mathbf{E} + D(\mathbf{B}, \mathbf{E}) + \mathbf{C}; \quad \mathbf{E}(0) = \mathbf{E}_0 \tag{5-10}$$

as found by differentiation of (5-7).

For the purposes of this book, linear stochastic state equations with $\mathbf{u(t)} = \mathbf{0}$ for all t are *asymptotically stable* [3] if and only if, given any $\mathbf{x}_0 \in \mathscr{R}^n$, $\lim_{t\to\infty} \|\mathbf{x}(t)\| = 0$ holds with probability one. Asymptotic stability of (5-1) of course implies that, given any positive semi-definite $\mathbf{E}_0$, $\lim_{t\to\infty} \|\mathbf{E}(t)\| = 0$ holds with $\mathbf{C} = \mathbf{0}$. Because both terms appearing in (5-7) are positive semi-definite when $\mathbf{E}_0$ is positive semi-definite, stochastic stability of (5-1) must also imply that $\mathbf{A}$ is asymptotically stable. If (5-1) is asymptotically stable, then the covariance of $\mathbf{x}(t)$ has a unique equilibrium value

$$\mathbf{X} = \lim_{t\to\infty} \mathbf{E}(t) \tag{5-11}$$

which is also the positive semi-definite solution of

$$\mathbf{0} = \mathbf{X}\mathbf{A}' + \mathbf{A}\mathbf{X} + D(\mathbf{B}, \mathbf{X}) + \mathbf{C} \tag{5-12}$$

because (5-10) is linear.

As in the case of deterministic design problems, a relative minimum of $f = f(\mathbf{c}, \mathbf{X})$ is sought subject to constraints

$$\mathbf{g}_1(\mathbf{c}, \mathbf{X}) = \mathbf{0}_1 \quad \text{and} \quad \mathbf{g}_2(\mathbf{c}, \mathbf{X}) \leqslant \mathbf{0}_2$$

as well as the constraint that $\mathbf{A}$ and $\mathbf{B}_k$ result in asymptotic stability. Functions f and $\mathbf{g}$ are assumed to be twice continuously differentiable throughout as are matrices $\mathbf{A}$, $\mathbf{B}_k$, and $\mathbf{C}$. The dependence of $\mathbf{X}$ on $\mathbf{c}$ of course poses all of the computational problems experienced with deterministic design problems. Moreover, the dependence of D on $\mathbf{X}$ as defined by (5-9) introduces new computational problems which are discussed in this chapter. However, the main thrust of this chapter is the formulation of meaningful control system design problems with stochastic signals so that many of the considerations pertinent to these problems are similar to those given in Chapter 4 for deterministic problems.

5.1 FORMULATION OF NECESSARY CONDITIONS

Dependence of f and $\mathbf{g}$ on $\mathbf{X}$ as well as on $\mathbf{c}$ is accounted for in an efficient fashion that is analogous to the formulation presented for deterministic design problems. Gradients and hessians are again expressed conveniently in terms of Lagrangian L given in (4-7) where F is defined by (4-9). Conditions for stationarity of Lagrangian L thus are again given by (4-10). However, matrix $\mathbf{Z}$ appearing in this Lagrangian is redefined as

$$\mathbf{Z} = \mathbf{X}\mathbf{A}' + \mathbf{A}\mathbf{X} + D(\mathbf{B}, \mathbf{X}) + \mathbf{C} \tag{5-13}$$

where D is defined by (5-9) for stochastic design problems.

All gradient methods are based on a direct calculation of gradients and hence require solution of the following equations. Equations

$$\mathbf{X}\mathbf{A}' + \mathbf{A}\mathbf{X} + D(\mathbf{B}, \mathbf{X}) + \mathbf{C} = \mathbf{0} \tag{5-14}$$

and

$$\mathbf{\Lambda}\mathbf{A} + \mathbf{A}'\mathbf{\Lambda} + D(\mathbf{B}', \mathbf{\Lambda}) + F_{\mathbf{X}} = \mathbf{0} \tag{5-15}$$

result from the first two conditions appearing in (4-10). Elements of gradient $L_{\mathbf{c}}$ also become

$$L_{c_i} = F_{c_i} + \mathrm{tr}\,\{\mathbf{\Lambda}'\mathbf{Z}_{c_i}\} \tag{5-16}$$

for each i where

$$\mathbf{Z}_{c_i} = \mathbf{X}\mathbf{A}'_{c_i} + \mathbf{A}_{c_i}\mathbf{X} + D_{c_i}(\mathbf{B}, \mathbf{X}) + \mathbf{C}_{c_i} \tag{5-17}$$

and

$$D_{c_i}(\mathbf{B}, \mathbf{X}) = \sum_{k=1}^{m} [(\mathbf{B}_k)_{c_i}\mathbf{X}\mathbf{B}'_k + \mathbf{B}_k\mathbf{X}(\mathbf{B}'_k)_{c_i}]. \tag{5-18}$$

These equations are of course solved on each trial in the selection of α for each iteration.

Use of directly calculated hessians as matrices $\mathbf{P}$ and $\mathbf{Q}$ which arise in the min–max method requires the solution of additional equations corresponding to (4-16). In particular, solution of

$$\mathbf{X}_j\mathbf{A}' + \mathbf{A}\mathbf{X}_j + D(\mathbf{B}, \mathbf{X}_j) + \mathbf{Z}_{cj} = \mathbf{0} \tag{5-19}$$

and

$$\boldsymbol{\Lambda}_j\mathbf{A} + \mathbf{A}'\boldsymbol{\Lambda}_j + D(\mathbf{B}', \boldsymbol{\Lambda}_j) + \left[\boldsymbol{\Lambda}\mathbf{A}_{cj} + \mathbf{A}'_{cj}\boldsymbol{\Lambda} + D_{cj}(\mathbf{B}', \boldsymbol{\Lambda}) + \frac{\mathrm{d}}{\mathrm{d}c_j}F_{\mathbf{X}}\right] = \mathbf{0} \tag{5-20}$$

is required for each j where matrices $\mathbf{X}_j$ and $\boldsymbol{\Lambda}_j$ are defined by the total derivatives of (5-14) and (5-15) with respect to c_j. These matrices then are used to compute

$$p_{ij} = \frac{\mathrm{d}}{\mathrm{d}c_j}F_{ci} + \operatorname{tr}\left\{\boldsymbol{\Lambda}'_j\mathbf{Z}_{ci} + \boldsymbol{\Lambda}'\left(\frac{\mathrm{d}}{\mathrm{d}c_j}\mathbf{Z}_{c_i}\right)\right\} \tag{5-21}$$

and

$$q_{ij} = \frac{\mathrm{d}}{\mathrm{d}c_j}F_{\lambda i} \tag{5-22}$$

where

$$\frac{\mathrm{d}}{\mathrm{d}c_j}\mathbf{Z}_{ci} = \mathbf{Z}_{c_ic_j} + \mathbf{X}_j\mathbf{A}'_{ci} + \mathbf{A}_{ci}\mathbf{X}_j + D_{ci}(\mathbf{B}, \mathbf{X}_j), \tag{5-23}$$

$$\mathbf{Z}_{c_ic_j} = \mathbf{X}\mathbf{A}'_{c_ic_j} + \mathbf{A}_{c_ic_j}\mathbf{X} + D_{c_ic_j}(\mathbf{B}, \mathbf{X}) + \mathbf{C}_{c_ic_j}, \tag{5-24}$$

and

$$D_{c_ic_j}(\mathbf{B}, \mathbf{X}) = \sum_{k=1}^{m} [(\mathbf{B}_k)_{c_ic_j}\mathbf{X}\mathbf{B}'_k + (\mathbf{B}_k)_{c_i}\mathbf{X}(\mathbf{B}'_k)_{c_j} + (\mathbf{B}_k)_{c_j}\mathbf{X}(\mathbf{B}'_k)_{c_i} + \mathbf{B}_k\mathbf{X}(\mathbf{B}'_k)_{c_ic_j}]. \tag{5-25}$$

Equations (5-19) through (5-25) are even more complex than the corresponding equations for deterministic design problems. The additional complexity, however, is due to nonzero matrices $\mathbf{B}_k$. Moreover, matrices $(\mathbf{B}_k)_{c_ic_j}$ may cause a significant increase in computer memory requirements. Hence, the Fletcher–Powell method is often used to compute **P** and **Q** indirectly when using the min–max method for problems with constraints.

5.2 METHODS OF SOLVING $\mathbf{XA}' + \mathbf{AX} + D(\mathbf{B}, \mathbf{X}) + \mathbf{C} = \mathbf{0}$

The complexity of computing gradients and hessians given in the previous section is of course primarily attributed to solving equations of the form appearing in

(5-12). If this equation has a solution, then this solution can be written in the form

$$\mathbf{X} = \int_0^\infty \{e^{\mathbf{A}t}[\mathbf{C} + D(\mathbf{B}, \mathbf{X})]\, e^{\mathbf{A}'t}\}\, dt \tag{5-26}$$

when **A** is asymptotically stable. Unfortunately, the integral appearing in (5-26) also depends on **X**, and hence this expression does not lead directly to computational methods of solution. Moreover, conditions for the existence of a unique positive semi-definite solution **X** depends not only on **C** but also on **A** and $\mathbf{B}_k$ in a very complicated way.

The primary result [4] which concerns the existence and uniqueness of positive semi-definite solutions to (5-12) is stated in terms of the following definitions. First, stochastic state vector $\mathbf{x}(t)$ is redefined by (5-1) assuming $\mathbf{u}(t) = \mathbf{0}$ for all t. Correspondingly, covariance matrix $\mathbf{E}(t)$ is redefined by (5-10) assuming $\mathbf{C} = \mathbf{0}$. Second, vector notation is defined by

$$\mathbf{e}_v' = [e_{11} \quad \sqrt{2}e_{12} \quad e_{22} \quad \sqrt{2}e_{13} \; . \; .] \tag{5-27}$$

for symmetric matrices so that only the upper triangular portion of **E** is used and hence dim $\mathbf{e}_v = n(n+1)/2$ when dim $\mathbf{E} = n \times n$. Then the matrix differential equation for **E** becomes

$$\dot{\mathbf{e}}_v = \mathbf{\Delta}\mathbf{e}_v; \quad \mathbf{e}_v(0) = \mathbf{e}_{v0} \tag{5-28}$$

in the same manner as introduced in Chapter 4 for the direct matrix inversion method of solving related algebraic equations. Third, symmetric matrices $\mathbf{\Phi}$ and $\mathbf{\Psi}$ are defined by the matrix algebraic equation

$$\mathbf{\Phi}\mathbf{A} + \mathbf{A}'\mathbf{\Phi} + D(\mathbf{B}', \mathbf{\Phi}) + \mathbf{\Psi} = \mathbf{0}. \tag{5-29}$$

This matrix equation can also be rewritten in the vector form

$$\mathbf{\Delta}'\boldsymbol{\phi}_v + \boldsymbol{\psi}_v = \mathbf{0} \tag{5-30}$$

with a coefficient matrix that is the transpose of the matrix $\mathbf{\Delta}$ defined by (5-28). The result is stated conveniently as

Theorem 5-1 *The following conditions are equivalent;*

i) Given any positive definite matrix $\mathbf{\Psi}$, *the unique solution* $\mathbf{\Phi}$ *is positive definite,*

ii) There exists a positive definite matrix $\mathbf{\Phi}$ *such that the corresponding matrix* $\mathbf{\Psi}$ *is positive definite, and*

iii) Matrix $\mathbf{\Delta}$ *is asymptotically stable.*

Proof Condition ii) follows directly from i) by construction. To prove that

iii) follows from ii), let $\mathbf{\Phi}$ be a positive definite matrix for which $\mathbf{\Psi}$ is also positive definite, and let V be the scalar function

$$V = E\{\mathbf{x}'\mathbf{\Phi}\mathbf{x}\} = E\{\mathrm{tr}[\mathbf{x}\mathbf{x}'\mathbf{\Phi}]\} = \mathrm{tr}\{\mathbf{E}\mathbf{\Phi}\}.$$

Because $\mathbf{E}_0$ is positive semi-definite, matrix $\mathbf{E}(t)$ is positive semi-definite on $[0, \infty)$ and elementary manipulations yield

$$\lambda_{\min}\{\mathbf{\Phi}\}\,\|\mathbf{E}\|_2 \leqslant V \leqslant n\lambda_{\max}\{\mathbf{\Phi}\}\,\|\mathbf{E}\|_2.$$

In addition, differentiation and elementary manipulations yield

$$\dot{V} = \mathrm{tr}\{\dot{\mathbf{E}}\mathbf{\Phi}\} = \mathrm{tr}\{[\mathbf{\Phi}\mathbf{A} + \mathbf{A}'\mathbf{\Phi} + D(\mathbf{B}',\mathbf{\Phi})]\mathbf{E}\} = -E\{\mathbf{x}'\mathbf{\Psi}\mathbf{x}\}$$

and hence

$$\dot{V} \leqslant -\lambda_{\min}\{\mathbf{\Psi}\}\,\|\mathbf{E}\|_2$$

holds given any positive semi-definite $\mathbf{E}_0$. Therefore, condition $\lim_{t\to\infty} \mathbf{E}(t) = \mathbf{0}$ holds given any positive semi-definite $\mathbf{E}_0$. Equivalently condition $\lim_{t\to\infty} \mathbf{e}_v(t) = \mathbf{0}$ holds given any $\mathbf{e}_{v0} \in \mathscr{R}^{n(n+1)/2}$ and hence $\mathbf{\Delta}$ is asymptotically stable. To prove that i) follows from iii), note first that (5-29) has the unique solution given by

$$\boldsymbol{\phi}_v = -(\mathbf{\Delta}')^{-1}\boldsymbol{\psi}_v = \int_0^\infty \{e^{\Delta' t}\boldsymbol{\psi}_v\}\,dt$$

when $\mathbf{\Delta}$ is asymptotically stable. Now let $\mathbf{\Psi}$ be some positive definite matrix and let

$$\mathbf{E}_0 = \mathbf{e}_0\mathbf{e}_0'$$

where $\mathbf{e}_0$ is any nonzero vector belonging to $\mathscr{R}^n$. Elementary manipulations then yield

$$\mathbf{e}_0'\mathbf{\Phi}\mathbf{e}_0 = \mathrm{tr}\{\mathbf{E}_0\mathbf{\Phi}\} = \mathbf{e}_{v0}'\boldsymbol{\phi}_v = \int_0^\infty \{\mathbf{e}_v'(t)\boldsymbol{\psi}_v\}\,dt.$$

Moreover, inequalities

$$\mathbf{e}_v'(t)\boldsymbol{\psi}_v = \mathrm{tr}\{\mathbf{E}(t)\mathbf{\Psi}\} \geqslant \lambda_{\min}\{\mathbf{\Psi}\}\,\|\mathbf{E}(t)\|_2 > 0 \tag{5-31}$$

are obtained for all $t \in [0, \infty)$ and hence

$$\mathbf{e}_0'\mathbf{\Phi}\mathbf{e}_0 > 0$$

holds for all nonzero $\mathbf{e}_0 \in \mathscr{R}^n$.

An immediate consequence of this theorem is

Corollary 5-1 *Suppose* $\mathbf{\Delta}$ *is asymptotically stable. Then, given any positive semi-definite matrix* $\mathbf{\Psi}$*, the unique solution* $\mathbf{\Phi}$ *is positive semi-definite.*

Proof The strict inequality appearing in (5-31) becomes an inequality whenever $\mathbf{\Psi}$ is positive semi-definite but singular.

Conditions appearing in Theorem 5-1 and Corollary 5-1 for matrices $\mathbf{\Phi}$ and $\mathbf{\Psi}$ also apply to matrices **X** and **C** respectively. This result follows from the vector form of (5-12), namely

$$\mathbf{\Delta x}_v + \mathbf{c}_v = \mathbf{0}, \tag{5-32}$$

and (5-30) because $\mathbf{\Delta}$ and $\mathbf{\Delta}'$ have identical eigenvalues.

Another result concerning the existence and uniqueness of solutions to (5-12) is stated as

Theorem 5-2 *Suppose* $\mathbf{\Delta}$ *is asymptotically stable. Then, given any symmetric matrix* **C**, $\mathbf{X} = \mathbf{X}_\infty$ *is the unique solution of (5-12) where* $\mathbf{X}_\infty$ *is the limit of sequence* $\{\mathbf{X}_j\}$ *defined by*

$$\mathbf{X}_{j+1}\mathbf{A}' + \mathbf{A}\mathbf{X}_{j+1} + [\mathbf{C} + D(\mathbf{B}, \mathbf{X}_j)] = \mathbf{0}; \quad \mathbf{X}_0 = \mathbf{0}. \tag{5-33}$$

Proof Because $\mathbf{\Delta}$ is asymptotically stable, matrix **A** is also asymptotically stable so let

$$\mathbf{Y} = \int_0^\infty \{e^{\mathbf{A}t}\mathbf{C}\, e^{\mathbf{A}'t}\}\, dt. \tag{5-34}$$

Also let $\mathscr{L}$ be the linear operator defined by

$$\mathscr{L}(\mathbf{X}) = \int_0^\infty \{e^{\mathbf{A}t} D(\mathbf{B}, \mathbf{X})\, e^{\mathbf{A}'t}\}\, dt. \tag{5-35}$$

Then equation (5-12) is equivalent to

$$\mathbf{X} = \mathscr{L}(\mathbf{X}) + \mathbf{Y}. \tag{5-36}$$

Moreover, repeated use of this equation in conjunction with (5-33) yields

$$\mathbf{X} = \mathscr{L}^j(\mathbf{X}) + \mathbf{X}_j \quad \text{for} \quad j = 1, 2, \ldots \tag{5-37}$$

and

$$\mathbf{X}_j = \sum_{i=0}^{j-1} \mathscr{L}^i(\mathbf{Y}), \tag{5-38}$$

where

$$\mathscr{L}^i(\mathbf{X}) = \mathscr{L}[\mathscr{L}^{i-1}(\mathbf{X})] \quad \text{and} \quad \mathscr{L}^0(\mathbf{X}) = \mathbf{X}.$$

If **C** is positive semi-definite, then **Y** is positive semi-definite and hence $(\mathbf{X} - \mathbf{X}_j)$ and $(\mathbf{X}_{j+1} - \mathbf{X}_j)$ are also positive semi-definite. Therefore, the sequence $\{\mathbf{X}_j\}$

converges monotonely with $\mathscr{L}^i(\mathbf{Y}) \to \mathbf{0}$ as $i \to \infty$ and $\mathbf{X}_\infty$ exists as a positive semi-definite matrix. In addition, linearity of $\mathscr{L}$ yields

$$\mathscr{L}(\mathbf{X}_\infty) + \mathbf{Y} = \sum_{i=0}^{\infty} \mathscr{L}^{i+1}(\mathbf{Y}) + \mathbf{Y} = \mathbf{X}_\infty$$

and hence

$$\mathbf{X}_\infty \mathbf{A}' + \mathbf{A}\mathbf{X}_\infty + D(\mathbf{B}, \mathbf{X}_\infty) + \mathbf{C} = \mathbf{0}. \tag{5-39}$$

Matrix $\mathbf{X}_\infty$ thus is the unique solution of (5-12) when $\mathbf{C}$ is positive semi-definite because $\mathbf{\Delta}$ is nonsingular. If $\mathbf{C}$ is positive definite, then there exists some $\alpha \in (0, \infty)$ for which $(\mathbf{Y} - \alpha\mathbf{I})$ is positive semi-definite. Moreover,

$$\mathscr{L}^i(\mathbf{I}) \quad \text{and} \quad [\mathscr{L}^i(\mathbf{Y}) - \alpha\mathscr{L}^i(\mathbf{I})]$$

are positive semi-definite for each i and hence $\mathscr{L}^i(\mathbf{I}) \to \mathbf{0}$ as $i \to \infty$. Now let $\mathbf{C}$ be any symmetric matrix and let $\mathbf{X}$ be the corresponding solution to (5-12). Then there exists some $\beta \in (0, \infty)$ for which $(\beta\mathbf{I} - \mathbf{X})$ and $(\mathbf{X} + \beta\mathbf{I})$ are positive semi-definite. Then

$$[\beta\mathscr{L}^j(\mathbf{I}) - \mathscr{L}^j(\mathbf{X})] \text{ and } [\mathscr{L}^j(\mathbf{X}) + \beta\mathscr{L}^j(\mathbf{I})]$$

are also positive semi-definite for each j and hence $\mathscr{L}^j(\mathbf{X}) \to \mathbf{0}$ as $j \to \infty$. Therefore, $\mathbf{X} = \mathbf{X}_\infty$ holds for all symmetric $\mathbf{C}$ by virtue of (5-37).

The sequence defined by (5-33) is useful as a procedure for computing the solution to (5-12) under circumstances where $\mathbf{\Delta}$ is asymptotically stable and only a few terms are required to approximate $\mathbf{X}_\infty$ with sufficient accuracy.

The condition that $\mathbf{\Delta}$ is asymptotically stable is particularly germane to stochastic design problems where $\mathbf{X}$ has the significance of a positive semi-definite covariance matrix. Unfortunately, however, a necessary and sufficient test for $\mathbf{A}$ and $\mathbf{B}_k$ to result in the asymptotic stability of $\mathbf{\Delta}$ is not forthcoming in a useful form for problems with large state-vector dimension. On the other hand, the following two sufficiency conditions are useful. The first [5] of these conditions is stated as

Theorem 5-3 *Suppose* $\mathbf{A}$ *is asymptotically stable and (5-12) has a positive definite solution* $\mathbf{X}$ *given some positive semi-definite* $\mathbf{C}$ *for which* $\mathbf{Y}$ *is positive definite. Then* $\mathbf{\Delta}$ *is asymptotically stable.*

Proof Because $\mathbf{X}$ and $\mathbf{Y}$ are positive definite, the sequence $\{\mathbf{X}_j\}$ corresponding to (5-34), (5-35), and (5-38) converges monotonely, and $\mathbf{X}_\infty$ is a positive definite solution of (5-12) by virtue of (5-39). Moreover, the construction given in the proof of Theorem 5-2 can now be used again to show that $\mathbf{X}_j$ converges to a positive definite solution of (5-12) given any positive definite $\mathbf{C}$. Therefore, matrix $\mathbf{\Delta}$ is nonsingular and (5-12) has a unique positive definite solution $\mathbf{X}$ given any positive definite $\mathbf{C}$. The desired result then follows from Theorem 5-1.

The second [2] of these sufficiency conditions is stated as

Theorem 5-4 *Suppose there exists a constant* $\beta < 1$ *such that*

$$\left\| \int_0^\infty \{e^{\mathbf{A}t} D(\mathbf{B}, \mathbf{X})\, e^{\mathbf{A}'t}\}\, dt \right\| \leqslant \beta \,\|\mathbf{X}\| \tag{5-40}$$

holds for some norm given any positive definite $\mathbf{X}$. *Then* $\mathbf{\Delta}$ *is asymptotically stable.*

Proof Given any positive definite $\mathbf{C}$ and corresponding $\mathbf{Y}$ which is also positive definite, inequality

$$\mathscr{L}^i(\mathbf{Y}) \leqslant \beta^i \,\|\mathbf{Y}\|$$

is found to hold for each i using (5-35) and (5-40). Inequality

$$\|\mathbf{X}_\infty\| \leqslant \frac{\|\mathbf{Y}\|}{1-\beta} \tag{5-41}$$

is also found to hold using (5-38). The construction given in the proof of Theorem 5-3 can now be used again to show that $\mathbf{\Delta}$ is asymptotically stable.

When (5-33) is used as an indirect method of computing $\mathbf{X}$, an estimate of the number of sequence terms required to obtain sufficient accuracy is needed. Such an estimate can be obtained when there exist constants $\Gamma, \gamma > 0$ such that

$$\|e^{\mathbf{A}t}\| \leqslant \Gamma\, e^{-\gamma t} \tag{5-42}$$

and

$$\frac{\Gamma^2}{2\gamma}\Delta < 1 \tag{5-43}$$

where

$$\Delta = \sum_{k=1}^{m} \|\mathbf{B}_k\|^2. \tag{5-44}$$

Then, by definition of $\mathbf{Y}$ and β, inequalities

$$\|\mathbf{Y}\| \leqslant \frac{\Gamma^2}{2\gamma}\|\mathbf{C}\| \tag{5-45}$$

and

$$\beta \leqslant \frac{\Gamma^2}{2\gamma}\Delta \tag{5-46}$$

are obtained so that

$$\|\mathbf{X}\| \leqslant \frac{\dfrac{\Gamma^2}{2\gamma}\|\mathbf{C}\|}{1 - \dfrac{\Gamma^2}{2\gamma}\Delta} \tag{5-47}$$

holds in accordance with (5-41).

An estimate of the required number of sequence terms is obtained by first selecting a norm so that Δ can be computed easily and yet is as small as possible. Typically either $\|\cdot\|_1$ or $\|\cdot\|_\infty$ is selected for this purpose. Then sequence member $\mathbf{X}_1$ is computed for some appropriate value of $\mathbf{c}$ so that $\mathbf{Y} = \mathbf{X}_1$ and approximation

$$\frac{\Gamma^2}{2\gamma} \cong \frac{\|\mathbf{X}_1\|}{\|\mathbf{C}\|} \tag{5-48}$$

are obtained from (5-45). Finally, convergence of the partial sum

$$B_j = \|\mathbf{X}_1\| \sum_{k=0}^{j-1} r^k \quad \text{for} \quad j = 1, 2, \ldots \tag{5-49}$$

where

$$r = \Delta \frac{\|\mathbf{X}_1\|}{\|\mathbf{C}\|} \tag{5-50}$$

to the upper bound appearing in (5-42) can be used to estimate j. For instance, index j can be defined as the smallest integer for which

$$R_j = \frac{B_\infty - B_j}{\|\mathbf{X}_1\|} = \frac{r^j}{1-r} < \epsilon \tag{5-51}$$

holds given some ϵ.

Use of (5-33) as an indirect method of computing solutions to equations of the form of (5-12) has two major difficulties. First, convergence of $\mathbf{X}_j$ to $\mathbf{X}$ is relatively slow unless $r \ll 1$ and in fact design problems arise for which (5-43) cannot be established. Second, numerical errors propagate so that $\|\mathbf{X} - \mathbf{X}_j\|$ may actually increase for suitably large j. The only known alternative to this indirect method is the direct matrix inversion method corresponding to (5-32) for symmetric $\mathbf{C}$. Elements of $\boldsymbol{\Delta}$ again are easily formed in terms of elements of $\mathbf{A}$ and $\mathbf{B}_k$ with a digital computer program, and matrix inversion then yields $\mathbf{x}_v$ directly. However, dimensionality continues to make this method unattractive when $n = \dim \mathbf{x}$ is relatively large because $\dim \mathbf{x}_v = \dim \mathbf{c}_v = n(n+1)/2$.

Design problems involving multiplicative disturbances typically exhibit structures that permit problem partitioning. Such partitioning is particularly desirable due to the computational difficulties of solving (5-12) with $\mathbf{B}_k \neq \mathbf{0}$. Partitioning associated with these structures corresponds to the largest dim $\mathbf{x}_1$ and dim $\mathbf{x}_3$ for which

$$\mathbf{A} = \begin{bmatrix} \mathbf{A}_{11} & \mathbf{0}_{12} & \mathbf{0}_{13} \\ \mathbf{A}_{21} & \mathbf{A}_{22} & \mathbf{0}_{23} \\ \mathbf{A}_{31} & \mathbf{A}_{32} & \mathbf{A}_{33} \end{bmatrix} \tag{5-52}$$

and

$$\mathbf{B}_k = \begin{bmatrix} \mathbf{0}_{11} & \mathbf{0}_{12} & \mathbf{0}_{13} \\ \mathbf{B}_{21} & \mathbf{B}_{22} & \mathbf{0}_{23} \\ \mathbf{0}_{31} & \mathbf{0}_{32} & \mathbf{0}_{33} \end{bmatrix}_k \tag{5-53}$$

hold for each k. If $\mathbf{C}$ is symmetric, then the distinct partitions of $\mathbf{X}$ corresponding to (5-12) are given by

$$\mathbf{X}_{11}\mathbf{A}'_{11} + \mathbf{A}_{11}\mathbf{X}_{11} + \mathbf{\Gamma}_{11} = \mathbf{0}_{11}, \tag{5-54}$$

$$\mathbf{X}_{12}\mathbf{A}'_{22} + \mathbf{A}_{11}\mathbf{X}_{12} + \mathbf{\Gamma}_{12} = \mathbf{0}_{12}, \tag{5-55}$$

$$\mathbf{X}_{22}\mathbf{A}'_{22} + \mathbf{A}_{22}\mathbf{X}_{22} + D(\mathbf{B}_{22}, \mathbf{X}_{22}) + \mathbf{\Gamma}_{22} = \mathbf{0}_{22}, \tag{5-56}$$

$$\mathbf{X}_{13}\mathbf{A}'_{33} + \mathbf{A}_{11}\mathbf{X}_{13} + \mathbf{\Gamma}_{13} = \mathbf{0}_{13}, \tag{5-57}$$

$$\mathbf{X}_{23}\mathbf{A}'_{33} + \mathbf{A}_{22}\mathbf{X}_{23} + \mathbf{\Gamma}_{23} = \mathbf{0}_{23}, \tag{5-58}$$

and

$$\mathbf{X}_{33}\mathbf{A}'_{33} + \mathbf{A}_{33}\mathbf{X}_{33} + \mathbf{\Gamma}_{33} = \mathbf{0}_{33} \tag{5-59}$$

in terms of a suitably defined matrix $\mathbf{\Gamma}$. These relationships reveal that only equation (5-56) is of the type that requires either (5-33) or direct matrix inversion for solution. Any of the methods given in Section 4.2 can be used to solve the remaining five equations. However, matrices

$$\mathbf{\Gamma}_{11} = \mathbf{C}_{11}, \tag{5-60}$$

$$\mathbf{\Gamma}_{12} = \mathbf{C}_{12} + \mathbf{X}_{11}\mathbf{A}'_{21}, \tag{5-61}$$

$$\mathbf{\Gamma}_{22} = \mathbf{C}_{22} + \mathbf{X}'_{12}\mathbf{A}'_{21} + \mathbf{A}_{21}\mathbf{X}_{12} + D\left([\mathbf{B}_{21} \quad \mathbf{B}_{22}], \begin{bmatrix} \mathbf{X}_{11} & \mathbf{X}_{12} \\ \mathbf{X}'_{12} & \mathbf{0}_{22} \end{bmatrix}\right), \tag{5-62}$$

$$\boldsymbol{\Gamma}_{13} = \mathbf{C}_{13} + \mathbf{X}_{11}\mathbf{A}'_{31} + \mathbf{X}_{12}\mathbf{A}'_{32}, \tag{5-63}$$

$$\boldsymbol{\Gamma}_{23} = \mathbf{C}_{23} + \mathbf{X}'_{12}\mathbf{A}'_{31} + \mathbf{X}_{22}\mathbf{A}'_{32} + \mathbf{A}_{21}\mathbf{X}_{13}, \tag{5-64}$$

and

$$\boldsymbol{\Gamma}_{33} = \mathbf{C}_{33} + \mathbf{X}'_{13}\mathbf{A}'_{31} + \mathbf{X}'_{23}\mathbf{A}'_{32} + \mathbf{A}_{31}\mathbf{X}_{13} + \mathbf{A}_{32}\mathbf{X}_{23} \tag{5-65}$$

reveal that (5-54) through (5-59) must be solved in order. If dim $\mathbf{x}_2$ is somewhat smaller than dim $\mathbf{x}$, then this method of solving the partitioned form of (5-12) results in a significant reduction in computations.

5.3 HUMAN CONTROLLER MODELS

Many significant problems which arise in control system design involve the human controller. Moreover, a number of these problems can be adequately represented in terms of a linear lumped-parameter system which is subjected to stationary Gaussian signals. For instance, such problems arise in compensatory tracking tasks.

Automated design of systems involving the human controller is complicated by the fact that the human controller exhibits adaptive characteristics which are very difficult to simulate. Recent literature [6] suggests that a deterministic model of the human operator which is adequate for many design purposes is linear but also includes both a remnant disturbance and adjustable parameters. Remnant disturbances in a sense model performance degradation in the human controller's response caused by an inability to estimate amplitudes of its forcing functions. Adjustable parameters in a sense model adaptation of the human controller to a particular tracking task so that a performance functional is minimized. Both amplitudes of the remnant disturbances and values of these adjustable parameters depend on system design.

No attempt is made here to justify a particular stochastic model of the human controller. Instead, typical characteristics of these models are introduced in this section by way of example. Similarly, no attempt is made here to justify the selection of a particular performance functional or the notion of a human controller minimizing a performance functional in any general context. Instead, an example is given in the next section for which the justification of this approach has already been established elsewhere.

A typical stochastic model [7, 8] of the human controller is depicted in Figure 5-1. This model is subjected to a forcing function $\mathbf{f}_r$, an observation disturbance $\mathbf{r}$, and a motor noise $\mathbf{s}$. The model is further characterized by a fixed delay-time matrix $\mathbf{T}$, a variable equalization transfer-function matrix $\mathbf{M}(s)$, and a fixed neuromuscular transfer-function matrix $\mathbf{N}(s)$. Elements of $\mathbf{M}(s)$ generally

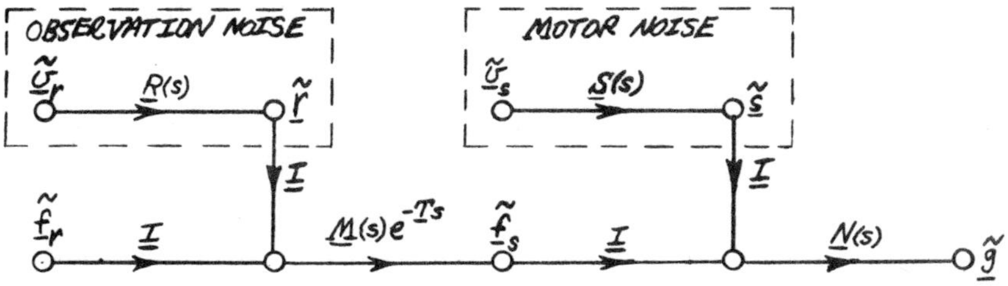

Figure 5-1 Flow graph of a typical human-controller model containing observation and motor noises.

depend upon parameters of the human controller that are adjusted in accordance with each tracking task.

Observation and motor noises typically are formulated in terms of statistically independent Brownian motions $\boldsymbol{\mu}_r$ and $\boldsymbol{\mu}_s$ respectively which define the corresponding white noise representations $\mathbf{v}_r$ and $\mathbf{v}_s$ appearing in Figure 5-1. In other words, these stochastic variables are related by

$$\boldsymbol{\mu}(t) = \int_0^t \mathrm{d}\mathbf{v}(\xi) \tag{5-66}$$

where subscripts r and s are omitted in the remainder of this section for convenience. Moreover, each element of $\mathrm{d}\boldsymbol{\mu}$ is given by

$$\mathrm{d}\mu_k(t) = f_k(t)\ \mathrm{d}w_k(t). \tag{5-67}$$

In other words, amplitude dependent observation and motor noises are generated by multiplicative disturbances.

Transfer function matrices $\mathbf{R}(s)$ and $\mathbf{S}(s)$ have the purpose of introducing appropriate statistical characteristics of observation and motor noises respectively. Typically these matrices are taken to be diagonal where for example

$$\mathbf{R}(s) = \begin{bmatrix} R_1(s) & 0 & \cdots \\ 0 & R_2(s) & \cdots \\ \cdots & \cdots & \cdots \end{bmatrix}$$

with

$$R_k(s) = K_k \frac{\sqrt{2\omega_k}}{s + \omega_k}.$$

In other words, the variance of the output of each filter would be equal to K_k^2 times the covariance coefficient of the filter input.

Fixed time-delay matrices depicted in Figure 5-1 are defined by the diagonal matrix

$$\mathbf{T} = \begin{bmatrix} T_1 & 0 & \cdots \\ 0 & T_2 & \cdots \\ \cdots & \cdots & \cdots \end{bmatrix}$$

so that

$$e^{-\mathbf{T}s} = \begin{bmatrix} e^{-T_1 s} & 0 & \cdots \\ 0 & e^{-T_2 s} & \cdots \\ \cdots & \cdots & \cdots \end{bmatrix}.$$

However, delay times T_k generally are small enough for most control applications so that low order Pade approximations of $e^{-T_k s}$ suffice. For instance, first order approximations

$$e^{-T_k s} \cong \frac{-T_k s + 2}{T_k s + 2}$$

for each k are frequently used in lumped-parameter models of the human controller.

Variable equalization transfer-function matrices depicted in Figure 5-1 take on many different forms and cannot be categorized in a simple fashion. The most elementary form of these matrices is the diagonal form

$$\mathbf{M}(s) = \begin{bmatrix} M_1(s) & 0 & \cdots \\ 0 & M_2(s) & \cdots \\ \cdots & \cdots & \cdots \end{bmatrix}$$

where for example

$$M_k(s) = \alpha_k \frac{\gamma_k \beta_k s + 1}{\beta_k s + 1} \quad \text{with} \quad \gamma_k \geqslant 1.$$

Gain α_k, time constant β_k, and lead ratio γ_k are not fixed but are in fact dependent upon the dynamics of each tracking task. In a similar fashion, fixed neuromuscular transfer-function matrices $\mathbf{N}(s)$ are taken to be diagonal in many elementary situations. Specifically, this matrix is

$$\mathbf{N}(s) = \begin{bmatrix} N_1(s) & 0 & \cdots \\ 0 & N_2(s) & \cdots \\ \cdots & \cdots & \cdots \end{bmatrix}$$

where for example

$$N_k(s) = \frac{1}{\tau_k s + 1}.$$

Time constant τ_k is relatively fixed over a variety of tracking-task dynamics.

Coefficient matrices $\mathbf{B}_k$ which appear in (5-1) are easily identified for observation and motor noises with lumped-parameter models of the human controller. In particular, lumped-parameter approximations permit expressing each element of $\mathbf{f}$ in state-vector form as

$$f_k(t) = \mathbf{h}'_k \mathbf{x}(t). \tag{5-68}$$

Now suppose state vector $\mathbf{x}$ is taken to be the solution of

$$\mathrm{d}\mathbf{x} = (\mathbf{A}\mathbf{x})\,\mathrm{d}t + \mathbf{E}\,\mathrm{d}\boldsymbol{\nu} + \sum_{k=1}^{m} \mathbf{g}_k\,\mathrm{d}\mu_k; \quad \mathbf{x}(0) = \mathbf{x}_0 \tag{5-69}$$

where $\boldsymbol{\nu}$ is independent Brownian motion with

$$E\{\boldsymbol{\nu}(t)\boldsymbol{\nu}'(\tau)\} = \mathbf{I} \min\{t, \tau\}. \tag{5-70}$$

Then this stochastic state equation is reduced to the form of (5-1) with

$$\mathrm{d}\mathbf{u} = \mathbf{E}\,\mathrm{d}\boldsymbol{\nu} \tag{5-71}$$

and

$$\mathbf{B}_k = \mathbf{g}_k \mathbf{h}'_k \tag{5-72}$$

after the elimination of $\mathrm{d}\mu_k$ and f_k. Finally, the covariance matrix $\mathbf{C}$ which appears in (5-4) becomes

$$\mathbf{C} = \mathbf{E}\mathbf{E}' \tag{5-73}$$

for this lumped-parameter system.

An elementary example is introduced here in order to illustrate the formulation of multiplicative disturbances and difficulties that arise with unrealistic models. The example is depicted in Figure 5-2 where independent disturbance v_1 is assumed to have a unit covariance coefficient and observation noise av_2 is assumed with the covariance coefficient of v_2 equal to the variance of forcing function x_1. Matrices which appear in (5-68) and (5-69) are easily identified to be

$$\mathbf{A} = \begin{bmatrix} 0 & -c \\ 1 & -\omega \end{bmatrix}, \quad \mathbf{E} = \begin{bmatrix} 1 \\ 0 \end{bmatrix}, \quad \mathbf{g}_1 = \begin{bmatrix} 0 \\ a \end{bmatrix}, \quad \text{and} \quad \mathbf{h}_1 = [1 \quad 0].$$

Equations (5-72) and (5-73) then yield

$$\mathbf{B}_1 = \begin{bmatrix} 0 & 0 \\ a & 0 \end{bmatrix} \quad \text{and} \quad \mathbf{C} = \begin{bmatrix} 1 & 0 \\ 0 & 0 \end{bmatrix}$$

for the system depicted in Figure 5-2.

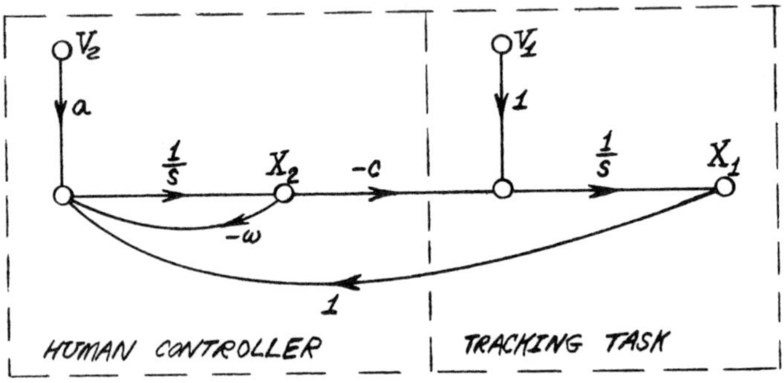

Figure 5-2 Simplified models of human controller and tracking task.

Equation (5-12) with D defined by (5-9) can now be solved directly using the equivalent vector form given in (5-32). Use of element ordering defined by (5-27) yields

$$\mathbf{\Delta} = \begin{bmatrix} 0 & -\sqrt{2}c & 0 \\ \sqrt{2} & -\omega & -\sqrt{2}c \\ a^2 & \sqrt{2} & -2\omega \end{bmatrix} \quad \text{and} \quad \mathbf{c}_v = \begin{bmatrix} 1 \\ 0 \\ 0 \end{bmatrix}.$$

Matrix $\mathbf{\Delta}$ is nonsingular if and only if $c \neq 0$ and $c \neq 2\omega/a^2$ hold, and the solution obtained under these conditions is

$$x_{11} = \frac{c + \omega^2}{c(2\omega - ca^2)}, \quad x_{12} = x_{21} = \frac{1}{2c}, \quad \text{and} \quad x_{22} = \frac{2 + \omega a^2}{2c(2\omega - ca^2)}.$$

Moreover, matrix $\mathbf{\Delta}$ is asymptotically stable if and only if $\omega > 0$ and $c \in (0, 2\omega/a^2)$, and solution $\mathbf{X}$ hence is positive semi-definite under these conditions. In fact, solution $\mathbf{X}$ is also nonsingular and hence positive definite under these conditions. Inspection of the solution reveals that $\mathbf{X}$ is not positive semi-definite given any $c \in (2\omega/a^2, \infty)$ even when $\omega > 0$ so that $\mathbf{A}$ is asymptotically stable. These conditions correspond to stochastic instability.

This elementary example also illustrates that meaningful automated design problems can be formulated with human controller observation noises even though no terms involving human controller forcing functions are included in the performance functional. Specifically, minimization of x_{11} with respect to $c \in (0, 2\omega/a^2)$ for $\omega > 0$ yields

$$c = \omega^2 \left[\left(1 + \frac{2}{\omega a^2} \right)^{1/2} - 1 \right] \leqslant \frac{\omega}{a^2}.$$

Finally, the example illustrates that the sequence defined by (5-33) converges to the solution $\mathbf{X}$ given here for all $c \in (0, 2\omega/a^2)$ even though (5-40) cannot be verified using (5-43). For example, the case defined by $a^2 = 2$ and $\omega = 1$ yields

$$\| \mathbf{X}_1 \| = \frac{c+2}{2c}, \quad \| \mathbf{C} \| = 1, \quad \text{and} \quad \Delta = 2$$

using either $\| \cdot \|_1$ or $\| \cdot \|_\infty$. The approximation given in (5-48) then yields

$$\frac{\Gamma^2}{2\gamma} \Delta = 1 + \frac{2}{c}$$

so that the sufficiency condition corresponding to (5-43) does not hold for any admissible c in this case.

5.4 EXAMPLE OF MANUAL CONTROL

A helicopter being piloted in a precision hover mode [9, 10] is presented here for the purpose of illustrating the simulation of manual control using a human-controller model with observation noise. Open-loop dynamics of the helicopter and pilot models are defined by the flow graph appearing in Figure 5-3, and the corresponding assignment of variables is summarized in Table 5-1. The feedback and observation-noise portions of the pilot model are defined by the flow graph appearing in Figure 5-4. In addition, the covariance coefficient of v_1 is taken to be unity, and the covariance coefficient of v_2 is determined in accordance with (5-66) and (5-67) by forcing function f_r and Brownian motion w having a unit covariance coefficient. Equivalent observation noise δ_r has a variance equal to k_r^2 times the covariance coefficient of v_2. Finally, the performance functional used for this model of the pilot is

$$f = 12.5 x_{11}^{1/2} + 1.25 x_{44}^{1/2} + 2.5 c_2 + c_4.$$

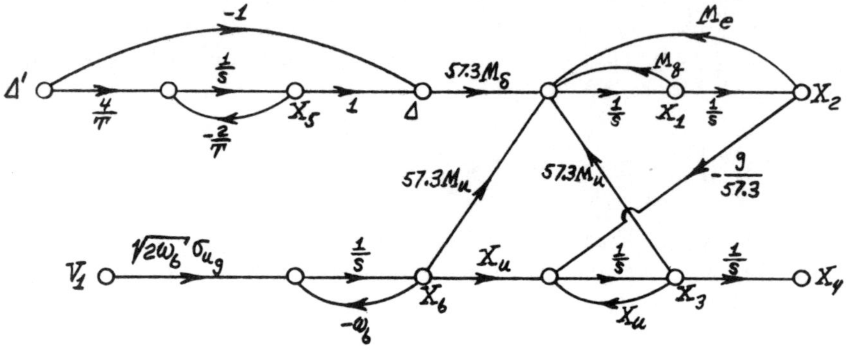

Figure 5-3 Flow graph of open-loop dynamics.

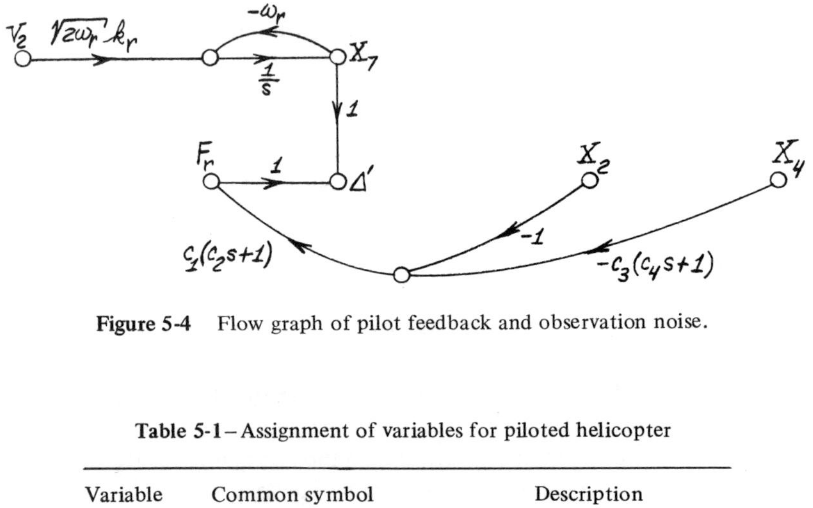

Figure 5-4 Flow graph of pilot feedback and observation noise.

Table 5-1 – Assignment of variables for piloted helicopter

Variable	Common symbol	Description
x_1	q (deg/sec)	Pitch rate
x_2	θ (deg)	Pitch angle
x_3	u (ft/sec)	Horizontal velocity
x_4	x (ft)	Horizontal displacement
x_5	—	—
x_6	u_g (ft/sec)	Wing gust velocity
x_7	δ_r (in)	Observation noise
v_1	—	—
v_2	—	—

Simulation of pilot flying capabilities is accomplished by parameter optimization corresponding to the minimization of f. Adjustable pilot parameters are gains c_1, c_3 and time constants c_2, c_4. All data presented here correspond to the fixed

Table 5-2—Values of fixed helicopter and pilot parameters

$M_u = 0.0281$	$\sigma_{ug} = 2.6$
$X_u = -0.1$	$\omega_b = 0.314$
$M_q = -3.0$	$g = 32.2$
$M_\theta = 0.0$	$T = 0.44$
$M_\delta = 0.396$	$\omega_r = 2.0$

parameter values summarized in Table 5-2. These parameter values yield open-loop helicopter transfer functions

$$\frac{\Theta}{\Delta} = \frac{22.7(s + 0.1)}{(s + 3.1075)[s^2 - 2(0.00663)(0.569)s + (0.569)^2]}$$

and

$$\frac{X}{\Delta} = \frac{-12.75}{s(s + 3.1075)[s^2 - 2(0.00663)(0.569)s + (0.569)^2]}.$$

Thus, the helicopter is unstable and the pilot is faced with a difficult control problem.

Typical convergence characteristics of the Newton–Raphson method of successive substitutions are illustrated in Table 5-3 for $k_r = 0.0$. At the solution

Table 5-3—Typical convergence of the Newton–Raphson method with $k_r = 0.0$

Iteration	Trial	α	c_1	c_2	c_3	c_4	f	$\lvert L_c \rvert$
0	1	0.00	0.3303	0.0748	−1.3051	0.4091	2.6005	3.60000
1	2	0.65	0.4130	0.1748	−1.7132	0.3446	2.2397	2.22000
2	1	1.00	0.5033	0.2430	−2.2132	0.2848	2.0898	0.33000
3	1	1.00	0.5112	0.2495	−2.4446	0.2762	2.0799	0.01950
4	1	1.00	0.5121	0.2517	−2.4756	0.2757	2.0798	0.00139
5	1	1.00	0.5121	0.2518	−2.4761	0.2757	2.0798	0.00000

point for this case, the hessian defined by (5-21) is

$$\mathbf{P} = \begin{bmatrix} 99.1 & -34.8 & -0.391 & 52.9 \\ -34.8 & 49.1 & 2.38 & -24.3 \\ -0.391 & 2.38 & 0.388 & -0.946 \\ 52.9 & -24.3 & -0.946 & 48.6 \end{bmatrix}$$

and is positive definite so that a relative minimum has been achieved. By way of comparison, typical convergence characteristics of the steepest-descent method of successive approximations are illustrated in Table 5-4 for $k_r = 0.0$.

Table 5-4—Typical convergence of the steepest-descent method with $k_r = 0.0$

Iteration	Trial	α	c_1	c_2	c_3	c_4	f	$\lvert L_c \rvert$
0	1	0.000	0.3303	0.0748	−1.3051	0.4091	2.6005	3.600
4	3	0.007	0.4040	0.1495	−1.3238	0.3924	2.3931	1.150
8	3	0.003	0.4598	0.1922	−1.3933	0.3510	2.3104	0.604
12	4	0.003	0.4779	0.2020	−1.5367	0.3081	2.2361	0.466
16	4	0.002	0.4787	0.2032	−1.5725	0.3062	2.2217	0.445

Excluding any consideration of computer memory requirements, the steepest-descent method is seen to be considerably less efficient than the Newton–Raphson method in this case. In addition, typical convergence characteristics of the conjugate gradient method of successive approximation are illustrated in Table 5-5 for $k_r = 0.0$. This method is seen to be somewhat more competitive with the Newton–Raphson method in this case even though convergence is not obtained. Typical four-step convergence is not exhibited in Table 5-5 because hessian $\mathbf{P}$ is a strong function of $\mathbf{c}$ except in a relatively small neighborhood of the minimum. For instance, this hessian is

$$\mathbf{P} = \begin{bmatrix} 158.4 & -75.6 & 2.019 & 40.2 \\ -75.6 & 86.5 & 8.05 & -28.8 \\ 2.019 & 8.05 & 3.062 & -2.118 \\ 40.2 & -28.8 & -2.118 & 21.1 \end{bmatrix}$$

at the initial point of these iterations and is seen to differ markedly from that at the solution point. In summary, these data illustrate typical advantages of the Newton–Raphson method when hessian $\mathbf{P}$ can be computed without

Table 5-5—Typical convergence of the conjugate gradient method with $k_r = 0.0$

Iteration	Trial	α	c_1	c_2	c_3	c_4	f	$\lvert L_c \rvert$
0	1	0.000	0.3303	0.0748	−1.3051	0.4091	2.6005	3.600
4	4	0.245	0.5032	0.2334	−1.4832	0.2730	2.2812	1.330
8	4	0.301	0.4653	0.1998	−2.0526	0.3082	2.1258	0.951
12	3	0.031	0.5083	0.2395	−2.2085	0.2731	2.0902	0.322
16	3	0.076	0.5029	0.2437	−2.2579	0.2887	2.0867	0.184

excessive memory requirements and is positive definite on a suitably large neighborhood of a relative minimum.

Optimal pilot performance is summarized in Table 5-6 for observation noises corresponding to various values of k_r. These data illustrate typical degradation in pilot performance caused by increasing k_r. Moreover, increasing k_r tends to result in optimal closed-loop systems that have increased damping as might be expected from the bound appearing in (5-47). By way of illustration, variable parameters appearing in the closed-loop characteristic polynomial

$$\det(s\mathbf{I} - \mathbf{A}) = (s + \omega_b)(s + \omega_r)(s + \omega_1)(s^2 + 2\zeta_2\omega_2 s + \omega_2^2) \times (s^2 + 2\zeta_3\omega_3 s + \omega_3^2)$$

are summarized in Table 5-7 for various observation noises. In particular, increasing k_r is seen to result primarily in increasing damping ratio ζ_2.

Optimal pilot performance given in Table 5-6 was computed by the direct matrix inversion method for solving equations of the form appearing in (5-12). By way of comparison, the indirect method of computing solutions to (5-12) by truncating the sequence defined in (5-33) can also be used. An estimate of the number of sequence terms required to compute $\mathbf{X}$ accurately can be established by the procedure presented in Section 5.3. This procedure is illustrated for the optimal value of $\mathbf{c}$ found for the case with $k_r = 0.0$. Elementary calculations for $k_r = 0.1$ then yield

$$\|\mathbf{X}_1\|_1 = 10.2, \quad \|\mathbf{C}\|_1 = 4.3, \quad \text{and} \quad \Delta = 0.0625$$

Table 5-6—Comparison of optimal pilot performance for various observation noises

k_r	c_1	c_2	c_3	c_4	f	x_{11}	x_{44}	x_{77}
0.00	0.5121	0.2518	−2.4761	0.2757	2.0798	6.6299	0.2404	0.00000
0.10	0.5089	0.2618	−2.4456	0.2803	2.1193	6.6491	0.2478	0.00227
0.20	0.4963	0.2908	−2.3565	0.2954	2.2469	6.8309	0.2740	0.00980
0.25	0.4843	0.3127	−2.2893	0.3081	2.3525	7.0572	0.2988	0.01633

Table 5-7–Variable closed-loop natural frequencies and damping ratios for various observation noises

k_r	ω_1	ζ_2	ω_2	ζ_3	ω_3
0.00	4.40	0.0624	3.16	0.105	1.35
0.10	3.90	0.0690	3.22	0.107	1.34
0.20	3.48	0.0877	3.38	0.112	1.30
0.25	3.21	0.1007	3.49	0.117	1.28

Table 5-8–Comparison of optimal solutions with truncated sequences for $k_r = 0.1$

j	c_1	c_2	c_3	c_4	f	x_{11}	x_{44}	x_{77}
1	0.5121	0.2518	−2.4761	0.2757	2.0798	6.6299	0.2404	0.00000
4	0.5089	0.2617	−2.4457	0.2803	2.1192	6.6491	0.2478	0.00227
∞	0.5089	0.2618	−2.4456	0.2803	2.1193	6.6491	0.2478	0.00227

so that $r = 0.15$ is obtained using (5-50). The ratio $R_4 = 0.0006$ suggests that at least three places of numerical accuracy in the computation of **X** will be obtained when the sequence is truncated at $j = 4$. This contention is substantiated by the data presented in Table 5-8 for this case.

5.5 OPTIMAL GAIN CONTROL

A number of design problems are formulated as the stochastic equivalent of optimal gain control. These problems are formulated with performance functionals

$$f = \lim_{t\to\infty} E\{\mathbf{e}'(t)\mathbf{e}(t)\} \tag{5-74}$$

under the assumption of stochastic stability. Error vector **e** is defined in terms of stochastic state and control vectors **x** and $\mathbf{u}_c$, respectively, as

$$\mathbf{e} = \mathbf{D}\mathbf{x} + \mathbf{E}\mathbf{u}_c. \tag{5-75}$$

Moreover, stochastic state and response equations are taken to be

$$d\mathbf{x} = (\mathbf{F}\mathbf{x} + \mathbf{G}\mathbf{u}_c)\,dt + \mathbf{B}\,d\mathbf{u} + \sum_{k=1}^{m} \mathbf{S}_k\mathbf{x}\,dw_k; \quad \mathbf{x}(0) = \mathbf{x}_0 \tag{5-76}$$

and

$$\mathbf{y} = \mathbf{H}\mathbf{x} \tag{5-77}$$

with $E\{\mathbf{x}_0\} = \mathbf{0}$. Forcing functions $\mathbf{u}$ and $\mathbf{w}$ are assumed to be independent Brownian motions with zero means and

$$E\{\mathbf{u}(t)\mathbf{u}'(\tau)\} = \mathbf{I} \min\{t, \tau\} \quad \text{and} \quad E\{\mathbf{w}(t)\mathbf{w}'(\tau)\} = \mathbf{I} \min\{t, \tau\}. \tag{5-78}$$

Finally, linear control equations employing response-variable feedback are taken to be

$$\mathbf{u}_c = -\mathbf{C}\mathbf{y} \tag{5-79}$$

where adjustable parameter vector $\mathbf{c}$ has been assigned to feedback gain matrix $\mathbf{C}$ by vector partitions as defined in (4-103).

Control and then response vectors can of course be eliminated from (5-75) and (5-76) using (5-79) and (5-77). Subsequently, under the assumption of stochastic stability, covariance matrix $\mathbf{X}$ which is defined by (5-8) and (5-11) is now found to be the positive semi-definite solution of

$$\mathbf{X}\mathbf{A}' + \mathbf{A}\mathbf{X} + D(\mathbf{S}, \mathbf{X}) + \mathbf{B}\mathbf{B}' = \mathbf{0}. \tag{5-80}$$

Function D is defined by (5-9) and $\mathbf{A}$ is again given by (4-105). In addition, performance functional f which is defined in (5-74) now becomes

$$f = \operatorname{tr}\{(\mathbf{D} - \mathbf{E}\mathbf{C}\mathbf{H})\mathbf{X}(\mathbf{D} - \mathbf{E}\mathbf{C}\mathbf{H})'\} \tag{5-81}$$

after the elimination of error vector $\mathbf{e}$.

The first necessary condition for stationarity of Lagrangian L, namely $L_\Lambda = \mathbf{0}$, is of course (5-80). The second necessary condition for stationarity of L is $L_\mathbf{X} = \mathbf{0}$ and becomes

$$\mathbf{\Lambda}\mathbf{A} + \mathbf{A}'\mathbf{\Lambda} + D(\mathbf{S}', \mathbf{\Lambda}) + \mathbf{W} = \mathbf{0} \tag{5-82}$$

where $\mathbf{W}$ is again given by (4-106). Finally, the third necessary condition for stationarity of L is $L_\mathbf{C} = \mathbf{0}$ and becomes

$$2[(\mathbf{E}'\mathbf{E})\mathbf{C}(\mathbf{H}\mathbf{X}\mathbf{H}') - (\mathbf{E}'\mathbf{D} + \mathbf{G}'\mathbf{\Lambda})(\mathbf{X}\mathbf{H}')] = \mathbf{0}.$$

The optimal feedback gain matrix then is given by

$$\mathbf{C} = (\mathbf{E}'\mathbf{E})^{-1}(\mathbf{E}'\mathbf{D} + \mathbf{G}'\mathbf{\Lambda})(\mathbf{X}\mathbf{H}')(\mathbf{H}\mathbf{X}\mathbf{H}')^{-1} \tag{5-83}$$

if $(\mathbf{E}'\mathbf{E})$ and $(\mathbf{H}\mathbf{X}\mathbf{H}')$ are nonsingular.

Optimal gains given in (5-83) are noted to bear a striking resemblance to optimal gains given in (4-107) for deterministic design problems. In fact, these gain matrices are identical when $l = 1$ is selected for the performance functional given in (4-100) for deterministic design problems and $\mathbf{S}_k = \mathbf{0}$ occurs in the state equations given in (5-76) for stochastic design problems. This identity follows from (4-107) and (5-83) and the fact that equations for $\mathbf{X}$ and $\mathbf{\Lambda}$ are interchanged in the two cases. In other words, covariance matrix $\mathbf{X}$ is identical to the Lagrange multiplier matrix $\mathbf{\Lambda}$ of the deterministic problem, and performance-integral matrix $\mathbf{X}$ appearing in (4-4) is equal to the Lagrange multiplier matrix $\mathbf{\Lambda}$ of the stochastic problem. This equivalence is not unexpected in light of equivalences between deterministic and stochastic design problems that were identified in Chapter 1.

The equivalence noted here is most useful. For example, synthesis techniques for transfer function matrices, which were developed in Chapter 4 for deterministic problems, thus can be used directly in the formulation of stochastic optimal-gain problems. In addition, essential aspects of the notion of complete responsiveness can be easily identified in terms of the matrix $\mathbf{\Delta}$ which appears in (5-28) and is defined by (5-27). These essential aspects are summarized by

Theorem 5-5 *Suppose* $\mathbf{\Delta}$ *is asymptotically stable and* $\{\mathbf{A}, \mathbf{B}, \mathbf{H}\}$ *is completely responsive. Then* $(\mathbf{HXH}')$ *is positive definite.*

Proof Asymptotic stability of $\mathbf{\Delta}$ implies $\mathbf{A}$ is asymptotically stable and hence

$$\mathbf{HXH}' = \int_0^\infty (\mathbf{H}\, e^{\mathbf{A}t}\mathbf{B})(\mathbf{H}\, e^{\mathbf{A}t}\mathbf{B})'\, dt + \int_0^\infty \{\mathbf{H}\, e^{\mathbf{A}t} D(\mathbf{S}, \mathbf{X})\, e^{\mathbf{A}'t}\mathbf{H}'\}\, dt.$$

The first integral in this expression is positive definite by Theorem 4-12, and the second integral is positive semi-definite because $\mathbf{X}$ is positive semi-definite by Corollary 5-1.

This sufficiency test for the invertability of $(\mathbf{HXH}')$ unfortunately cannot be applied in advance of problem solution because $\mathbf{A}$ depends on $\mathbf{X}$ and $\mathbf{\Lambda}$.

5.6 EXAMPLE OF OPTIMAL GAIN CONTROL

Design of control systems for flexible launch vehicles is a classical problem in which stochastic disturbances arise. Specifically, body bending modes are induced by air turbulence, and hence these modes must be controlled by feedback. An elementary version of this design problem is presented here as an illustration of stochastic optimal-gain control.

Linearized equations of motion [11], which include actuator dynamics and displacement of the 1st body bending mode, are given here in accordance with state and control variable assignments summarized in Table 5-9. These equations

Table 5-9– Assignment of variables for flexible launch vehicles

Variable	Common symbol	Description
x_1	β (rad)	Control deflection angle
x_2	$\dot{\phi}$ (rad/sec)	Attitude angle rate
x_3	ϕ (rad)	Attitude angle
x_4	α (rad)	Angle of attack
x_5	$\dot{\eta}_1$ (meters/sec)	Displacement rate of 1st mode
x_6	η_1 (meters)	Displacement of 1st mode
u_{1c}	β_c (rad)	Control deflection command
v_1	–	–

correspond to coefficient matrices

$$\mathbf{F} = \begin{bmatrix} -18.0 & 0 & 0 & 0 & 0 & 0 \\ -0.5 & 0 & 0 & 0.08 & 0 & 0 \\ 0 & 1.0 & 0 & 0 & 0 & 0 \\ -0.02 & 1.0 & -0.04 & -0.01 & 0 & 0 \\ 16.0 & 0 & 0 & 5.0 & -0.02 & -5.0 \\ 0 & 0 & 0 & 0 & 1.0 & 0 \end{bmatrix},$$

$$\mathbf{G} = \begin{bmatrix} 18.0 \\ 0 \\ 0 \\ 0 \\ 0 \\ 0 \end{bmatrix}, \quad \mathbf{B} = \begin{bmatrix} 0 \\ 1.0 \\ 0 \\ 0 \\ 0 \\ 0 \end{bmatrix},$$

and $\mathbf{S}_k = \mathbf{0}$ in the notation of (5-76). In addition, a flow graph of these open-loop equations is shown in Figure 5-5 for convenience where v_1 is the white noise representation of an equivalent stochastic disturbance due to air turbulence. This flow graph illustrates that equivalent rigid body motion induces bending mode displacement but bending mode displacement does not in turn induce equivalent rigid body motion.

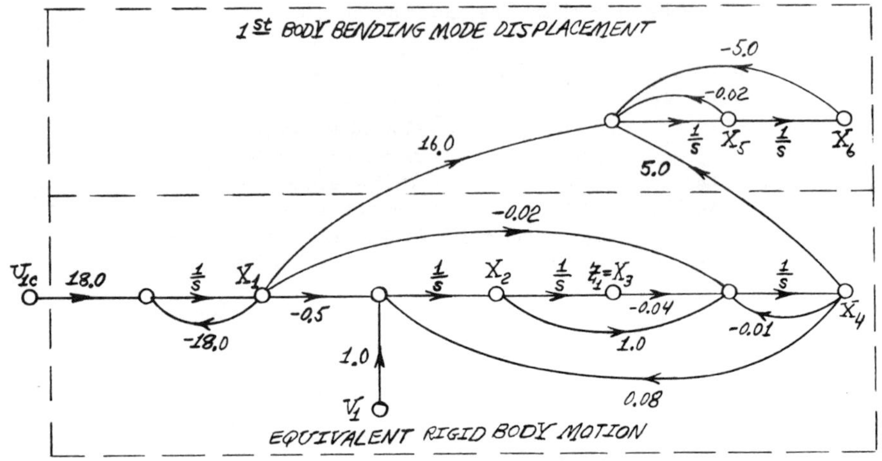

Figure 5-5 Flow graph of open-loop dynamics.

Primary design requirements for feedback control of a launch vehicle can be stated in terms of the desired transfer function of a command augmentation system. In particular, open-loop equations of motion yield the transfer function

$$\frac{Z_1}{U_{1c}} = \frac{-9.00(s + 0.0132)}{(s + 18.0)(s + 0.306)(s - 0.0411)(s - 0.255)}$$

corresponding to pitch attitude angle which is the primary response variable. Thus, every desired transfer function of the form

$$\frac{Z_1}{Z_{1c}} = \frac{-K}{(s + \alpha)(s^2 + 2\zeta\omega s + \omega^2)}$$

results in compatible design specifications. Typical design requirements correspond to $K = 10.0$, $\zeta = 0.8$, and $\omega = 1.0$. In addition, parameter α can then be selected so that the feedback gain around the actuator is suppressed using the technique described in Section 4.5 but without state augmentation. This procedure yields $\alpha = 16.4$ to three places of numerical accuracy.

Synthesis techniques given in Chapter 4 for transfer function matrices can now be applied directly. If feedback control is expressed as

$$u_{1c} = \tfrac{10}{9} z_{1c} - \mathbf{k}'\mathbf{x} \tag{5-84}$$

for these command augmentation systems, then feedback gains

$$\mathbf{k}' = [0.0001777 \quad -3.0356 \quad -1.8219 \quad -0.1599 \quad 0.0 \quad 0.0]$$

are obtained to four places of numerical accuracy. In addition, command augmentation systems corresponding to (5-84) result in transfer function

$$\frac{Z_1}{Z_{1c}} = \frac{-10.0(s + 0.0132)[s^2 + 2(0.00447)(2.24)s + (2.24)^2]}{(s + \alpha_1)(s^2 + 2\zeta_1\omega_1 s + \omega_1^2)(s + \alpha_2)(s^2 + 2\zeta_2\omega_2 s + \omega_2^2)},$$

and hence $\alpha_1 = 16.4$, $\zeta_1 = 0.8$, $\omega_1 = 1.0$ and $\alpha_2 = 0.0132$, $\zeta_2 = 0.00447$, $\omega_2 = 2.24$ are obtained for these design specifications.

The command augmentation system derived previously is not acceptable because the body bending mode has not been suitably damped. Lack of suitable damping gives rise to unacceptably large body bending mode displacements when the vehicle is subjected to air turbulence. Specifically, the standard deviation of x_6 for the previous design is 52.7 meters per unit covariance coefficient of v_1 in rad/sec^2 or equivalently 0.920 meters per unit covariance coefficient of v_1 in deg/sec^2. This measure of sensitivity to air turbulence can be reduced significantly in a direct fashion by introducing the following stochastic optimization problem.

A feedback gain matrix is introduced in accordance with (5-77) and 5-79), and matrix

$$\mathbf{H} = \begin{bmatrix} 0 & 1 & 0 & 0 & 0 & 0 \\ 0 & 0 & 1 & 0 & 0 & 0 \\ 0 & 0 & 0 & 1 & 0 & 0 \\ 0 & 0 & 0 & 0 & 1 & 0 \\ 0 & 0 & 0 & 0 & 0 & 1 \end{bmatrix}$$

is selected so that the feedback gain around the actuator is suppressed. In addition, a performance functional is introduced in accordance with (5-74) and (5-75), and matrices

$$\mathbf{E} = [0.9] \quad \text{and} \quad \mathbf{D} = [0.0 \quad -2.7320 \quad -1.6397 \quad -0.1439 \quad 0.0 \quad 0.0]$$

are selected so that feedback gain matrix

$$\mathbf{C} = [-3.0356 \quad -1.8219 \quad -0.1599 \quad 0.0 \quad 0.0]$$

is optimal in the absence of constraints. Finally, Cases 1 through 4 of optimal gain control are introduced by inequality constraints $x_{66} \leqslant 2772.72$, $x_{66} \leqslant 328.30$, $x_{66} \leqslant 94.56$, and $x_{66} \leqslant 47.28$ respectively.

An additional case of optimal gain control is introduced for the purposes of comparison. Specifically, Case 5 is defined as the linear optimal control system

Table 5-10–Optimal feedback gains for cases 1 through 5

Case	k_1	k_2	k_3	k_4	k_5	k_6	λ_1
1	0.0000	−3.0356	−1.8219	−0.1599	0.0000	0.0000	0.00000
2	0.0000	−3.0915	−1.8042	−0.1795	0.0083	0.0139	0.00067
3	0.0000	−3.4954	−1.7619	−0.2884	0.0457	0.1052	0.02440
4	0.0000	−10.0513	−3.4944	−1.2148	0.1486	1.3021	2.96559
5	0.1403	−4.8180	−3.0023	−0.4554	0.0178	0.0882	–

resulting from the performance integral

$$I = \int_0^\infty \{0.01(30x_3 + x_6)^2 + u_{1c}^2\}\, dt$$

which is used for deterministic design problems. This performance integral has been suggested [11] in the design of feedback controls for a similar flexible launch vehicle.

Feedback gains of the command augmentation systems corresponding to Cases 1 through 5 are summarized in Table 5-10 using notation defined by (5-84). These data illustrate that feedback gains k_5 and k_6 for Cases 1 through 4 are nonzero when the inequality constraint is effective and increase as λ_1 increases. Of course, no inequality constraint is imposed in Case 5. Table 5-11 summarizes the performance of these five cases in terms of variances x_{33} and x_{66} and in terms of previously defined transfer-function parameters. These data illustrate that x_{33} increases as x_{66} decreases for Cases 1 through 4. In addition, pole placement and pole suppression become increasingly less accurate as x_{66} decreases for Cases 1 through 4. Cases 3 and 5 are roughly equivalent in terms

Table 5-11–Performance of cases 1 through 5

Case	x_{33}	x_{66}	α_1	ζ_1	ω_1	α_2	ζ_2	ω_2
1	0.3761	2772.72	16.40	0.800	1.001	0.132	0.0045	2.24
2	0.4206	328.30	16.22	0.801	0.999	0.0132	0.0442	2.24
3	0.6896	94.56	15.30	0.806	0.949	0.0141	0.2494	2.38
4	2.2210	47.28	7.51	0.987	0.538	0.0346	0.6697	7.04
5	0.3017	175.39	18.00	0.809	1.192	0.0108	0.1310	2.36

Table 5-12–Convergence of the min–max method for Case 2 with $k_1 = 0.0$

Iteration	$-k_2$	$-k_3$	$-k_4$	$k_5 \times 10^2$	$k_6 \times 10$	$\lambda_1 \times 10^3$	x_{66}	f
0	3.0356	1.8219	0.1599	0.0000	0.0000	0.0000	2772.72	0.0000000
1	3.0409	1.8199	0.1619	0.0869	0.0135	0.0125	1485.05	0.0224693
2	3.0540	1.8182	0.1652	0.2348	0.0367	0.0495	845.59	0.0581526
3	3.0718	1.8152	0.1700	0.4491	0.0712	0.1486	532.66	0.1031015
4	3.0876	1.8108	0.1754	0.6753	0.1094	0.3496	389.12	0.1387731
5	3.0927	1.8062	0.1788	0.8054	0.1334	0.5776	337.42	0.1501867
6	3.0917	1.8044	0.1795	0.8299	0.1386	0.6657	328.56	0.1508315
7	3.0915	1.8042	0.1795	0.8305	0.1388	0.6709	328.30	0.1508317

of system performance. However, Case 5 has the principal disadvantage of a nonzero feedback gain around the actuator.

As a matter of interest, data appearing in Tables 5-10 and 5-11 were computed using the Newton–Raphson method of successive substitutions. Typical convergence characteristics of the min–max method for inequality constraints are illustrated in Table 5-12 for Case 2. Iterations given in Table 5-12 are initialized with the optimal feedback gains of Case 1. The constraint function corresponding to Cases 2 through 4 has only an indirect dependence on feedback gains, and this dependence is very nonlinear thereby causing rather slow convergence. Typical convergence characteristics of the method for solving linear optimal control problems defined by (4-118) through (4-121) are illustrated in Table 5-13. Iterations given in Table 5-13 are initialized with the optimal feedback gains of Case 4, and this initial point is a very poor approximation to the solution so that a relatively large number of iterations is required. Values of performance functional $f = \text{tr}\{\mathbf{X}\}$ where $\mathbf{X}$ is defined by (4-4) for Case 5 are

Table 5-13–Convergence of (4-118) through (4-121) for Case 5

Iteration	k_1	$-k_2$	$-k_3$	$-k_4$	k_5	k_6	f
0	0.0000	10.0513	3.4944	1.2148	0.14863	1.30214	112.49059
1	0.3625	7.8520	3.3032	1.0113	0.16117	0.60331	42.90454
2	0.2305	6.0808	3.0335	0.7455	0.08838	0.29773	32.37511
3	0.1747	5.2639	3.0007	0.5673	0.04670	0.15989	29.46692
4	0.1493	4.9261	3.0016	0.4828	0.02560	0.10549	28.65686
5	0.1413	4.8290	3.0022	0.4582	0.01867	0.08998	28.56549
6	0.1404	4.8182	3.0023	0.4555	0.01785	0.08824	28.56424
7	0.1403	4.8180	3.0023	0.4554	0.01784	0.08822	28.56426

given in Table 5-13 for the purposes of illustration. Tables 5-12 and 5-13 illustrate quadratic convergence of the Newton–Raphson method of successive substitutions in a suitably small neighborhood of the solution.

5.7 LINEAR OPTIMAL CONTROL

As in the case of deterministic design problems, perhaps the most significant case of optimal gain control occurs with $(\mathbf{E}'\mathbf{E})$ and $(\mathbf{HXH}')$ nonsingular for $\mathbf{H} = \mathbf{I}$. This special case of optimal gain control is called linear optimal control for stochastic design problems because optimal gains no longer depend on $\mathbf{B}$. Feedback gains are given by

$$\mathbf{C} = (\mathbf{E}'\mathbf{E})^{-1}(\mathbf{E}'\mathbf{D} + \mathbf{G}'\boldsymbol{\Lambda}) \tag{5-85}$$

for linear optimal control of stochastic systems where Lagrange multiplier matrix $\boldsymbol{\Lambda}$ is defined by

$$\mathbf{0} = (\mathbf{D}'\mathbf{D}) + \boldsymbol{\Lambda}\mathbf{F} + \mathbf{F}'\boldsymbol{\Lambda} + D(\mathbf{S}', \boldsymbol{\Lambda}) - \mathbf{C}'(\mathbf{E}'\mathbf{E})\mathbf{C}' \tag{5-86}$$

Except for the linear term $D(\mathbf{S}', \boldsymbol{\Lambda})$, this quadratic matrix equation is identical in form to (4-113).

In order to investigate existence and uniqueness of solutions to (5-85) and (5-86), these equations are further reduced to the form

$$\mathbf{0} = \mathbf{P} + \boldsymbol{\Lambda}\mathbf{Q} + \mathbf{Q}'\boldsymbol{\Lambda} + D(\mathbf{S}', \boldsymbol{\Lambda}) - \boldsymbol{\Lambda}\mathbf{R}\boldsymbol{\Lambda} \tag{5-87}$$

or equivalently to the form

$$\mathbf{0} = \mathbf{P} + \boldsymbol{\Lambda}\mathbf{A} + \mathbf{A}'\boldsymbol{\Lambda} + D(\mathbf{S}', \boldsymbol{\Lambda}) + \boldsymbol{\Lambda}\mathbf{R}\boldsymbol{\Lambda} \tag{5-88}$$

where

$$\mathbf{A} = \mathbf{Q} - \mathbf{R}\boldsymbol{\Lambda}. \tag{5-89}$$

Coefficient matrices appearing in (5-87) are taken to be arbitrary real matrices except that $\mathbf{P}$ and $\mathbf{R}$ are nominally assumed to be positive semi-definite and $\mathbf{S}_k \neq \mathbf{0}$ is assumed. Subsequent results given in this section are similar to results given in Reference [5].

Properties of the quadratic matrix equations given in (5-87) are stated here in terms of notation introduced in Section 4.9 for eigenvalue real parts and complex null spaces of matrices. In addition, matrix $\boldsymbol{\Delta}'$ denotes the coefficient matrix that corresponds to $[\boldsymbol{\Lambda}\mathbf{A} + \mathbf{A}'\boldsymbol{\Lambda} + D(\mathbf{S}', \boldsymbol{\Lambda})]$ when equations and variables are ordered as

$$\boldsymbol{\lambda}_v' = [\lambda_{11} \quad \sqrt{2}\lambda_{12} \quad \lambda_{22} \quad \sqrt{2}\lambda_{13} \quad \ldots].$$

For the purpose of stating pertinent conditions in a convenient fashion, the following definition also is introduced as

Definition 5-1 The ordered pair $\{\mathbf{\Pi}, \mathbf{Q}\}$ is *completely observable* if and only if

$$I(\mathbf{\Pi}, \mathbf{Q}) = \int_0^T (\mathbf{\Pi} e^{\mathbf{Q}t})'(\mathbf{\Pi} e^{\mathbf{Q}t})\, dt \tag{5-90}$$

is nonsingular for some $T > 0$.

For instance, the requirement that $\{\mathbf{\Pi}, \mathbf{Q}\}$ be completely observable is imposed subsequently and implies that no eigenvector of $\mathbf{Q}$ belongs to $N(\mathbf{P})$ where

$$\mathbf{P} = \mathbf{\Pi}'\mathbf{\Pi}. \tag{5-91}$$

This property follows from

$$\mathbf{m}^{\dagger} I(\mathbf{\Pi}, \mathbf{Q})\mathbf{m} = (\lambda\lambda^*)\mathbf{m}^{\dagger}\mathbf{P}\mathbf{m} \int_0^T e^{(\lambda + \lambda^*)t}\, dt$$

where $\mathbf{m}$ is an eigenvector of $\mathbf{Q}$ and λ is the corresponding eigenvalue of $\mathbf{Q}$.

Complete observability is of course a well known property as is the equivalence

Proposition 5-1 *Integral* $I(\mathbf{\Pi}, \mathbf{Q})$ *is nonsingular if and only if matrix*

$$\Gamma(\mathbf{Q}', \mathbf{\Pi}') = [\mathbf{\Pi}' \quad (\mathbf{\Pi}\mathbf{Q})' \ldots (\mathbf{\Pi}\mathbf{Q}^{n-1})'] \tag{5-92}$$

has rank n where dim $\mathbf{Q} = n \times n$.

On the other hand, the subsequent result [5] is useful and is not encountered frequently. This result is stated as

Lemma 5-1 *Let* $\mathbf{D}$ *and* $\mathbf{E}$ *be arbitrary matrices of suitable dimensions and let*

$$\mathbf{H}\mathbf{H}' = \mathbf{D}\mathbf{D}' + \mathbf{G}\mathbf{G}'.$$

Then the rank of $\Gamma(\mathbf{F} + \mathbf{D}\mathbf{E}, \mathbf{H})$ *is not less than the rank of* $\Gamma(\mathbf{F}, \mathbf{G})$.

Proof First note that $\mathbf{f} \in N(\mathbf{H}')$ implies that $\mathbf{f} \in N(\mathbf{D}')$ and $\mathbf{f} \in N(\mathbf{G}')$ and hence

$$N(\mathbf{H}') \subset N(\mathbf{D}') \cap N(\mathbf{G}'). \tag{5-93}$$

If relationships of the form $N_{\perp}(\mathbf{H}') = R(\mathbf{H})$ are used for range spaces in accordance with Lemma 2-2, then the orthogonal complement of (5-93) yields

$$R(\mathbf{H}) \supset R(\mathbf{D}) \cup R(\mathbf{G}).$$

Therefore, the relationship $R(\mathbf{H}) \supset R(\mathbf{G})$ holds and hence

$$R[\Gamma(\mathbf{F}, \mathbf{H})] \supset R[\Gamma(\mathbf{F}, \mathbf{G})] \tag{5-94}$$

also holds. Moreover, relationships

$$R(\mathbf{H}) \supset R(\mathbf{D}) \supset R(\mathbf{DE})$$

hold so that, given any $\mathbf{h} \in \mathscr{R}^n$, there exists a vector $\mathbf{j} \in \mathscr{R}^m$ such that $\mathbf{Hj} = (\mathbf{DE})\mathbf{h}$. In other words, a matrix $\mathbf{J}$ exists such that

$$\mathbf{HJ} = (\mathbf{DE})\mathbf{H},$$

and hence elementary column operations on $\Gamma(\mathbf{F} + \mathbf{DE}, \mathbf{H})$ can be used to demonstrate that

$$R[\Gamma(\mathbf{F} + \mathbf{DE}, \mathbf{H})] = R[\Gamma(\mathbf{F}, \mathbf{H})]. \tag{5-95}$$

Relationships (5-94) and (5-95) combine to yield

$$R[\Gamma(\mathbf{F} + \mathbf{DE}, \mathbf{H})] \supset R[\Gamma(\mathbf{F}, \mathbf{G})]$$

and hence the rank of $\Gamma(\mathbf{F} + \mathbf{DE}, \mathbf{H})$ cannot be less than the rank of $\Gamma(\mathbf{F}, \mathbf{G})$.

A corollary result of Lemma 5-1, which is used frequently in the sequel, is stated as

Lemma 5-2 *Let* $\mathbf{P}$ *and* $\mathbf{R}$ *be positive semi-definite and let* $\mathbf{\Pi}$ *be defined by* (*5-91*). *Then the matrix integral*

$$J(\mathbf{\Lambda}) = \int_0^T \{e^{(\mathbf{Q}-\mathbf{R\Lambda})'t}[\mathbf{P} + \mathbf{\Lambda R \Lambda}]\, e^{(\mathbf{Q}-\mathbf{R\Lambda})t}\}\, dt \tag{5-96}$$

with $T > 0$ *is nonsingular for all real symmetric* $\mathbf{\Lambda}$ *when* $\{\mathbf{\Pi}, \mathbf{Q}\}$ *is completely observable.*

Proof The result follows directly from Proposition 5-1 and Lemma 5-1 using $\mathbf{D} = \mathbf{\Lambda R}^{1/2}$, $\mathbf{F} = -\mathbf{R}^{1/2}$, $\mathbf{F} = \mathbf{Q}'$, and $\mathbf{G} = \mathbf{\Pi}'$.

In other words, if matrix $\mathbf{\Phi}$ is defined by

$$\mathbf{\Phi}'\mathbf{\Phi} = \mathbf{P} + \mathbf{\Lambda R \Lambda},$$

then $\{\mathbf{\Phi}, \mathbf{Q} - \mathbf{R\Lambda}\}$ is completely observable for all real symmetric $\mathbf{\Lambda}$ when $\{\mathbf{\Pi}, \mathbf{Q}\}$ is completely observable.

The main result given in this section concerning solutions to (5-87) can now be proved in a form which is similar to Lemma 4-7. This result is stated as

Theorem 5-6 *Suppose*

i) $\mathbf{P}$ *and* $\mathbf{R}$ *are positive semi-definite,*

ii) $\{\mathbf{\Pi}, \mathbf{Q}\}$ *is completely observable,*
and

iii) there exists a positive definite matrix $\mathbf{Y}$ *for which the equivalent coefficient matrix of*

$$[\mathbf{\Lambda}(\mathbf{Q}-\mathbf{RY})+(\mathbf{Q}-\mathbf{RY})'\mathbf{\Lambda}+D(\mathbf{S}',\mathbf{\Lambda})]$$

is asymptotically stable.
Then there exists one and only one positive definite solution to (5-87).

Proof Existence of a positive definite solution is proved by introducing sequences defined by

$$\mathbf{0}=\mathbf{T}_i+\mathbf{P}+\mathbf{\Lambda}_i\mathbf{Q}+\mathbf{Q}'\mathbf{\Lambda}_i+D(\mathbf{S}',\mathbf{\Lambda}_i)-\mathbf{\Lambda}_i\mathbf{R}\mathbf{\Lambda}_i \tag{5-97}$$

$$\mathbf{A}_i=\mathbf{Q}-\mathbf{R}\mathbf{\Lambda}_i \tag{5-98}$$

$$\mathbf{\Sigma}_i\mathbf{A}_i+\mathbf{A}_i'\mathbf{\Sigma}_i+D(\mathbf{S}',\mathbf{\Sigma}_i)=\mathbf{T}_i \tag{5-99}$$

and

$$\mathbf{\Lambda}_{i+1}=\mathbf{\Lambda}_i+\mathbf{\Sigma}_i \tag{5-100}$$

for each i. In addition, matrix $\mathbf{\Delta}_i'$ denotes the equivalent coefficient matrix of (5-99). First, assume there exists a positive definite $\mathbf{\Lambda}_i$ for which $\mathbf{T}_i$ is positive semi-definite. Because $\{\mathbf{\Pi}, \mathbf{Q}\}$ is completely observable, no eigenvector of $\mathbf{Q}$ belongs to $N(\mathbf{P})$ and hence no eigenvector of $\mathbf{Q}$ belongs to $N[\mathbf{P}+\mathbf{T}_i+D(\mathbf{S}',\mathbf{\Lambda}_i)]$. Matrix $\mathbf{A}_i$ thus is asymptotically stable by Lemma 4-5. Moreover, matrix $\mathbf{\Delta}_i$ is also asymptotically stable by Theorem 5-3 because complete observability of $\{\mathbf{\Pi}, \mathbf{Q}\}$ implies

$$\int_0^\infty \{e^{\mathbf{A}_i't}(\mathbf{P}+\mathbf{T}_i+\mathbf{\Lambda}_i\mathbf{R}\mathbf{\Lambda}_i)\,e^{\mathbf{A}_it}\}\,dt$$

is nonsingular by Lemma 5-2. Elimination of $\mathbf{T}_i$ from (5-97) and (5-99) now yields

$$\mathbf{\Lambda}_{i+1}\mathbf{A}_i+\mathbf{A}_i'\mathbf{\Lambda}_{i+1}+D(\mathbf{S}',\mathbf{\Lambda}_{i+1})+\mathbf{P}+\mathbf{\Lambda}_i\mathbf{R}\mathbf{\Lambda}_i=\mathbf{0}.$$

Matrices $\mathbf{\Sigma}_i$ and $\mathbf{\Lambda}_{i+1}$ then are seen to be negative and positive semi-definite respectively by Corollary 5-1. Complete observability of $\{\mathbf{\Pi}, \mathbf{Q}\}$ and Lemma 5-2 now imply that

$$\mathbf{\Lambda}_{i+1}=\int_0^\infty \{e^{\mathbf{A}_i't}[\mathbf{P}+\mathbf{\Lambda}_i\mathbf{R}\mathbf{\Lambda}_i+D(\mathbf{S}',\mathbf{\Lambda}_{i+1})]\,e^{\mathbf{A}_it}\}\,dt$$

is nonsingular. Manipulation of (5-97) through (5-100) finally yields (4-122) and hence $\mathbf{T}_{i+1}$ is positive semi-definite. In other words, limit

$$\mathbf{\Lambda} = \lim_{i\to\infty} \mathbf{\Lambda}_i$$

exists by Proposition 4-4 under the previous assumption, and this limit is a positive definite solution of (5-88) for which the corresponding coefficient matrix $\mathbf{\Delta}$ is asymptotically stable. Now, given any positive definite $\mathbf{Y}$ for which the equivalent coefficient matrix of $[\mathbf{\Lambda}(\mathbf{Q} - \mathbf{RY}) + (\mathbf{Q} - \mathbf{RY})'\mathbf{\Lambda} + D(\mathbf{S}', \mathbf{\Lambda})]$ is asymptotically stable, let $\mathbf{W}$ be any positive semi-definite matrix such that

$$[\mathbf{W} - (\mathbf{P} + \mathbf{YQ} + \mathbf{Q}'\mathbf{Y} + D(\mathbf{S}', \mathbf{Y}) - \mathbf{YRY})]$$

is also positive semi-definite. In addition, let $\mathbf{Z}$ be the unique positive semi-definite solution to

$$\mathbf{Z}(\mathbf{Q} - \mathbf{RY}) + (\mathbf{Q} - \mathbf{RY})'\mathbf{Z} + D(\mathbf{S}', \mathbf{Z}) + \mathbf{W} = \mathbf{0}$$

in accordance with Corollary 5-1. Finally, let

$$\mathbf{\Lambda}_0 = \mathbf{Y} + \mathbf{Z}$$

so that (5-97) yields

$$\mathbf{T}_0 = [\mathbf{W} - (\mathbf{P} + \mathbf{YQ} + \mathbf{Q}'\mathbf{Y} + D(\mathbf{S}', \mathbf{Y}) - \mathbf{YRY})] + \mathbf{ZRZ}.$$

Matrices $\mathbf{\Lambda}_0$ and $\mathbf{T}_0$ thus are positive definite and positive semi-definite respectively, thereby establishing existence. In order to establish uniqueness, let $\mathbf{\Lambda}_1$ and $\mathbf{\Lambda}_2$ be any two distinct positive definite solutions of (5-88). Then $\mathbf{A}_1$ and $\mathbf{A}_2$ are asymptotically stable by Lemma 4-5, and equivalent coefficient matrices $\mathbf{\Delta}_1$ and $\mathbf{\Delta}_2$ are also asymptotically stable by Theorem 5-3. Now let

$$\mathbf{Z} = \mathbf{\Lambda}_2 - \mathbf{\Lambda}_1$$

so that (5-88) can be used to eliminate $\mathbf{P}$, thereby obtaining

$$\mathbf{ZA}_1 + \mathbf{A}_1'\mathbf{Z} + D(\mathbf{S}', \mathbf{Z}) - \mathbf{W} = \mathbf{ZA}_2 + \mathbf{A}_2'\mathbf{Z} + D(\mathbf{S}', \mathbf{Z}) + \mathbf{W} = \mathbf{0}$$

where

$$\mathbf{W} = \mathbf{ZRZ}.$$

Because $\mathbf{W}$ is positive semi-definite, matrix $\mathbf{Z}$ is both positive and negative semi-definite by Corollary 5-1 and hence $\mathbf{Z} = \mathbf{0}$.

Assumptions imposed in Theorem 5-6 should be noted to imply the assumptions imposed in Theorem 4-13. Specifically, a positive semi-definite solution to (5-87) exists by the theorem just proved, and complete observability of $\{\mathbf{\Pi}, \mathbf{Q}\}$ of course implies that no eigenvector of $\mathbf{Q}$ belongs to $N(\mathbf{P})$. Lemma 4-5 therefore holds, and a consequence of this lemma was found to be that eigenvectors of $\mathbf{Q}'$ belonging to $N(\mathbf{R})$ correspond to eigenvalues with negative real parts. In other words, the assumptions of Theorem 5-6 imply $\{\mathbf{P}, \mathbf{Q}, \mathbf{R}\}$ is regular.

Equations (5-97) through (5-100) which are used in the proof of Theorem 5-6 of course constitute a method of solving (5-87) by successive substitutions. These equations in fact correspond to the Newton method of solving nonlinear algebraic equations and hence exhibit quadratic convergence in a suitably small neighborhood of a solution to (5-87). The primary difficulties of this computational method stem from the solution of (5-99) on each iteration and from the selection of a positive semi-definite $\mathbf{\Lambda}_0$ such that $\mathbf{\Delta}_0$ is asymptotically stable. An alternate method is based on the repetitive application of (4-129) and is stated as

Theorem 5-7 *Suppose the assumptions of Theorem 5-6 hold. Then* $\mathbf{\Lambda} = \mathbf{\Lambda}_\infty$ *is the unique positive definite solution of (5-87) where* $\mathbf{\Lambda}_\infty$ *is the limit of the positive semi-definite sequence* $\{\mathbf{\Lambda}_i\}$ *defined by*

$$\mathbf{0} = \mathbf{P} + D(\mathbf{S}', \mathbf{\Lambda}_i) + \mathbf{\Lambda}_{i+1}\mathbf{Q} + \mathbf{Q}'\mathbf{\Lambda}_{i+1} - \mathbf{\Lambda}_{i+1}\mathbf{R}\mathbf{\Lambda}_{i+1}; \quad \mathbf{\Lambda}_0 = \mathbf{0}. \tag{5-101}$$

Proof Let $\mathbf{\Lambda}$ be the positive definite solution of (5-88) so that correspondingly $\mathbf{A}$ is asymptotically stable. Manipulation of (5-88) and (5-101) then yields

$$\mathbf{\Lambda} - \mathbf{\Lambda}_{i+1} = \int_0^\infty \{e^{\mathbf{A}'t}[D(\mathbf{S}', \mathbf{\Lambda} - \mathbf{\Lambda}_i) + (\mathbf{\Lambda} - \mathbf{\Lambda}_{i+1})\mathbf{R}(\mathbf{\Lambda} - \mathbf{\Lambda}_{i+1})]\, e^{\mathbf{A}t}\}\, dt$$

and hence $(\mathbf{\Lambda} - \mathbf{\Lambda}_{i+1})$ is positive semi-definite when $(\mathbf{\Lambda} - \mathbf{\Lambda}_i)$ is positive semi-definite. In addition, if $\mathbf{\Lambda}_i$ is positive semi-definite, then $\{\mathbf{P} + D(\mathbf{S}', \mathbf{\Lambda}_i), \mathbf{Q}, \mathbf{R}\}$ is regular so that $\mathbf{\Lambda}_{i+1}$ is positive semi-definite and

$$\mathbf{A}_{i+1} = \mathbf{Q} - \mathbf{R}\mathbf{\Lambda}_{i+1}$$

is asymptotically stable. Manipulation of (5-101) now yields

$$\mathbf{\Lambda}_{i+1} - \mathbf{\Lambda}_i = \int_0^\infty \{e^{\mathbf{A}'_{i+1}t}[D(\mathbf{S}', \mathbf{\Lambda}_i - \mathbf{\Lambda}_{i-1})$$
$$+ (\mathbf{\Lambda}_{i+1} - \mathbf{\Lambda}_i)\mathbf{R}(\mathbf{\Lambda}_{i+1} - \mathbf{\Lambda}_i)]\, e^{\mathbf{A}_{i+1}t}\}\, dt$$

so that $(\mathbf{\Lambda}_{i+1} - \mathbf{\Lambda}_i)$ is positive semi-definite when $(\mathbf{\Lambda}_i - \mathbf{\Lambda}_{i-1})$ is positive semi-definite. Thus, by Theorems 4-13 and 5-6, initialization $\mathbf{\Lambda}_0 = \mathbf{0}$ results in monotone convergence of the positive semi-definite solution to (5-101) in accordance with Proposition 4-4. In addition, complete observability of $\{\mathbf{\Pi}, \mathbf{Q}\}$ implies that no eigenvector of $\mathbf{Q}$ belongs to $N[\mathbf{P} + D(\mathbf{S}', \mathbf{\Lambda}_i)]$ and hence $\mathbf{\Lambda}_i$ is positive definite for each $i > 0$ by Lemma 4-5. Limit $\mathbf{\Lambda}_\infty$ thus is also a positive definite solution to (5-87) and $\mathbf{\Lambda} = \mathbf{\Lambda}_\infty$ by Theorem 5-6.

Convergence of the positive definite solutions of (5-101) to the positive definite solution of (5-87) unfortunately is not quadratic. Convergence characteristics of (5-101), however, resemble those of (5-33) as might be expected because these equations are identical when $\mathbf{R} = \mathbf{0}$.

5.8 OPTIMAL LINEAR FILTERS AND PREDICTORS

Design problems involving the optimal filtering of noise corrupted data often arise in the context of control. In addition, linear filters are of course frequently used for this purpose. The formulation of optimal linear filters thus is a significant problem to be considered and also constitutes a comprehensive illustration of many topics presented in Chapters 4 and 5. The formulation presented here is based on Reference [12] and results in the so-called Wiener–Kalman–Bucy filter [13, 14] which can be derived in a much more general context.

Stochastic signals considered here are represented in linear constant-coefficient state-equation form and are given the following interpretation. Stochastic state equations

$$d\mathbf{x} = (\mathbf{F}\mathbf{x})\,dt + \mathbf{G}\,d\mathbf{u}; \quad \mathbf{x}(0) = \mathbf{x}_0 \tag{5-102}$$

represent a *plant* which has a Gaussian initial state $\mathbf{x}_0$ and a Brownian-motion forcing function $\mathbf{u}$. An optimal estimate of plant state vector $\mathbf{x}$ is sought as a function of time by observing response vector $\mathbf{y}$. Stochastic response equations

$$\mathbf{y}\,dt = (\mathbf{H}\mathbf{x})\,dt + \mathbf{J}\,d\boldsymbol{\mu} \tag{5-103}$$

represent an *observer* which includes an additive Brownian-motion noise $\boldsymbol{\mu}$. Plant and observer equations are also depicted in Figure 5-6 in terms of white-noise representations $\mathbf{v}$ and $\boldsymbol{\nu}$ which are defined by

$$\mathbf{u}(t) = \int_0^t d\mathbf{v}(\xi) \quad \text{and} \quad \boldsymbol{\mu}(t) = \int_0^t d\boldsymbol{\nu}(\xi)$$

respectively.

A linear *filter* is introduced in terms of stochastic state equations

$$d\mathbf{z} = (\mathbf{A}\mathbf{z} + \mathbf{C}\mathbf{y})\,dt; \quad \mathbf{z}(0) = \mathbf{z}_0. \tag{5-104}$$

This filter is constrained by $\dim \mathbf{z} = \dim \mathbf{x}$ so that error vector

$$\mathbf{e} = \mathbf{x} - \mathbf{z} \tag{5-105}$$

can be introduced. In other words, each element of filter state vector $\mathbf{z}$ is intended to be an optimal estimate of the corresponding element of plant state vector $\mathbf{x}$.

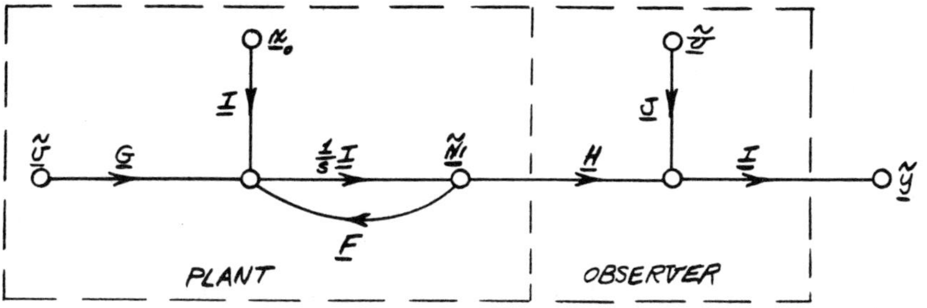

Figure 5-6 Flow graph of plant and observer equations.

The filter is further constrained by requiring that the optimal estimate be unbiased. That is, condition $E\{\mathbf{z}(t)\} = E\{\mathbf{x}(t)\}$ is imposed for all t assuming that

$$E\{\mathbf{u}(t)\} = \mathbf{0} \quad \text{and} \quad E\{\boldsymbol{\mu}(t)\} = \mathbf{0} \tag{5-106}$$

hold for all t. Notation

$$\bar{\mathbf{x}}(t) = E\{\mathbf{x}(t)\} \quad \text{and} \quad \bar{\mathbf{e}}(t) = E\{\mathbf{e}(t)\} \tag{5-107}$$

is introduced first for convenience in determining conditions which result in an unbiased estimate. Use of (5-102) then yields

$$\mathrm{d}\bar{\mathbf{x}} = E\{\mathrm{d}\mathbf{x}\} = (\mathbf{F}\bar{\mathbf{x}})\,\mathrm{d}t$$

or rather

$$\frac{\mathrm{d}\bar{\mathbf{x}}}{\mathrm{d}t} = \mathbf{F}\bar{\mathbf{x}}; \quad \bar{\mathbf{x}}(0) = E\{\mathbf{x}_0\}$$

so that

$$\bar{\mathbf{x}}(t) = \mathrm{e}^{\mathbf{F}t}E\{\mathbf{x}_0\}. \tag{5-108}$$

In a similar fashion, use of (5-102), (5-103), and (5-104) yields

$$\mathrm{d}\bar{\mathbf{e}} = E\{\mathrm{d}\mathbf{e}\} = [\mathbf{A}\bar{\mathbf{e}} + (\mathbf{F} - \mathbf{CH} - \mathbf{A})\bar{\mathbf{x}}]\,\mathrm{d}t$$

or rather

$$\frac{\mathrm{d}\bar{\mathbf{e}}}{\mathrm{d}t} = \mathbf{A}\bar{\mathbf{e}} + (\mathbf{F} - \mathbf{CH} - \mathbf{A})\,\mathrm{e}^{\mathbf{F}t}E\{\mathbf{x}_0\}; \quad \bar{\mathbf{e}}(\mathbf{0}) = E\{\mathbf{x}_0\} - \mathbf{z}_0. \tag{5-109}$$

Because the filter must be initialized for each application, initial state $\mathbf{z}_0$ is deterministic and hence $E\{\mathbf{z}_0\} = \mathbf{z}_0$ has been used in the derivation of (5-109). Conditions

$$\mathbf{A} = \mathbf{F} - \mathbf{CH} \tag{5-110}$$

and

$$\mathbf{z}_0 = E\{\mathbf{x}_0\} \tag{5-111}$$

thus are seen to result in $\bar{\mathbf{e}} = \mathbf{0}$ and hence $E\{\mathbf{z}\} = \bar{\mathbf{x}}$ for all t given any $E\{\mathbf{x}_0\}$. The class of filters corresponding to these relationships is depicted in Figure 5-7. Gain matrix $\mathbf{C}$ must now be specified so that $\mathbf{z}$ is the optimal unbiased estimate of $\mathbf{x}$.

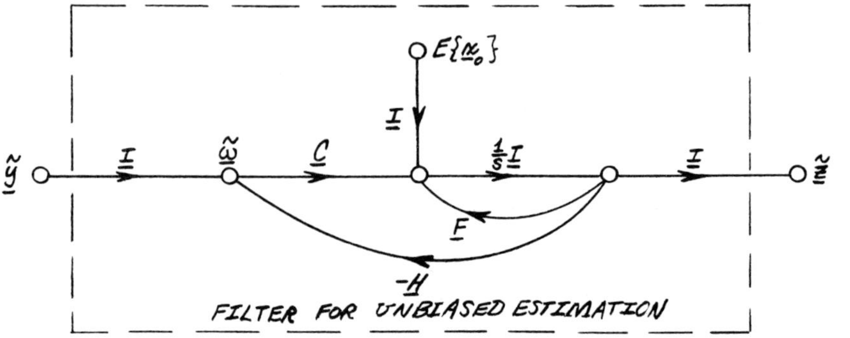

Figure 5-7 Flow graph of filter for unbiased estimation.

The class of linear filters depicted in Figure 5-7 results in a zero-mean error vector $\mathbf{e}$ which is seen to satisfy the stochastic state equation

$$\mathrm{d}\mathbf{e} = (\mathbf{A}\mathbf{e})\,\mathrm{d}t + [\mathbf{G} \quad -\mathbf{CJ}] \begin{bmatrix} \mathrm{d}\mathbf{u} \\ \mathrm{d}\boldsymbol{\mu} \end{bmatrix}; \quad \mathbf{e}(0) = \mathbf{x}_0 - E\{\mathbf{x}_0\}. \tag{5-112}$$

Brownian motion forcing functions are now assumed to have the positive semi-definite covariance matrix

$$E\left\{\begin{bmatrix} \mathbf{u}(t) \\ \boldsymbol{\mu}(t) \end{bmatrix} \begin{bmatrix} \mathbf{u}(\tau) \\ \boldsymbol{\mu}(\tau) \end{bmatrix}'\right\} = \begin{bmatrix} \mathbf{I} & \mathbf{K}' \\ \mathbf{K} & \mathbf{I} \end{bmatrix} \min\,\{t, \tau\}. \tag{5-113}$$

In addition, the performance functional adopted here for optimal filter design is taken to be

$$f = \lim_{t \to \infty} E\{\mathbf{e}'(t)\mathbf{e}(t)\}$$

under the assumption that $\mathbf{A}$ is asymptotically stable. A more convenient form of this performance functional is

$$f = \mathrm{tr}\{\mathbf{\Sigma}\} \tag{5-114}$$

where

$$\mathbf{\Sigma} = \lim_{t\to\infty} E\{\mathbf{e}(t)\mathbf{e}'(t)\}. \tag{5-115}$$

Covariance matrix $\mathbf{\Sigma}$ then is found to satisfy

$$\mathbf{\Sigma A}' + \mathbf{A\Sigma} + [\mathbf{G} \quad -\mathbf{CJ}] \begin{bmatrix} \mathbf{I} & \mathbf{K}' \\ \mathbf{K} & \mathbf{I} \end{bmatrix} [\mathbf{G} \quad -\mathbf{CJ}]' = \mathbf{0} \tag{5-116}$$

by the methods used to derive (5-12).

Optimal gains can now be found in a direct fashion by the introduction of Lagrangian

$$L = \mathrm{tr}\{\mathbf{\Sigma}\} + \mathrm{tr}\{\mathbf{\Lambda}'\mathbf{Z}\}$$

where

$$\mathbf{Z} = \mathbf{\Sigma A}' + \mathbf{A\Sigma} + [\mathbf{G} \quad -\mathbf{CJ}] \begin{bmatrix} \mathbf{I} & \mathbf{K}' \\ \mathbf{K} & \mathbf{I} \end{bmatrix} [\mathbf{G} \quad -\mathbf{CJ}]'.$$

Stationarity conditions $L_{\mathbf{\Lambda}} = \mathbf{0}$, $L_{\mathbf{\Sigma}} = \mathbf{0}$ and $L_{\mathbf{C}} = \mathbf{0}$ then yield (5-116),

$$\mathbf{\Lambda A} + \mathbf{A}'\mathbf{\Lambda} + \mathbf{I} = \mathbf{0},$$

and

$$2\mathbf{\Lambda}[\mathbf{C}(\mathbf{JJ}') - (\mathbf{GK}'\mathbf{J}' + \mathbf{\Sigma H}')] = \mathbf{0}$$

respectively. Lagrange multiplier matrix $\mathbf{\Lambda}$ is seen to be nonsingular when $\mathbf{A}$ is assumed to be asymptotically stable so that optimal gains are given by

$$\mathbf{C} = [\mathbf{G}(\mathbf{JK})' + \mathbf{\Sigma H}'](\mathbf{JJ}')^{-1} \tag{5-117}$$

when $(\mathbf{JJ}')$ is also assumed to be nonsingular. In addition, equation (5-116) becomes

$$\mathbf{0} = \mathbf{GG}' + \mathbf{\Sigma F}' + \mathbf{F\Sigma} - \mathbf{C}(\mathbf{JJ}')\mathbf{C}' \tag{5-118}$$

when $\mathbf{C}$ is optimal.

Existence and uniqueness of a positive semi-definite solution $\mathbf{\Sigma}$ to (5-117) and (5-118) can now be established directly from the results given in Section 4-9. Substitution of (5-117) into (5-118) yields

$$\mathbf{0} = \mathbf{R} + \mathbf{\Sigma Q}' + \mathbf{Q\Sigma} - \mathbf{\Sigma P\Sigma} \tag{5-119}$$

where

$$\mathbf{R} = \mathbf{G}[\mathbf{I} - (\mathbf{JK})'(\mathbf{JJ}')^{-1}(\mathbf{JK})]\mathbf{G}', \tag{5-120}$$

$$\mathbf{Q} = \mathbf{F} - \mathbf{G}(\mathbf{JK})'(\mathbf{JJ}')^{-1}\mathbf{H}, \tag{5-121}$$

and

$$\mathbf{P} = \mathbf{H}'(\mathbf{JJ}')^{-1}\mathbf{H}. \tag{5-122}$$

In accordance with Theorem 4-13, a unique positive semi-definite solution to (5-119) exists when $\{\mathbf{R}, \mathbf{Q}', \mathbf{P}\}$ is regular. In other words, gains given in (5-117) are optimal and result in a positive semi-definite covariance matrix $\boldsymbol{\Sigma}$ when $\{\mathbf{P}, \mathbf{Q}, \mathbf{R}\}$ is regular. Matrices $\mathbf{P}$ and $\mathbf{R}$ exist as positive semi-definite matrices when $(\mathbf{JJ}')$ is nonsingular and the coefficient matrix appearing in (5-113) is positive semi-definite. Nevertheless, regularity may not be achieved even if $\mathbf{F}$ is assumed to be asymptotically stable. The special and frequently encountered case with $\mathbf{K} = \mathbf{0}$ and $\mathbf{F}$ asymptotically stable results in a regular $\{\mathbf{P}, \mathbf{Q}, \mathbf{R}\}$. Optimality of $\mathbf{C}$ is then assured and covariance matrix $\boldsymbol{\Sigma}$ is positive definite. Computational methods given in Section 4.9 also apply directly to (5-119) under these conditions.

The optimal filter depicted in Figure 5-7 is the Kalman–Bucy realization of the Wiener filter corresponding to assumptions imposed here. The Kalman–Bucy realization is often associated with the alternate performance functional

$$f = \lim_{t\to\infty} E\{[\mathbf{He}(t)]'[\mathbf{He}(t)]\}. \tag{5-123}$$

Minimization of this performance functional also results in (5-117) and (5-118) providing that the corresponding Lagrange multiplier matrix $\boldsymbol{\Lambda}$ is nonsingular. This matrix is now given by

$$\boldsymbol{\Lambda}\mathbf{A} + \mathbf{A}'\boldsymbol{\Lambda} + \mathbf{H}'\mathbf{H} = \mathbf{0}$$

and hence is nonsingular if and only if $\{\mathbf{H}, \mathbf{A}\}$ is completely observable. Moreover, use of (5-110) and elementary column operations reveal the range space equivalence $R[\Gamma(\mathbf{F}', \mathbf{H}')] = R[\Gamma(\mathbf{A}', \mathbf{H}')]$, and hence $\boldsymbol{\Lambda}$ is nonsingular if and only if $\{\mathbf{H}, \mathbf{F}\}$ is completely observable. Complete observability of the plant is assumed in most formulations of the optimal filtering problem.

Signal $\boldsymbol{\omega}$, defined by Figure 5-7 for unbiased estimators, is found to satisfy

$$\mathbf{He} = \boldsymbol{\omega} - \mathbf{J}\boldsymbol{\upsilon}. \tag{5-124}$$

When $\{\mathbf{H}, \mathbf{F}\}$ is completely observable, optimal gains result from the minimization of (5-123) and hence $\boldsymbol{\omega}$ is often regarded to be the optimal unbiased estimate of observer noise $\mathbf{J}\boldsymbol{\upsilon}$.

Optimal linear predictors can also be deduced in an elementary fashion from the previous derivation of optimal filters. Given some $T \geqslant 0$ and any value of real time t, suppose an optimal estimate $\boldsymbol{\zeta}(t)$ of stochastic state $\mathbf{x}(t+T)$ is sought. If $\dim \boldsymbol{\zeta} = \dim \mathbf{x}$ is assumed, then error vector

$$\boldsymbol{\epsilon}(t) = \mathbf{x}(t+T) - \boldsymbol{\zeta}(t) \tag{5-125}$$

can be defined and optimal prediction corresponds to the minimization of performance functional

$$f = \lim_{t \to \infty} E\{\boldsymbol{\epsilon}'(t)\boldsymbol{\epsilon}(t)\}. \tag{5-126}$$

In addition, if only unbiased estimators are considered, the condition $E\{\boldsymbol{\epsilon}(t)\} = \mathbf{0}$ implies that $E\{\boldsymbol{\zeta}(t)\} = E\{\mathbf{x}(t+T)\}$ must hold. Moreover, stochastic state vector $\mathbf{x}(t+T)$ is given by

$$\mathbf{x}(t+T) = \mathrm{e}^{\mathbf{F}T}\mathbf{x}(t) + \int_t^{t+T} \mathrm{e}^{(t+T-\xi)\mathbf{F}}\mathbf{G}\, \mathrm{d}\mathbf{u}(\xi),$$

and evaluation of the expectation $E\{\mathbf{x}(t+T)\}$ yields

$$E\{\boldsymbol{\zeta}(t)\} = \mathrm{e}^{\mathbf{F}T}E\{\mathbf{x}(t)\}$$

because Brownian-motion increments have zero mean. Linear filters constructed so that (5-110) and (5-111) hold have the property $E\{\mathbf{z}(t)\} = E\{\mathbf{x}(t)\}$. Therefore, no further generality is lost by selecting

$$\boldsymbol{\zeta}(t) = \mathrm{e}^{\mathbf{F}T}\mathbf{z}(t) \tag{5-127}$$

as the unbiased estimator of $\mathbf{x}(t+T)$, and the corresponding predictor is depicted in Figure 5-8.

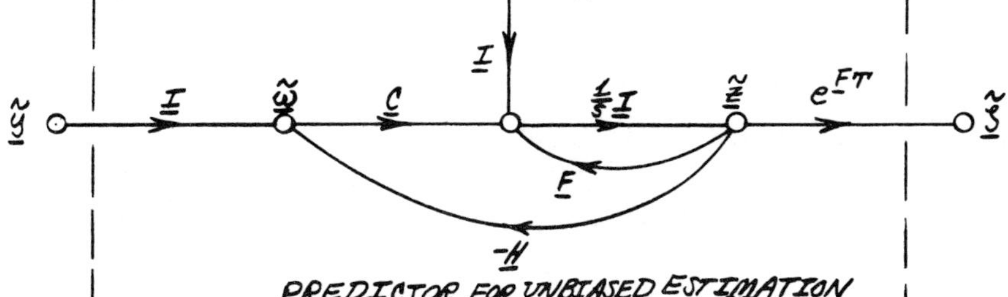

Figure 5-8 Flow graph of predictor for unbiased estimation.

The performance functional given in (5-126) can now be put in a more useful form noting error vector $\boldsymbol{\epsilon}$ becomes

$$\boldsymbol{\epsilon}(t) = e^{\mathbf{F}T}\mathbf{e}(t) + \boldsymbol{\chi}(t + T) \tag{5-128}$$

where stochastic state vector $\boldsymbol{\chi}$ satisfies

$$d\boldsymbol{\chi} = (\mathbf{F}\boldsymbol{\chi})\,dt + \mathbf{G}\,d\mathbf{u}; \quad \boldsymbol{\chi}(t) = \mathbf{0}. \tag{5-129}$$

Moreover, manipulation of (5-128) yields

$$E\{\boldsymbol{\epsilon}'(t)\boldsymbol{\epsilon}(t)\} = \mathrm{tr}[e^{\mathbf{F}T}E\{\mathbf{e}'(t)\}\, e^{\mathbf{F}'T} + 2e^{\mathbf{F}T}E\{\mathbf{e}(t)\boldsymbol{\chi}'(t + T)\} + \mathbf{E}(T)] \tag{5-130}$$

where definition

$$\mathbf{E}(T) = E\{\boldsymbol{\chi}(t + T)\boldsymbol{\chi}'(t + T)\}$$

has been introduced for convenience. This definition combines with (5-129) to yield

$$\mathbf{E}(T) = \int_0^T \{e^{\mathbf{F}\xi}\mathbf{G}\mathbf{G}'\, e^{\mathbf{F}'\xi}\}\, d\xi. \tag{5-131}$$

In addition, because $\mathbf{u}$ and $\boldsymbol{\mu}$ are zero-mean Brownian motions, increments $d\mathbf{u}(\xi)$ and $d\boldsymbol{\mu}(\xi)$ on $[t, t + T]$ are statistically independent of $\mathbf{e}(t)$ so that $E\{\mathbf{e}(t)\boldsymbol{\chi}'(t + T)\} = \mathbf{0}$ holds. The performance functional given in (5-126) thus becomes

$$f = \mathrm{tr}\{e^{\mathbf{F}T}\,\boldsymbol{\Sigma} e^{\mathbf{F}'T} + \mathbf{E}(T)\} \tag{5-132}$$

where covariance matrix $\boldsymbol{\Sigma}$ is defined by (5-115) and satisfies (5-116).

Minimization of the performance functional given in (5-132) again results in optimal gains being given by (5-117) and (5-118). However, the Lagrange multiplier matrix $\boldsymbol{\Lambda}$ is now given by

$$\boldsymbol{\Lambda}\mathbf{A} + \mathbf{A}'\boldsymbol{\Lambda} + e^{\mathbf{F}'T}e^{\mathbf{F}T} = \mathbf{0}$$

and hence is nonsingular for all finite $T \geqslant 0$. Differentiation of (5-132) with respect to T also yields

$$\frac{\partial f}{\partial T} = \mathrm{tr}\{e^{\mathbf{F}T}(\mathbf{F}\boldsymbol{\Sigma} + \boldsymbol{\Sigma}\mathbf{F}' + \mathbf{G}\mathbf{G}')\, e^{\mathbf{F}'T}\}.$$

Moreover, if gains are optimal in accordance with (5-117), then (5-118) can be used to obtain

$$\frac{\partial f}{\partial T} = \mathrm{tr}\,\{(e^{\mathbf{F}T}\mathbf{C}\mathbf{J})(e^{\mathbf{F}T}\mathbf{C}\mathbf{J})'\} \geqslant 0. \tag{5-133}$$

In other words, the minimum value of performance functional f is a non-decreasing function of prediction interval $T \geqslant 0$.

5.9 AIRCRAFT TRACKING EXAMPLE

An illustration of optimal filtering and prediction is given here in terms of an idealized example of aircraft tracking. Dynamics of the pilot, aircraft, and trajectory geometry are depicted in Figure 5-9 in linearized and normalized form. An optimal predictor of future aircraft position, namely $x_1(t+T)$ with $T > 0$, is sought. Design of this predictor can be accomplished using any combination of tracking radar, doppler radar, and accelerometer observations.

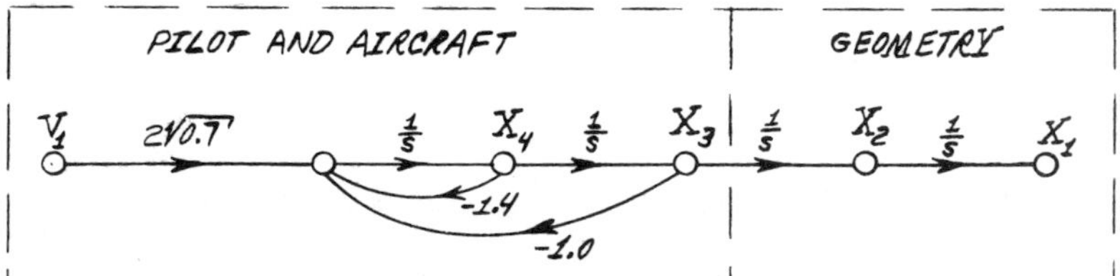

Figure 5-9 Flow graph of simplified dynamics for aircraft tracking.

Use of a tracking radar observation only results in the simplest predictor and hence is considered first as Case 1. Simplified dynamics of a tracking radar observation are depicted in Figure 5-10. Coefficient matrices corresponding to this plant and observer are

$$\mathbf{F} = \begin{bmatrix} 0 & 1 & 0 & 0 \\ 0 & 0 & 1 & 0 \\ 0 & 0 & 0 & 1 \\ 0 & 0 & -1 & -1.4 \end{bmatrix}, \quad \mathbf{G} = \begin{bmatrix} 0 \\ 0 \\ 0 \\ 2\sqrt{0.7} \end{bmatrix}, \quad \begin{array}{l} \mathbf{H} = [1 \quad 0 \quad 0 \quad 0], \\ \text{and} \quad \mathbf{J} = [10]. \end{array}$$

In addition, Brownian motions u_1 and u_2 are independent so that $\mathbf{K} = [0]$ and the matrix of covariance coefficients appearing in (5-113) becomes the identity matrix.

Gains given in (5-117) for Case 1 yield an optimal estimate to the additive noise of the tracking radar because $\{\mathbf{H}, \mathbf{F}\}$ is completely observable. Moreover, use of (5-120), (5-121), and (5-122) demonstrate that $\{\mathbf{P}, \mathbf{Q}, \mathbf{R}\}$ is regular. Therefore, covariance matrix $\mathbf{\Sigma}$, which satisfies (5-119), can be computed

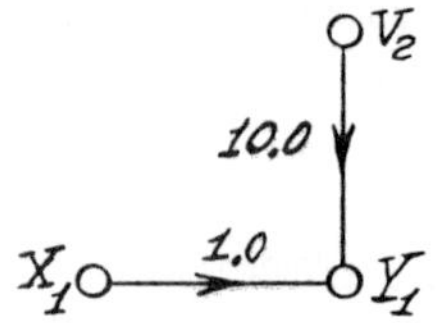

Figure 5-10 Flow graph of simplified dynamics for tracking radar observation.

directly using the method defined by Theorem 4-15. Optimal gains then are found from (5-117) to be

$$\mathbf{C} = \begin{bmatrix} 0.5730 \\ 0.1641 \\ 0.008446 \\ -0.008632 \end{bmatrix},$$

and the optimal filter for tracking-radar observations is depicted in Figure 5-11. This filter results in variances

$$\sigma_{11} = 57.30, \quad \sigma_{22} = 8.560, \quad \sigma_{33} = 0.9971, \quad \text{and} \quad \sigma_{44} = 0.9948$$

which are the major-diagonal elements of $\mathbf{\Sigma}$. The characteristic polynomial of this filter is also determined to be

$$\det(s\mathbf{I} - \mathbf{A}) = [s^2 + 2(0.708)(0.412)s + (0.412)^2]$$
$$\times [s^2 + 2(0.700)(0.993)s + (0.993)^2]$$

during the computation of $\mathbf{\Sigma}$ using the method defined by Theorem 4-15.

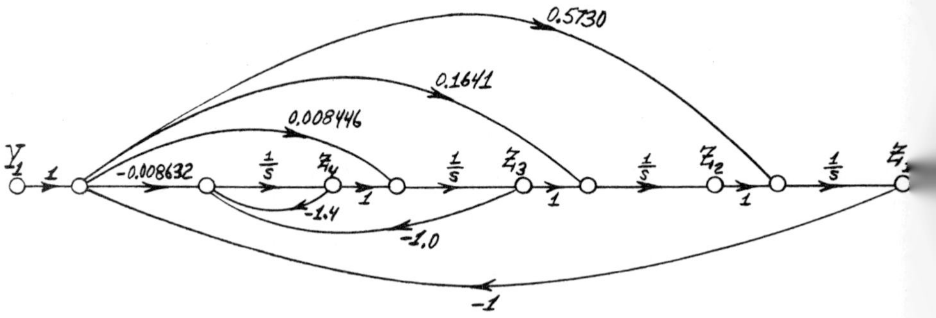

Figure 5-11 Flow graph of the optimal filter for Case 1 with $E\{\mathbf{x}_0\} = \mathbf{0}$.

Use of both tracking radar and doppler radar observations results in a somewhat more complicated predictor and is referred to as Case 2. Simplified dynamics of a doppler radar observation are depicted in Figure 5-12 so that coefficient matrices corresponding to Case 2 are

$$\mathbf{F} = \begin{bmatrix} 0 & 1 & 0 & 0 & 0 \\ 0 & 0 & 1 & 0 & 0 \\ 0 & 0 & 0 & 1 & 0 \\ 0 & 0 & -1 & -1.4 & 0 \\ 0 & 0.5 & 0 & 0 & -0.5 \end{bmatrix}, \quad \mathbf{G} = \begin{bmatrix} 0 \\ 0 \\ 0 \\ 2\sqrt{0.7} \\ 0 \end{bmatrix},$$

$$\mathbf{H} = \begin{bmatrix} 1 & 0 & 0 & 0 & 0 \\ 0 & 0 & 0 & 0 & 1 \end{bmatrix}, \quad \mathbf{J} = \begin{bmatrix} 10 & 0 \\ 0 & 2 \end{bmatrix}, \quad \text{and} \quad \mathbf{K} = \begin{bmatrix} 0 \\ 0 \end{bmatrix}.$$

Optimal gains then are found, by the method used for Case 1, to be

$$\mathbf{C} = \begin{bmatrix} 0.2842 & 1.3687 \\ 0.07785 & 0.6855 \\ 0.006834 & 0.08222 \\ -0.005986 & -0.06975 \\ 0.05475 & 0.4278 \end{bmatrix}$$

and result in variances

$$\sigma_{11} = 28.42, \quad \sigma_{22} = 5.282, \quad \sigma_{33} = 0.9853, \quad \text{and} \quad \sigma_{44} = 0.9804.$$

In addition, the characteristic polynomial found for the optimal filter of Case 2 is

$$\det(s\mathbf{I} - \mathbf{A}) = [s^2 + 2(0.813)(0.707)s + (0.707)^2] \times [s^2 + 2(0.677)(0.940)s + (0.940)^2](s + 0.189).$$

The optimal filter for Case 2 is seen to be somewhat more complicated than that of Case 1 mainly because filter state vector dimension has been increased by one. However, variances of position and velocity errors have been moderately reduced by the addition of a doppler radar observation.

Use of tracking radar, doppler radar, and accelerometer observations results in the most complicated predictor considered here and is referred to as Case 3.

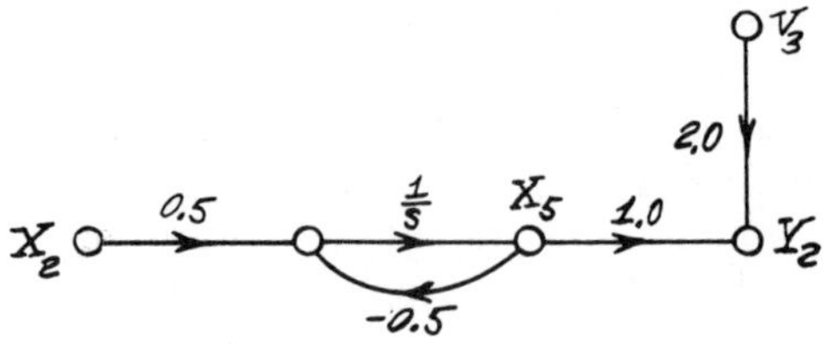

Figure 5-12 Flow graph of simplified dynamics for doppler radar observation.

Simplified dynamics of an accelerometer observation are depicted in Figure 5-13 so that coefficient matrices corresponding to Case 3 are

$$\mathbf{F} = \begin{bmatrix} 0 & 1 & 0 & 0 & 0 & 0 & 0 \\ 0 & 0 & 1 & 0 & 0 & 0 & 0 \\ 0 & 0 & 0 & 1 & 0 & 0 & 0 \\ 0 & 0 & -1 & -1.4 & 0 & 0 & 0 \\ 0 & 0.5 & 0 & 0 & -0.5 & 0 & 0 \\ 0 & 0 & 9 & 0 & 0 & -6 & -9 \\ 0 & 0 & 0 & 0 & 0 & 1 & 0 \end{bmatrix}, \quad \mathbf{G} = \begin{bmatrix} 0 \\ 0 \\ 0 \\ 2\sqrt{0.7} \\ 0 \\ 0 \\ 0 \end{bmatrix},$$

$$\mathbf{H} = \begin{bmatrix} 1 & 0 & 0 & 0 & 0 & 0 & 0 \\ 0 & 0 & 0 & 0 & 1 & 0 & 0 \\ 0 & 0 & 0 & 0 & 0 & 0 & 1 \end{bmatrix}, \quad \mathbf{J} = \begin{bmatrix} 10 & 0 & 0 \\ 0 & 2 & 0 \\ 0 & 0 & 0.4 \end{bmatrix}, \quad \text{and} \quad \mathbf{K} = \begin{bmatrix} 0 \\ 0 \\ 0 \end{bmatrix}.$$

Optimal gains then are found, by the method used for previous cases, to be

$$\mathbf{C} = \begin{bmatrix} 0.1962 & 0.5793 & 0.7627 \\ 0.02643 & 0.1539 & 1.6965 \\ 0.0008300 & 0.01055 & 1.7198 \\ -0.0006194 & -0.007058 & -0.05555 \\ 0.02317 & 0.1207 & 0.3632 \\ -0.0005712 & -0.005554 & 0.8276 \\ 0.001220 & 0.01452 & 1.2841 \end{bmatrix}$$

and result in variances

$$\sigma_{11} = 19.62, \quad \sigma_{22} = 0.9994, \quad \sigma_{33} = 0.5145, \quad \text{and} \quad \sigma_{44} = 0.8305.$$

In addition, the characteristic polynomial found for Case 3 is

$$\begin{aligned}\det(s\mathbf{I} - \mathbf{A}) = {} & [s^2 + 2(0.872)(0.207)s + (0.207)^2] \\ & \times [s^2 + 2(0.589)(1.91)s + (1.91)^2](s + 0.456) \\ & \times [s^2 + 2(0.987)(3.26)s + (3.26)^2].\end{aligned}$$

The optimal filter for Case 3 is of course relatively complicated with dim $\mathbf{z} = 7$. On the other hand, variances of position, velocity, and acceleration errors have been further reduced significantly by the addition of an accelerometer observation.

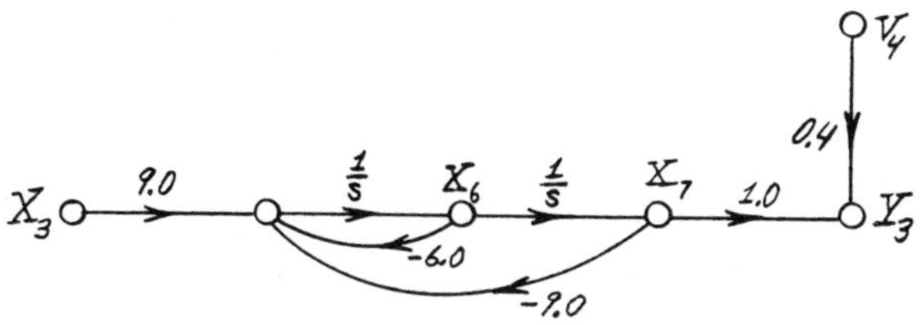

Figure 5-13 Flow graph of simplified dynamics for accelerometer observation.

The three examples given here of various combinations of observers and resulting optimal filters are intended to illustrate a typical compromise between complexity and performance. A number of approximations is also possible. For example, suppose only an estimate of $x_1(t)$ is sought. Then Case 1 results in a filter with one input and one output. The corresponding transfer function is found to be

$$\frac{Z_1}{Y_1} = \frac{0.573(s + 0.288)[s^2 + 2(0.695)(1.01)s + (1.01)^2]}{[s^2 + 2(0.708)(0.412)s + (0.412)^2][s^2 + 2(0.700)(0.993)s + (0.993)^2]}$$

and is approximated rather well by

$$\frac{Z_1}{Y_1} \cong \frac{0.573(s + 0.288)}{s^2 + 2(0.708)(0.412)s + (0.412)^2}.$$

This approximation is depicted in Figure 5-14 and is seen to result in a significant simplification. On the other hand, direct implementation of the optimal filter generally is required when prediction with $T > 0$ is sought.

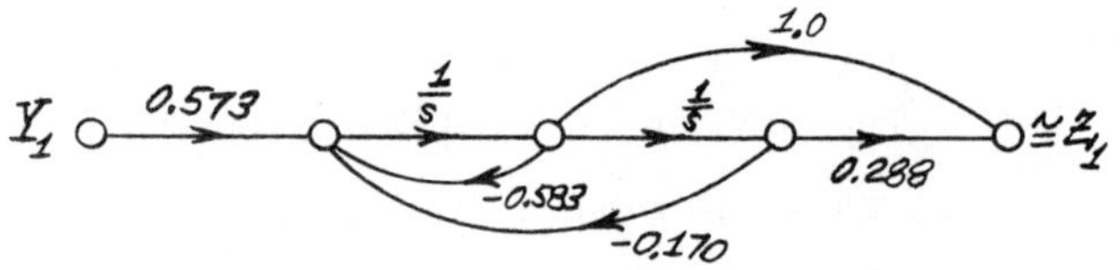

Figure 5-14 Flow graph of filter approximation for Case 1.

In order to predict the trajectory $x_1(t + \tau)$ of the aircraft on some nonzero interval $[t, t + T]$, the first row of $e^{\mathbf{F}\tau}$ must be implemented as a function of τ. Unfortunately, implementation of relatively complex functions would be required unless T is relatively small. An approximation which is frequently employed for T relatively small is based on a truncated Taylor series of $e^{\mathbf{F}\tau}$. For example, the approximation

$$e^{\mathbf{F}\tau} \cong \mathbf{I} + \mathbf{F}\tau + \mathbf{F}^2 \frac{\tau^2}{2} + \mathbf{F}^3 \frac{\tau^3}{6}$$

yields

$$\zeta_1(t + \tau) \cong z_1(t) + z_2(t)\tau + z_3(t) \frac{\tau^2}{2} + z_4(t) \frac{\tau^3}{6}$$

for all three optimal filters presented here. In other words, the first four filter state variables form the coefficients of a cubic extrapolation of aircraft trajectory on some suitably small interval.

5.10 SUMMARY

A variety of control system design problems with stochastic signals have been formulated here. These design problems include both additive and multiplicative Brownian-motion forcing functions and necessitate introduction of the concept of stochastic stability. Selection of suitable performance functionals for these design problems is seen to permit a parameter optimization formulation that is entirely analogous to the formulation for parameter optimization with deterministic signals.

Linear stochastic models of the human controller are discussed in some detail. Observation noise and motor noise are often included in these models and are represented by multiplicative Brownian motion. In addition, variable parameters are often included in these models so that simulation of the human controller can be accomplished for a variety of design conditions. These variable parameters are adjusted for this purpose by optimization techniques. An extensive example of human controller simulation using parameter optimization techniques is included.

Optimal gain control is revisited with stochastic signals and is found to be analogous to optimal gain control with deterministic signals. Therefore, methods of design problem formulation which are presented in Chapter 4 such as the synthesis of transfer function matrices also apply to optimal gain control problems with stochastic signals. An extensive example of optimal gain control with stochastic signals is also included.

Optimal filtering and prediction problems are introduced and are seen to be analogous to linear optimal control problems with deterministic signals. A relatively elaborate example is also introduced primarily for the purpose of illustrating the origin of these problems.

Computational aspects of design problems formulated in this chapter are also discussed. Computation of gradients and hessians which appear in necessary conditions for relative minima is formulated in the same manner as for design problems with deterministic signals. However, the solution of attendant linear and quadratic matrix equations requires somewhat more complicated but related computational methods for design problems which include multiplicative Brownian motion.

BIBLIOGRAPHY

1. Bucy, R. S. and P. D. Joseph: *Filtering for Stochastic Processes with Applications to Guidance*, Interscience Publishers, New York, 1968.
2. Wonham, W. M.: "Optimal stationary control of a linear system with state dependent noise", *SIAM Journal on Control*, **5**, No. 3, August 1967.
3. Kushner, H. J.: *Stochastic Stability and Control*, Academic Press, New York, 1967.
4. Kleinman, D. L.: "On the stability of linear stochastic systems", *IEEE Transactions on Automatic Control* (*Correspondence*), **AC-14**, No. 6, August 1969.
5. Wonham, W. M.: "On a matrix Riccati equation of stochastic control", *SIAM Journal on Control*, **6**, No. 4, 1968.
6. McRuer, D. T. *et al.*: "Human pilot dynamics in compensatory systems, theory, model, and experiments with controlled element and forcing function variations", Report No. AFFDL-TR-65-16, Flight Dynamics Laboratory, Wright-Patterson Air Force Base, Ohio, 1965.
7. Levison, W. H. *et al.*: "A model for human controller remnant", *IEEE Transactions on Man-Machine Systems*, **MMS-10**, No. 4, 1969.
8. Kleinman, D. L. *et al.*: *An Optimal Control Model of Human Behavior*, Proceedings of the Fifth Annual NASA–University Conference on Manual Control, M.I.T., Cambridge, March 1969.

9. Dillow, J. D.: "The paper pilot: A digital computer program to predict pilot rating for the hover task", Report No. AFFDL-TM-69-3, Flight Dynamics Laboratory, Wright-Patterson Air Force Base, Ohio, January 1970.
10. Anderson, R. O.: "An additional or alternate method of specifying hover flying qualities", Report No. AFFDL-TR-69-120, Flight Dynamics Laboratory, Wright-Patterson Air Force Base, Ohio, May 1970.
11. Rynaski, E. G. *et al.*: "Optimal control of a flexible launch vehicle", Report No. IH-2089-F-1, Cornell Aeronautical Laboratory, Inc., Buffalo, July 1966.
12. Athans, M. and E. Tse: "A direct derivation of the optimal linear filter using the maximum principle", *IEEE Transactions on Automatic Control*, **AC-12**, No. 6, December 1967.
13. Lee, Y. W.: *Statistical Theory of Communication*, John Wiley and Sons, Inc., New York, 1960.
14. Kalman, R. E. and R. S. Bucy: "New results in linear filtering and prediction theory", *Transactions of ASME, Journal of Basic Engineering, Ser. D,* **83**, March 1961.

PROBLEMS

5-1 Consider the stochastic state equation

$$d\mathbf{x} = \mathbf{f}(\mathbf{x})\, dt + \mathbf{G}(\mathbf{x})\, d\mathbf{u}; \quad \mathbf{x}(0) = \mathbf{x}_0$$

where $\mathbf{x}_0$ is a zero-mean Gaussian vector and $\mathbf{u}$ is independent zero-mean Brownian motion with

$$E\{\mathbf{u}(t)\mathbf{u}'(\tau)\} = \mathbf{I} \min\{\mathbf{t}, \tau\}.$$

Also consider a twice continuously differentiable scalar function $V = V(\mathbf{x})$. Then, under some mild assumptions concerning $\mathbf{f}$, $\mathbf{G}$, and V, stochastic calculus [1] yields

$$dV = [V_{\mathbf{x}}'\mathbf{f} + \tfrac{1}{2}\operatorname{tr}(\mathbf{G}\mathbf{G}'V_{\mathbf{xx}})]\, dt + (V_{\mathbf{x}}'\mathbf{G})\, d\mathbf{u}$$

which is the stochastic equivalent of a total derivative.

a) Derive an ordinary differential equation for $E\{V(\mathbf{x})\}$ noting that

i) $\mathbf{x}(t)$ and $d\mathbf{u}(t)$ are statistically independent for all $t \geqslant 0$

and

ii) the expectation of a product of statistically independent functions is equal to the product of expectations of the functions.

b) Verify (5-10).

5-2 Verify that (5-10) holds for

$$d\mathbf{x} = (\mathbf{A}\mathbf{x})\, dt + d\mathbf{u} + \sum_{i=1}^{n} x_i \mathbf{D}_i\, d\mathbf{w}$$

and

$$E\left\{\begin{bmatrix}\mathbf{u}(t)\\ \mathbf{w}(t)\end{bmatrix}\begin{bmatrix}\mathbf{u}(\tau)\\ \mathbf{w}(\tau)\end{bmatrix}'\right\} = \begin{bmatrix}\bar{\mathbf{C}} & \mathbf{0}\\ \mathbf{0} & \mathbf{I}\end{bmatrix}\min\{t, \tau\}$$

by identifying **C** and $D(\mathbf{B}, \mathbf{E})$. Also verify that

$$D(\mathbf{B}', \mathbf{E}) = [\mathrm{tr}\{\mathbf{D}_i'\mathbf{E}\mathbf{D}_j\}].$$

5-3 Write a computer program for solving (5-14) by direct matrix inversion.

5-4 Demonstrate that the equivalent coefficient matrices of

$$[\mathbf{XA}' + \mathbf{AX} + D(\mathbf{B}, \mathbf{X})] \quad \text{and} \quad [\mathbf{XA} + \mathbf{A}'\mathbf{X} + D(\mathbf{B}'\ \mathbf{X})]$$

are the transposes of each other.

5-5 Write a computer program for solving (5-14) by (5-33) and (4-46).

5-6 Write a computer program based on Problem 3-28 for solving stochastic parameter optimization problems.

5-7 Solve (5-10) for

$$\mathbf{A} = \begin{bmatrix} -1 & 0 \\ 1 & -1 \end{bmatrix}, \quad \mathbf{B}_1 = \begin{bmatrix} \sqrt{2} & 0 \\ 0 & 0 \end{bmatrix}, \quad \text{and} \quad \mathbf{C} = \begin{bmatrix} 0 & 0 \\ 0 & 0 \end{bmatrix}$$

assuming $\mathbf{E}_0$ is symmetric. Also find all symmetric $\mathbf{E}_0$ that are eigenmatrices.

5-8 Repeat Problem 5-7 assuming $\mathbf{E}_0$ is not symmetric.

5-9 Explain the relationship between eigenvalues of equivalent coefficient matrices formed from $[\mathbf{XA}' + \mathbf{AX} + D(\mathbf{B}, \mathbf{X})]$ when **X** is taken to be symmetric and when **X** is taken to be arbitrary.

5-10 Draw a flow graph that depicts system coupling corresponding to (5-52) and (5-53).

5-11 Dynamics of a compensatory tracking task are depicted in Figure 5-15. Brownian

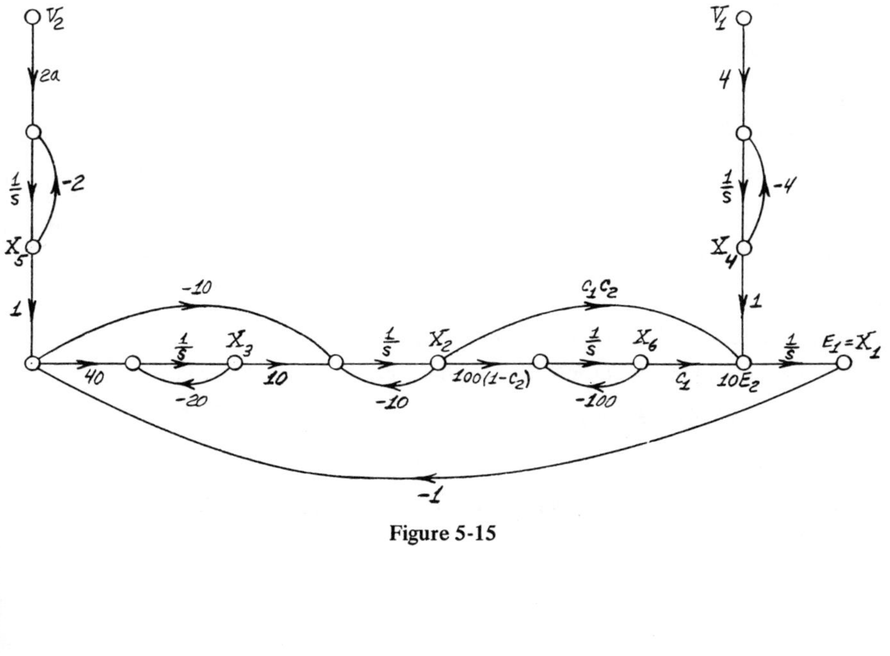

Figure 5-15

motion u_1 has a unit convariance coefficient, and Brownian motion u_2 has a covariance coefficient that is equal to the variance of x_1. The performance functional given in (5-74) is formed with the error vector defined by Figure 5-15. Find the optimal values of controller gain c_1 and lead ratio c_2 with $a = 0.0$.

5-12 Repeat Problem 5-11 with $a = 0.31623$.

5-13 Repeat the pilot simulation presented in Section 5.4 with $k_r = 0.0$ and helicopter dynamics corresponding to

$$M_u = 0.03106, \quad X_u = -0.2, \quad M_q = -1.0,$$
$$M_\theta = 0.0, \quad M_\delta = 0.483, \quad \text{and} \quad \sigma_g = 5.1.$$

Use

$$\mathbf{c}' = [0.170 \quad 0.634 \quad -1.88 \quad 0.240]$$

for the purpose of initialization.

5-14 Repeat Problem 5-13 for $k_r = 0.1$.

5-15 Repeat Problem 5-13 for

$$M_u = 0.03106, \quad X_u = -0.05, \quad M_q = -1.0$$
$$M_\theta = -3.0, \quad M_\delta = 0.385, \quad \text{and} \quad \sigma_g = 5.1.$$

Use

$$\mathbf{c}' = [0.133 \quad 0.57 \quad -1.17 \quad 0.430]$$

for the purpose of initialization.

5-16 Repeat Problem 5-15 with $k_r = 0.1$.

5-17 Derive conditions for stationarity in the form of (5-80), (5-82), and (5-83) for (5-76), (5-77), (5-79), and

$$f = \lim_{t \to \infty} E\{\mathbf{x}'\mathbf{\Phi}\mathbf{x} + 2\mathbf{u}_c'\mathbf{\Psi}\mathbf{x} + \mathbf{u}_c'\mathbf{\Omega}\mathbf{u}_c\}.$$

5-18 Derive conditions for stationarity in the form of (5-80), (5-82), and (5-83) for (5-76), (5-77), (5-79), and

$$d\mathbf{x} = (\mathbf{F}\mathbf{x} + \mathbf{G}\mathbf{u}_c)\,dt + \mathbf{B}\,d\mathbf{u} + \sum_{k=1}^{m} \mathbf{G}_k\mathbf{u}_c\,dw_k; \quad \mathbf{x}(0) = \mathbf{x}_0.$$

5-19 Repeat the design of a control system for flexible launch vehicles given in Section 5.6 when the second body bending mode is included. This mode is defined by

$$\ddot{\eta}_2 = -0.06\dot{\eta}_2 - 32.0\eta_2 + 23.0\beta + 2.0\alpha,$$

and η_2 must satisfy the constraint imposed on η_1.

5-20 Repeat Problem 5-19 when the third body bending mode defined by

$$\ddot{\eta}_3 = -0.09\dot{\eta}_3 - 84.0\eta_3 + 26.0\beta + 12.0\alpha$$

is also included.

5-21 Derive coefficient matrices **P**, **Q**, **R**, and $\mathbf{S}_k$ which appear in (5-87) for optimal gain control.

5-22 Repeat Problem 5-21 for the performance functional introduced in Problem 5-17.

5-23 Specialize the results of Problem 5-18 to the case of linear optimal control, and specify coefficient matrices of the equation that is analogous to (5-87).

5-24 Use the results of Problems 4-39 and 4-40 to state the relationship between complete controllability and complete observability.

5-25 Consider matrices

$$\mathbf{P} = \begin{bmatrix} 1 & 0 & 0 \\ 0 & 0 & 0 \\ 0 & 0 & 0 \end{bmatrix} \quad \text{and} \quad \mathbf{Q} = \tfrac{1}{2}\begin{bmatrix} -3 & 1 & -2 \\ 1 & -3 & 2 \\ 1 & -1 & 0 \end{bmatrix}.$$

a) Determine whether condition ii) appearing in the definition of regularity is satisfied.
b) Determine whether $\{\mathbf{\Pi}, \mathbf{Q}\}$ is completely observable where $\mathbf{\Pi}$ is defined by (5-91).
c) Summarize the relationship between condition ii) appearing in the definition of regularity and complete observability of $\{\mathbf{\Pi}, \mathbf{Q}\}$.

5-26 Demonstrate that (5-97) through (5-100) form the Newton method of successive substitutions for solving (5-87).

5-27 Derive the Newton method of successive substitutions for solving the matrix algebraic equation derived in Problem 5-23.

5-28 Use the results of Problem 5-27 to determine whether Theorem 5-6 also applies to the matrix algebraic equation derived in Problem 5-23.

5-29 Write a computer program for solving (5-87) using (5-97) through (5-100) and the results of Problem 5-5.

5-30 Write a computer program for solving (5-87) using (5-101) and the results of Problem 4-53.

5-31 A plant and observer are depicted in Figure 5-16 where corresponding Brownian motions u_1 and u_2 are assumed to be statistically independent.
a) Determine the optimal filter for this system.
b) Determine the optimal predictor for $x_1(t + \tau)$ with $\tau \in [0, T]$.
c) Find the transfer function of the optimal predictor determined in part b).

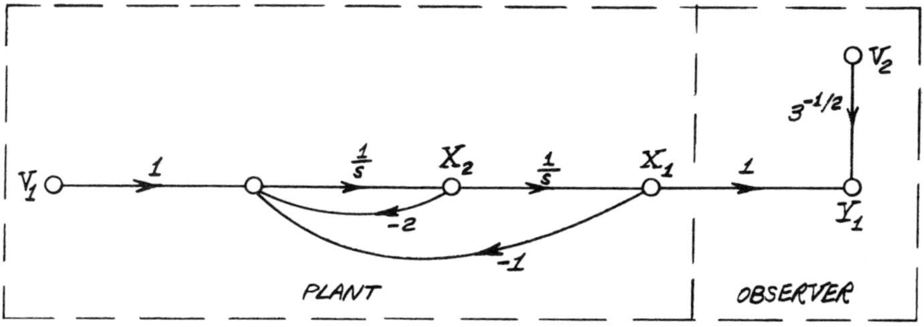

Figure 5-16

5-32 Derive stationarity conditions for the optimal filter corresponding to performance functional

$$f = \lim_{t \to \infty} E\{\mathbf{e}'(t)\mathbf{E}\mathbf{e}(t)\}$$

where **E** is positive semi-definite. Give a sufficient condition for Lagrange multiplier matrix $\boldsymbol{\Lambda}$ to be nonsingular and determine the optimal gain matrix **C** under this condition.

5-33 Modify the results of Problem 5-53 so that a computer program for solving (5-119) is obtained.

5-34 Repeat the aircraft tracking example given in Section 5.9 using only tracking radar and accelerometer observations.

CHAPTER 6

Finite-time design problems

All design problems considered thus far have been formulated with performance functionals of the form $f(\mathbf{c}, \mathbf{X})$. Matrix $\mathbf{X}$ is defined by (4-5) for deterministic problems and is defined by (5-7) and (5-11) for stochastic design problems. In other words, all design problems considered thus far are characterized by infinite time intervals. This class of design problems of course arises naturally in many situations and has the advantage that differential equations need not be solved. On the other hand, this class of design problems has the disadvantage that only values of $\mathbf{c}$ which result in an asymptotically stable system are permitted.

There are many other significant design problems, however, that are formulated in terms of finite time intervals. Solutions of differential equations generally are required for finite-time design problems, but the corresponding systems generally need not be asymptotically stable. In addition, finite-time approximations to design problems with infinite time intervals are useful in determining an admissible value of $\mathbf{c}$ for initializing previously discussed computational methods.

Finite-time design problems with optimal parameters are formulated here first. The primary purpose of this formulation is the development of a convenient method for introducing constraints in the form of differential equations. Moreover, this formulation facilitates the investigation of optimal parameters as a function of interval length for design problems such as optimal gain control.

Finite-time design problems with optimal forcing functions are also formulated in this chapter. Necessary and sufficient conditions for optimal forcing functions can now be derived by methods similar to those presented in Chapter 2. In addition, this formulation facilitates the investigation of optimal forcing functions as functions of interval length for design problems such as optimal gain control.

6.1 FORMULATION OF NECESSARY CONDITIONS FOR OPTIMAL PARAMETERS

Finite-time design problems with adjustable parameters are introduced here as follows. Nonlinear state equations are taken to be

$$\dot{\mathbf{x}} = \mathbf{f}(\mathbf{x}); \quad \mathbf{x}(0) = \boldsymbol{\gamma} + \boldsymbol{\Gamma}\mathbf{c} \tag{6-1}$$

where $\mathbf{c}$ denotes adjustable parameters. In addition, the performance functional is taken to be

$$J = \int_0^{t_f} F(\mathbf{x})\, \mathrm{d}t + G(\mathbf{x}_f) \tag{6-2}$$

where final time t_f is fixed and $\mathbf{x}_f = \mathbf{x}(t_f)$. Finally, equality constraints are introduced as

$$\mathbf{g}(\mathbf{x}_f) = \mathbf{0}. \tag{6-3}$$

The finite-time design problem being posed in this section is the minimization of J with respect to an admissible $\mathbf{c}$.

In typical situations, the state vector defined by (6-1) can be expressed as the function $\mathbf{x}(t) = \boldsymbol{\chi}(t, \mathbf{c})$ so that correspondingly $\dot{\mathbf{x}}(t) = \boldsymbol{\chi}_t(t, \mathbf{c})$. Therefore, the performance functional defined by (6-2) can be expressed as the function $J = f(\mathbf{c})$, and the finite-time design problem being posed here is then amenable to parameter optimization techniques given in Chapters 2 and 3. For instance, a Lagrange multiplier vector $\boldsymbol{\lambda}$, which does not depend on t, is introduced for the formulation of Lagrangian $L = L(\mathbf{x}_f, \boldsymbol{\lambda})$ defined as

$$L = G(\mathbf{x}_f) + \boldsymbol{\lambda}'\mathbf{g}(\mathbf{x}_f) \tag{6-4}$$

and Lagrangian $\mathscr{L} = \mathscr{L}(\mathbf{c}, \boldsymbol{\lambda})$ defined as

$$\mathscr{L} = J + \boldsymbol{\lambda}'\mathbf{g}(\mathbf{x}_f). \tag{6-5}$$

Gradients $\mathscr{L}_{\mathbf{c}}$, $\mathscr{L}_{\boldsymbol{\lambda}}$ and hessian $\mathscr{L}_{\boldsymbol{\lambda}\mathbf{c}}$ can now be formulated explicitly in a relatively straight-forward fashion. Direct formulation of hessian $\mathscr{L}_{\mathbf{cc}}$, however, proves to be unwieldy due to the dependence of $\mathbf{x}$ on $\mathbf{c}$. This dependence is analogous to the dependence of $\mathbf{X}$ on $\mathbf{c}$ and is again accounted for in a Lagrange multiplier formulation.

Lagrange multipliers which are introduced to account for the dependence of $\mathbf{x}$ on $\mathbf{c}$ are defined in accordance with Section 2.8. However, these Lagrange multipliers now depend on t because state equations represent constraints that must be satisfied for all $t \in [0, t_f]$. Specifically, *co-state* vector $\mathbf{p}$ is introduced for this purpose and is used to form *Hamiltonian* $H = H(\mathbf{x}, \mathbf{p})$ as

$$H = F(\mathbf{x}) + \mathbf{p}'\mathbf{f}(\mathbf{x}). \tag{6-6}$$

Lagrangian $\mathscr{L}$ can now be rewritten in the convenient form

$$\mathscr{L} = \int_0^{t_f} \{H - \mathbf{p}'\dot{\mathbf{x}}\}\, \mathrm{d}t + L, \tag{6-7}$$

and the state equation given in (6-1) can also be rewritten as

$$\dot{\mathbf{x}} = H_{\mathbf{p}}; \quad \mathbf{x}(0) = \boldsymbol{\Upsilon} + \boldsymbol{\Gamma}\mathbf{c}. \tag{6-8}$$

In addition, co-state vector $\mathbf{p}$ is defined by

$$-\dot{\mathbf{p}} = H_{\mathbf{x}}; \quad \mathbf{p}(t_f) = L_{\mathbf{x}_f} \tag{6-9}$$

and hence, under suitable assumptions, becomes the function $\mathbf{p}(t) = \boldsymbol{\pi}(t, \mathbf{c}, \boldsymbol{\lambda})$ so that correspondingly $\dot{\mathbf{p}}(t) = \boldsymbol{\pi}_t(t, \mathbf{c}, \boldsymbol{\lambda})$. State and co-state equations are expressed here in a form resembling that of Hamiltonian dynamics and are analogous to $\mathbf{0} = L_{\Lambda}$ and $\mathbf{0} = L_{\mathbf{X}}$, respectively, for the algebraic constraints introduced in Section 4.1.

Gradients and hessians of $\mathscr{L}$ can now be derived in a convenient manner under the following assumptions which are introduced for simplicity and assure that $\mathscr{L}$ is twice continuously differentiable. Specifically, let S be some open subset of $\mathscr{R}^m$ and assume

i) $\mathbf{x}(t)$ exists on $[0, t_f]$ given any $\mathbf{c} \in S$,

ii) H is twice continuously differentiable with respect to $\mathbf{x}$ on $\mathscr{R}^n$ given any $\mathbf{p} \in \mathscr{R}^n$,

and

iii) L is twice continuously differentiable with respect to $\mathbf{x}_f$ on $\mathscr{R}^n$ given any $\boldsymbol{\lambda} \in \mathscr{R}^l$.

These assumptions assure not only the existence and uniqueness of $\mathbf{x}(t) = \boldsymbol{\chi}(t, \mathbf{c})$ and $\mathbf{p}(t) = \boldsymbol{\pi}(t, \mathbf{c}, \boldsymbol{\lambda})$ but also the existence and continuity of

$$\frac{\partial^2 \boldsymbol{\chi}}{\partial t \partial c_i}, \quad \frac{\partial^2 \boldsymbol{\pi}}{\partial t \partial c_i}, \quad \text{and} \quad \frac{\partial^2 \boldsymbol{\pi}}{\partial t \partial \lambda_j}$$

for each i and j [1]. For notational simplicity, pertinent derivatives are denoted as

$$\mathbf{x_c}(t) = \boldsymbol{\chi}_{\mathbf{c}}(t, \mathbf{c}), \quad \mathbf{p_c}(t) = \boldsymbol{\pi}_{\mathbf{c}}(t, \mathbf{c}, \boldsymbol{\lambda}), \quad \mathbf{p}_{\boldsymbol{\lambda}}(t) = \boldsymbol{\pi}_{\boldsymbol{\lambda}}(t, \mathbf{c}, \boldsymbol{\lambda})$$

and

$$\dot{\mathbf{x}}_{\mathbf{c}}(t) = [\boldsymbol{\chi}_{\mathbf{c}}(t, \mathbf{c})]_t, \quad \dot{\mathbf{p}}_{\mathbf{c}}(t) = [\boldsymbol{\pi}_{\mathbf{c}}(t, \mathbf{c}, \boldsymbol{\lambda})]_t, \quad \dot{\mathbf{p}}_{\boldsymbol{\lambda}}(t) = [\boldsymbol{\pi}_{\boldsymbol{\lambda}}(t, \mathbf{c}, \boldsymbol{\lambda})]_t.$$

In addition, these second partial derivatives can also be expressed as

$$\dot{\mathbf{x}}_{\mathbf{c}}(t) = [\boldsymbol{\chi}_t(t, \mathbf{c})]_{\mathbf{c}}, \quad \dot{\mathbf{p}}_{\mathbf{c}}(t) = [\boldsymbol{\pi}_t(t, \mathbf{c}, \boldsymbol{\lambda})]_{\mathbf{c}} \quad \text{and} \quad \dot{\mathbf{p}}_{\boldsymbol{\lambda}}(t) = [\boldsymbol{\pi}_t(t, \mathbf{c}, \boldsymbol{\lambda})]_{c}$$

by Proposition 2-3 and interchanging the order of differentiation.

Gradient $\mathscr{L}_{\mathbf{c}}$ is now obtained by first expanding partial derivatives to obtain

$$\mathscr{L}_{\mathbf{c}} = \int_0^{t_f} \{\mathbf{x}_{\mathbf{c}}' H_{\mathbf{x}} + \mathbf{p}_{\mathbf{c}}' H_{\mathbf{p}} - \mathbf{p}_{\mathbf{c}}' \dot{\mathbf{x}} - \dot{\mathbf{x}}_{\mathbf{c}}' \mathbf{p}\} \, \mathrm{d}t + \mathbf{x}_{\mathbf{c}}'(t_f) L_{\mathbf{x}_f}.$$

Equation (6-8) then is used to eliminate the second and third terms appearing in the previous integrand. Integration by parts using the fourth term appearing in this integrand now results in

$$\mathscr{L}_{\mathbf{c}} = \int_0^{t_f} \{\mathbf{x}_{\mathbf{c}}' H_{\mathbf{x}} + \mathbf{x}_{\mathbf{c}}'\dot{\mathbf{p}}\}\, dt - \mathbf{x}_{\mathbf{c}}'\mathbf{p}\Big|_{t=0}^{t=t_f} + \mathbf{x}_{\mathbf{c}}'(t_f)L_{\mathbf{x}_f}.$$

Finally, use of (6-9) and $\mathbf{x}_{\mathbf{c}}(0) = \mathbf{\Gamma}$ yields the simple form

$$\mathscr{L}_{\mathbf{c}} = \mathbf{\Gamma}'\mathbf{p}(0). \tag{6-10}$$

Gradient $\mathscr{L}_{\mathbf{c}}$ and gradient

$$\mathscr{L}_{\boldsymbol{\lambda}} = L_{\boldsymbol{\lambda}} \tag{6-11}$$

of course appear in the first necessary condition for a constrained relative minimum of J in S.

Hessians $\mathscr{L}_{\mathbf{cc}}$ and $\mathscr{L}_{\boldsymbol{\lambda}\mathbf{c}}$, which appear in additional necessary and sufficient conditions for constrained relative minima, are readily found from (6-10) and (6-11) to be

$$\mathscr{L}_{\mathbf{cc}} = \mathbf{\Gamma}'\mathbf{p}_{\mathbf{c}}(0) \tag{6-12}$$

and

$$\mathscr{L}_{\boldsymbol{\lambda}\mathbf{c}} = L_{\boldsymbol{\lambda}\mathbf{x}_f}\mathbf{x}_{\mathbf{c}}(t_f) \tag{6-13}$$

by differentiation. In addition, equations (6-8) and (6-9) are differentiated with respect to elements of $\mathbf{c}$ and the order of differentiation is interposed. These steps yield linear differential equations

$$\dot{\mathbf{x}}_{\mathbf{c}} = H_{\mathbf{px}}\mathbf{x}_{\mathbf{c}}; \quad \mathbf{x}_{\mathbf{c}}(0) = \mathbf{\Gamma} \tag{6-14}$$

and

$$-\dot{\mathbf{p}}_{\mathbf{c}} = H_{\mathbf{xx}}\mathbf{x}_{\mathbf{c}} + H_{\mathbf{xp}}\mathbf{p}_{\mathbf{c}}; \quad \mathbf{p}_{\mathbf{c}}(t_f) = L_{\mathbf{x}_f\mathbf{x}_f}\mathbf{x}_{\mathbf{c}}(t_f) \tag{6-15}$$

respectively.

Although hessians $\mathscr{L}_{\mathbf{cc}}$ and $\mathscr{L}_{\boldsymbol{\lambda}\mathbf{c}}$, as given in (6-12) and (6-13) respectively, are acceptable for computational purposes, symmetry properties $\mathscr{L}_{\mathbf{cc}} = \mathscr{L}_{\mathbf{cc}}'$ and $\mathscr{L}_{\boldsymbol{\lambda}\mathbf{c}} = \mathscr{L}_{\mathbf{c}\boldsymbol{\lambda}}'$ have not been put in evidence. To this end, a state transition matrix $\mathbf{\Phi}$ is defined by

$$\dot{\mathbf{\Phi}} = H_{\mathbf{px}}\mathbf{\Phi}; \quad \mathbf{\Phi}(0) = \mathbf{I} \tag{6-16}$$

so that postmultiplication by $\mathbf{\Gamma}$ and comparison with (6-14) reveals

$$\mathbf{x}_{\mathbf{c}}(t) = \mathbf{\Phi}(t)\mathbf{\Gamma}. \tag{6-17}$$

In a similar fashion, matrix $\boldsymbol{\Psi}$ is defined by

$$-\dot{\boldsymbol{\Psi}} = H_{\mathbf{xx}}\boldsymbol{\Phi} + H_{\mathbf{xp}}\boldsymbol{\Psi}; \quad \boldsymbol{\Psi}(t_f) = L_{\mathbf{x}_f\mathbf{x}_f}\boldsymbol{\Phi}(t_f) \tag{6-18}$$

so that postmultiplication by $\boldsymbol{\Gamma}$ and comparison with (6-15) now reveals

$$\mathbf{p_c}(t) = \boldsymbol{\Psi}(t)\boldsymbol{\Gamma}. \tag{6-19}$$

Because state transition matrix $\boldsymbol{\Phi}$ is nonsingular on $[0, t_f]$ by the uniqueness theory of solutions to ordinary differential equations, matrix $\boldsymbol{\Gamma}$ can be eliminated from (6-17) and (6-19) to obtain

$$\mathbf{p_c}(t) = \boldsymbol{\Omega}(t)\mathbf{x_c}(t) \tag{6-20}$$

where

$$\boldsymbol{\Omega} = \boldsymbol{\Psi}\boldsymbol{\Phi}^{-1}. \tag{6-21}$$

Differentiation of (6-21) then yields

$$-\dot{\boldsymbol{\Omega}} = H_{\mathbf{xx}} + \boldsymbol{\Omega}H_{\mathbf{px}} + H'_{\mathbf{px}}\boldsymbol{\Omega}; \quad \boldsymbol{\Omega}(t_f) = L_{\mathbf{x}_f\mathbf{x}_f} \tag{6-22}$$

and hence $\boldsymbol{\Omega}$ is seen to be symmetric on $[0, t_f]$. Moreover, substitution of (6-20) into (6-12) yields

$$\mathscr{L}_{\mathbf{cc}} = \boldsymbol{\Gamma}'\boldsymbol{\Omega}(0)\boldsymbol{\Gamma} \tag{6-23}$$

which is also seen to be symmetric.

An expression for $\mathscr{L}_{\mathbf{c}\boldsymbol{\lambda}}$ is readily found from (6-9) and (6-10) by differentiation to be

$$\mathscr{L}_{\mathbf{c}\boldsymbol{\lambda}} = \boldsymbol{\Gamma}'\mathbf{p}_{\boldsymbol{\lambda}}(0) \tag{6-24}$$

where

$$-\dot{\mathbf{p}}_{\boldsymbol{\lambda}} = H_{\mathbf{xp}}\mathbf{p}_{\boldsymbol{\lambda}}; \quad \mathbf{p}_{\boldsymbol{\lambda}}(t_f) = L_{\mathbf{x}_f\boldsymbol{\lambda}}. \tag{6-25}$$

Equations (6-14) and (6-24) then are used to show that

$$\frac{\mathrm{d}}{\mathrm{d}t}(\mathbf{p}'_{\boldsymbol{\lambda}}\mathbf{x_c}) = \mathbf{0}$$

and hence

$$\mathbf{p}'_{\boldsymbol{\lambda}}(0)\mathbf{x_c}(0) = \mathbf{p}'_{\boldsymbol{\lambda}}(t_f)\mathbf{x_c}(t_f) \tag{6-26}$$

hold. Elimination of $\mathbf{x_c}(0)$ and $\mathbf{p}'_{\boldsymbol{\lambda}}(t_f)$ finally demonstrates that $\mathscr{L}_{\boldsymbol{\lambda}\mathbf{c}} = \mathscr{L}'_{\mathbf{c}\boldsymbol{\lambda}}$.

Finite-time problems defined by (6-1), (6-2), and (6-3) include a number of important special cases. For example, suppose

$$J = \int_0^{t_f} F(\mathbf{x}, \mathbf{c})\,\mathrm{d}t + G(\mathbf{x}_f, \mathbf{c}) \tag{6-27}$$

is to be minimized with respect to **c** where

$$\dot{\mathbf{x}} = \mathbf{f}(\mathbf{x}, \mathbf{c}); \quad \mathbf{x}(0) = \mathbf{x}_0. \tag{6-28}$$

Previous results are immediately applicable to this problem when state and co-state assignments

$$\mathbf{y} = \begin{bmatrix} \mathbf{x} \\ \mathbf{c} \end{bmatrix} \quad \text{and} \quad \mathbf{q} = \begin{bmatrix} \mathbf{p} \\ \mathbf{r} \end{bmatrix}$$

are introduced. State equations now become

$$\dot{\mathbf{y}} = \begin{bmatrix} \mathbf{f} \\ \mathbf{0} \end{bmatrix}; \quad \mathbf{y}(0) = \begin{bmatrix} \mathbf{x}_0 \\ \mathbf{0} \end{bmatrix} + \begin{bmatrix} \mathbf{0} \\ \mathbf{I} \end{bmatrix} \mathbf{c},$$

and hence the Hamiltonian for (6-27) and (6-28) becomes

$$H = F(\mathbf{x}, \mathbf{c}) + \mathbf{p}'\mathbf{f}(\mathbf{x}, \mathbf{c}). \tag{6-29}$$

In addition, the set of co-state equations corresponding to **c** are found to be

$$-\dot{\mathbf{r}} = H_{\mathbf{c}}; \quad \mathbf{r}(t_f) = G_{\mathbf{c}}$$

and hence

$$\mathbf{r}(0) = \int_0^{t_f} H_{\mathbf{c}} \, \mathrm{d}t + G_{\mathbf{c}}.$$

Stationarity conditions corresponding to (6-27) and (6-28) thus reduce to the *canonical* form

$$\dot{\mathbf{x}} = H_{\mathbf{p}}; \quad \mathbf{x}(0) = \mathbf{x}_0, \tag{6-30}$$

$$-\dot{\mathbf{p}} = H_{\mathbf{x}}; \quad \mathbf{p}(t_f) = G_{\mathbf{x}_f}, \tag{6-31}$$

and

$$\mathbf{0} = \int_0^{t_f} H_{\mathbf{c}} \, \mathrm{d}t + G_{\mathbf{c}}. \tag{6-32}$$

Hessian $\mathscr{L}_{\mathbf{cc}}$ can also be simplified somewhat for this example.

Use of the min–max method for determining optimal parameters by successive approximations of course requires the computation of $\mathscr{L}_{\mathbf{c}}$ and $\mathscr{L}_{\lambda}$. Computation of these gradients for finite-time problems, however, involve difficulties not previously encountered in design problems with infinite intervals. First, the performance functional given in (6-2) and state equations given in (6-8) must be solved by numerical integration from $t = 0$ to $t = t_f$ on each trial of each iteration. Moreover, solution $\mathbf{x}$ must be stored in tabulated form on the last trial of each iteration so that co-state equations given in (6-9) can be solved by numerical integration from $t = t_f$ to $t = 0$. Computation of gradients $\mathscr{L}_{\mathbf{c}}$ and $\mathscr{L}_{\boldsymbol{\lambda}}$ thus may require large amounts of both computer time and computer memory.

In order to avoid the high cost of the trials that are generally required to locate a relative minimum along each direction vector, the method of successive substitutions corresponding to (3-89) is often employed for finite-time problems instead of the min–max method of successive approximations. Computation of $\mathscr{L}_{\mathbf{cc}}$ and $\mathscr{L}_{\boldsymbol{\lambda}\mathbf{c}}$ for this purpose requires the numerical solution of additional differential equations and perhaps the storage of additional tabulated data. For example, use of (6-12) and (6-13) requires the numerical integration of (6-14) from $t = 0$ to $t = t_f$. Moreover, solution $\mathbf{x_c}$ must be stored in tabulated form so that (6-15) can be numerically integrated from $t = t_f$ to $t = 0$. In other words, each iteration of the Newton–Raphson method using (6-12) and (6-13) requires the numerical integration of a total of $2n(1 + m)$ first order differential equations where $n = \dim \mathbf{x}$ and $m = \dim \mathbf{c}$. Also, computer memory must be provided for the storage in tabulated form of a total of $n(1 + m)$ time functions. On the other hand, use of (6-23) and (6-24) in conjunction with the Newton–Raphson method requires the numerical integration of $2n\left(1 + \frac{l}{2} + \frac{n+1}{4}\right)$ first order differential equations on each iteration where $l = \dim \mathbf{g}$. Also, computer memory must be provided for the storage of n time variables. In either case, computation of Hessians $\mathscr{L}_{\mathbf{cc}}$ and $\mathscr{L}_{\boldsymbol{\lambda}\mathbf{c}}$ may be very costly for relatively large t_f.

6.2 OPTIMAL GAIN CONTROL

Typical behavior of finite-time problems with $t_f \to \infty$ is investigated here by way of example for optimal gain-control problems. This investigation proceeds conveniently using the matrix form of the canonical equations. Specifically, if matrices

$$\mathbf{C} = [\mathbf{c}_1 \quad \mathbf{c}_2 \quad \ldots], \quad \mathbf{F} = [\mathbf{f}_1 \quad \mathbf{f}_2 \quad \ldots], \quad \text{and} \quad \mathbf{X} = [\mathbf{x}_1 \quad \mathbf{x}_2 \quad \ldots]$$

are introduced in terms of vector partitions of $\mathbf{c}$, $\mathbf{f}$, and $\mathbf{x}$ respectively, then equations (6-27) and (6-28) are rewritten as

$$J = \int_0^{t_f} F(\mathbf{X}, \mathbf{C})\, \mathrm{d}t + \mathbf{G}(\mathbf{X}_f, \mathbf{C}) \tag{6-33}$$

and

$$\dot{\mathbf{X}} = \mathbf{F}(\mathbf{X}, \mathbf{C}); \quad \mathbf{X}(0) = \mathbf{X}_0. \tag{6-34}$$

In addition, if co-state matrix

$$\mathbf{P} = [\mathbf{p}_1 \quad \mathbf{p}_2 \quad \ldots]$$

is introduced similarly, then the Hamiltonian given in (6-29) becomes

$$H = F(\mathbf{X}, \mathbf{C}) + \operatorname{tr}\{\mathbf{P}'\mathbf{F}(\mathbf{X}, \mathbf{C})\}. \tag{6-35}$$

Finally, canonical equations corresponding to this class of finite-time problems then become

$$\dot{\mathbf{X}} = H_{\mathbf{P}}; \quad \mathbf{X}(0) = \mathbf{X}_0, \tag{6-36}$$

$$-\dot{\mathbf{P}} = H_{\mathbf{X}}; \quad \mathbf{P}(t_f) = G_{\mathbf{X}_f}, \tag{6-37}$$

and

$$\mathbf{0} = \int_0^{t_f} H_{\mathbf{C}}\, dt + G_{\mathbf{C}} \tag{6-38}$$

using the matrix form of (6-30), (6-31), and (6-32) respectively.

Optimal gain control for deterministic finite-time design problems is based on a performance functional formed with $G = 0$ and on integrand

$$F = \sum_{j=1}^{k} \mathbf{e}_j'\mathbf{e}_j. \tag{6-39}$$

Error vector $\mathbf{e}_j(t)$ is defined for each j by

$$\mathbf{e}_j = (\mathbf{D} - \mathbf{ECH})\mathbf{x}_j \tag{6-40}$$

and

$$\dot{\mathbf{x}}_j = \mathbf{A}\mathbf{x}_j; \quad \mathbf{x}(0) = \mathbf{b}_j \tag{6-41}$$

where $\mathbf{A}$ is given in (4-105). This integrand can also be written as

$$F = \operatorname{tr}\{\mathbf{D} - \mathbf{ECH})\mathbf{X}(\mathbf{D} - \mathbf{ECH})'\} \tag{6-42}$$

where

$$\mathbf{X} = \sum_{j=1}^{k} \mathbf{x}_j\mathbf{x}_j'.$$

In addition, differentiation of $\mathbf{X}$ and use of notation

$$\mathbf{B} = [\mathbf{b}_1 \quad \mathbf{b}_2 \quad \ldots \quad \mathbf{b}_k]$$

yield

$$\dot{\mathbf{X}} = \mathbf{XA}' + \mathbf{AX}; \quad \mathbf{X}(0) = \mathbf{BB}' \tag{6-43}$$

which is the matrix equivalent form of linear closed-loop state equations.

Canonical equations given in (6-36), (6-37), and (6-38) can now be applied directly to optimal gain control on a finite time interval. The first canonical equation of course results in (6-43) which has the solution

$$\mathbf{X}(t) = e^{\mathbf{A}t}\mathbf{B}\mathbf{B}' e^{\mathbf{A}'t}. \tag{6-44}$$

In addition, the second canonical equation results in

$$-\dot{\mathbf{P}} = \mathbf{P}\mathbf{A} + \mathbf{A}'\mathbf{P} + \mathbf{W}; \quad \mathbf{P}(t_f) = \mathbf{0} \tag{6-45}$$

where $\mathbf{W}$ is given by (4-106). If $\lambda_i\{\mathbf{A}\} + \lambda_j\{\mathbf{A}\} \neq 0$ is assumed to hold for all i and j, then

$$\mathbf{0} = \overline{\mathbf{X}}\mathbf{A} + \mathbf{A}'\overline{\mathbf{X}} + \mathbf{W} \tag{6-46}$$

has a unique solution $\overline{\mathbf{X}}$ by Lemma 4-1 and the solution of (6-45) can be written as

$$\mathbf{P} = \overline{\mathbf{X}} - e^{(t_f - t)\mathbf{A}'}\overline{\mathbf{X}}\, e^{(t_f - t)\mathbf{A}}. \tag{6-47}$$

The third canonical equation now becomes

$$\mathbf{0} = 2[(\mathbf{E}'\mathbf{E}\mathbf{C}\mathbf{H} - \mathbf{E}'\mathbf{D} - \mathbf{G}'\overline{\mathbf{X}})\mathbf{Y}(t_f) + \mathbf{G}'\mathbf{Z}(t_f)]\,\mathbf{H}' \tag{6-48}$$

where

$$\mathbf{Y}(t_f) = \int_0^{t_f} \{e^{\mathbf{A}t}\mathbf{B}\mathbf{B}' e^{\mathbf{A}'t}\}\, dt \tag{6-49}$$

and

$$\mathbf{Z}(t_f) = \int_0^{t_f} \{e^{(t_f - t)\mathbf{A}'}[\overline{\mathbf{X}}\, e^{\mathbf{A}t_f}\mathbf{B}\mathbf{B}']\, e^{\mathbf{A}'t}\}\, dt. \tag{6-50}$$

Optimal gains are finally given by

$$\mathbf{C} = [\mathbf{E}'\mathbf{E}]^{-1}[(\mathbf{E}'\mathbf{D} + \mathbf{G}'\overline{\mathbf{X}})\mathbf{Y}(t_f)\mathbf{H}' - \mathbf{G}'\mathbf{Z}(t_f)\mathbf{H}']\,[\mathbf{H}\mathbf{Y}(t_f)\mathbf{H}']^{-1} \tag{6-51}$$

when $\mathbf{H}\mathbf{Y}(t_f)\mathbf{H}'$ is assumed to be nonsingular for some $t_f > 0$.

Optimal gains given in (6-51) are easily seen to limit to the optimal gains given in (4-107) when $\mathbf{A}$ is asymptotically stable. Specifically, equivalences $\mathbf{X} = \overline{\mathbf{X}}$, $\mathbf{\Lambda} = \mathbf{Y}(\infty)$, and $\mathbf{0} = \mathbf{Z}(\infty)$ are easily verified. In addition, matrices given by (6-49) and (6-50) can be used as follows to establish a rationale for selecting t_f in order to use the optimal-gain formulation presented in this section as a method of initializing the corresponding infinite-time problem. Suppose t_f is selected to be sufficiently large so that $\mathbf{A}$ is asymptotically stable and hence there exist $\gamma, \Gamma > 0$ for which

$$\| e^{\mathbf{A}t} \| \leqslant \Gamma\, e^{-\gamma t}.$$

Then norm bounding and integration yield

$$\|\mathbf{Y}(t_f)\| \leqslant \frac{\Gamma^2}{2\gamma} \|\mathbf{BB}'\| (1 - \mathrm{e}^{-2\gamma t_f})$$

and

$$\|\mathbf{Z}(t_f)\| \leqslant \frac{\Gamma^3}{2\gamma} \|\overline{\mathbf{X}}\| \, \|\mathbf{BB}'\| (2\gamma t_f)\, \mathrm{e}^{-2\gamma t_f}.$$

Selection of $t_f \cong 3/\gamma$, for example, is based on these bounds and can be used to specify a finite-time problem for the purpose of approximating optimal gains given in (4-107). An estimate of γ usually can be obtained from system design specifications.

6.3 FORMULATION OF NECESSARY CONDITIONS FOR OPTIMAL FORCING FUNCTIONS

Finite-time design problems with adjustable forcing functions are introduced here as follows. Nonlinear state equations are taken to be

$$\dot{\mathbf{x}} = \mathbf{f}(\mathbf{x}, \mathbf{u}); \quad \mathbf{x}(0) = \mathbf{x}_0 \tag{6-52}$$

where $\mathbf{u}$ denotes adjustable forcing functions. In addition, the performance functional $J = J[\mathbf{u}]$ is taken to be

$$J = \int_0^{t_f} F(\mathbf{x}, \mathbf{u})\, \mathrm{d}t + G(\mathbf{x}_f) \tag{6-53}$$

where final time t_f is fixed and $\mathbf{x}_f = \mathbf{x}(t_f)$. The finite-time design problem being posed in this section is the minimization of J with respect to time functions $\mathbf{u}$. For simplicity, only continuous time functions are considered in this chapter.

As in the case of adjustable parameters, introduction of a co-state vector and Hamiltonian simplifies the formulation of necessary conditions. Hamiltonian $H = H(\mathbf{x}, \mathbf{u}, \mathbf{p})$ is defined as

$$H = F(\mathbf{x}, \mathbf{u}) + \mathbf{p}'\mathbf{f}(\mathbf{x}, \mathbf{u}) \tag{6-54}$$

so that the performance functional can be rewritten as

$$J = \int_0^{t_f} \{H - \mathbf{p}'\dot{\mathbf{x}}\}\, \mathrm{d}t + G \tag{6-55}$$

and state equations can be rewritten as

$$\dot{\mathbf{x}} = H_{\mathbf{p}}; \quad \mathbf{x}(0) = \mathbf{x}_0. \tag{6-56}$$

In addition, co-state vector $\mathbf{p}$ is defined by

$$-\dot{\mathbf{p}} = H_{\mathbf{x}}; \quad \mathbf{p}(t_f) = G_{\mathbf{x}_f} \tag{6-57}$$

under suitable assumptions.

Conditions presented in this chapter for optimal forcing functions are similar to those presented previously for optimal parameters, and these conditions are also based on similar assumptions. In order to simplify the sequel, only continuous forcing functions $\mathbf{u}(t)$ are considered, and hence these functions belong to $C^m[0, t_f]$ which denotes the space of all continuous $\mathbf{u}(t)$ with dim $\mathbf{u} = m$ and $t \in [0, t_f]$. In addition, the only performance functionals considered here have a property that is similar to functions which are twice continuously differentiable on $\mathscr{R}^m$. Specifically,

Definition 6-1 Performance functional J is *twice continuously differentiable* on $C^m[0, t_f]$ for some $\mathbf{x}_0$ if and only if

i) $\mathbf{x}(t)$ exists on $[0, t_f]$ given any $\mathbf{u}(t) \in C^m[0, t_f]$,
ii) H is twice continuously differentiable with respect to $[\mathbf{u}' \;\; \mathbf{x}']'$ on $\mathscr{R}^{m+n}$ given any $\mathbf{p} \in \mathscr{R}^n$,
and
iii) G is twice continuously differentiable with respect to $\mathbf{x}_f$ on $\mathscr{R}^n$.

Behavior of performance functional J is now investigated by evaluating (6-55), (6-56), and (6-57) with forcing functions

$$\mathbf{u}(t, \alpha) = \mathbf{u}(t) + \alpha\boldsymbol{\mu}(t) \tag{6-58}$$

where $\mathbf{u}(t), \boldsymbol{\mu}(t) \in C^m[0, t_f]$ and $\alpha \in (-\epsilon, \epsilon)$ for some $\epsilon > 0$. Conditions appearing in Definition 6-1 assure that the corresponding solutions $\mathbf{x}(t, \alpha)$ and $\mathbf{p}(t, \alpha)$ exist and are unique [1] on $[0, t_f] \times (-\epsilon, \epsilon)$ with t_f finite. Existence, uniqueness, and continuity of partial derivatives $[\mathbf{x}_t(t, \alpha)]_\alpha$ and $[\mathbf{p}_t(t, \alpha)]_\alpha$ are also assured [1] so that

$$[\mathbf{x}_\alpha(t, \alpha)]_t = [\mathbf{x}_t(t, \alpha)]_\alpha \quad \text{and} \quad [\mathbf{p}_\alpha(t, \alpha)]_t = [\mathbf{p}_\alpha(t, \alpha)]_t$$

hold on $[0, t_f] \times (-\epsilon, \epsilon)$ with t_f finite. Function $f(\alpha) = J[\mathbf{u} + \alpha\boldsymbol{\mu}]$ therefore exists and is twice continuously differentiable with respect to α on $(-\epsilon, \epsilon)$.

The notion of a relative minimum for performance functional J can now be introduced in terms of any norm of $\boldsymbol{\mu}(t) \in C^m[0, t_f]$. Norms of such time functions are denoted as $\|\boldsymbol{\mu}(t)\|$. For example, a suitable norm for $\boldsymbol{\mu}(t) \in C^m[0, t_f]$ is

$$\|\boldsymbol{\mu}(t)\| = \sum_{i=1}^{m} \Big[\sup_{t \in [0, t_f]} |\mu_i(t)| \Big]. \tag{6-59}$$

A weak relative minimum [2] is then defined as

Definition 6-2 Performance functional J has a *weak relative minimum* at $\mathbf{u}(t)$ if and only if there exists a $\delta > 0$ corresponding to some $\epsilon > 0$ such that

$$J[\mathbf{u}] \leqslant J[\mathbf{u} + \epsilon \boldsymbol{\mu}]$$

holds for all admissible $\boldsymbol{\mu}(t) \in C^m [0, t_f]$ with $\|\boldsymbol{\mu}(t)\| < \delta$. If in addition the equality holds only when $\|\boldsymbol{\mu}(t)\| = 0$, then the weak relative minimum at $\mathbf{u}(t)$ is called *strict.*

Throughout this chapter, only relative minima at $\mathbf{u}(t) \in C^m [0, t_f]$ are considered.

Necessary conditions for relative minima are stated first in terms of the *first variation* $V_1 = V_1 [\boldsymbol{\mu}]$ and the *second variation* $V_2 = V_2 [\boldsymbol{\mu}]$ which are defined as

$$V_1 = J_\alpha [\mathbf{u} + \alpha \boldsymbol{\mu}] \,|_{\alpha = 0} \tag{6-60}$$

and

$$V_2 = \tfrac{1}{2} J_{\alpha\alpha} [\mathbf{u} + \alpha \boldsymbol{\mu}] \,|_{\alpha = 0}. \tag{6-61}$$

Specifically, the first and second necessary conditions are stated as

Lemma 6-1 *Suppose performance functional J is twice continuously differentiable. Then J has a weak relative minimum at* $\mathbf{u}(t) \in C^m [0, t_f]$ *only if* $V_1 = 0$ *and* $V_2 \geqslant 0$ *hold for all admissible* $\boldsymbol{\mu}(t) \in C^m [0, t_f]$.

Proof The results follow in the order stated from Theorems 2-1 and 2-2 noting that $f(\alpha) = J[\mathbf{u} + \alpha \boldsymbol{\mu}]$ is twice continuously differentiable at $\alpha = 0$.

Requirements imposed by *stationarity* $V_1 = 0$ and *convexity* $V_2 \geqslant 0$ conditions must now be identified in terms of the Hamiltonian.

Convenient expressions for the first and second variations of J can be derived from (6-55) as follows. Differentiation of $J[\mathbf{u} + \alpha \boldsymbol{\mu}]$ with respect to α yields

$$J_\alpha [\mathbf{u} + \alpha \boldsymbol{\mu}] = \int_0^{t_f} \{\mathbf{x}'_\alpha \mathbf{H}_\mathbf{x} + \boldsymbol{\mu}' H_\mathbf{u} + \mathbf{p}'_\alpha H_\mathbf{p} - \mathbf{p}'_\alpha \dot{\mathbf{x}} - \mathbf{p}' \dot{\mathbf{x}}_\alpha\} \, dt + \mathbf{x}'_\alpha(t_f, \alpha) G_{\mathbf{x}_f}.$$

Equation (6-56) then is used to eliminate the third and fourth terms appearing in the previous integrand. Integration by parts using the fifth term appearing in this integrand now results in

$$J_\alpha [\mathbf{u} + \alpha \boldsymbol{\mu}] = \int_0^{t_f} \{\mathbf{x}'_\alpha H_\mathbf{x} + \boldsymbol{\mu}' H_\mathbf{u} + \mathbf{x}'_\alpha \dot{\mathbf{p}}\} \, dt - \mathbf{x}'_\alpha \mathbf{p} \Big|_{t=0}^{t=t_f} + \mathbf{x}'_\alpha(t_f, \alpha) G_{\mathbf{x}_f}.$$

Finally, use of (6-57) and $\mathbf{x}_\alpha(0, \alpha) = \mathbf{0}$ yields

$$J_\alpha[\mathbf{u} + \alpha\boldsymbol{\mu}] = \int_0^{t_f} \{\boldsymbol{\mu}' H_\mathbf{u}\}\, dt \tag{6-62}$$

where $H_\mathbf{u}$ is evaluated with $\mathbf{x}(t, \alpha)$, $\mathbf{u}(t, \alpha)$, and $\mathbf{p}(t, \alpha)$. The first variation of J at $\mathbf{u}(t) \in C^m[0, t_f]$ is found to be

$$V_1 = \int_0^{t_f} \{\boldsymbol{\mu}' H_\mathbf{u}\}\, dt \tag{6-63}$$

by evaluating (6-62) at $\alpha = 0$. In addition, the second variation of J at $\mathbf{u}(t) \in C^m[0, t_f]$ is found by differentiating (6-62) with respect to α and evaluating the result at $\alpha = 0$. These steps and use of notation

$$\mathbf{x}(t) = \mathbf{x}(t, 0), \quad \mathbf{p}(t) = \mathbf{p}(t, 0)$$

and

$$\boldsymbol{\chi}(t) = \mathbf{x}_\alpha(t, \alpha)\big|_{\alpha=0}, \quad \boldsymbol{\pi}(t) = \mathbf{p}_\alpha(t, \alpha)\big|_{\alpha=0} \tag{6-64}$$

then result in

$$V_2 = \tfrac{1}{2} \int_0^{t_f} \{\boldsymbol{\mu}' H_{\mathbf{uu}}\boldsymbol{\mu} + \boldsymbol{\mu}' H_{\mathbf{ux}}\boldsymbol{\chi} + \boldsymbol{\mu}' H_{\mathbf{up}}\boldsymbol{\pi}\}\, dt. \tag{6-65}$$

Furthermore, differentiation of (6-56) and (6-57) with respect to α, interchange of the order of differentiation, and evaluation at $\alpha = 0$ result in

$$\dot{\boldsymbol{\chi}} = H_{\mathbf{px}}\boldsymbol{\chi} + H_{\mathbf{pu}}\boldsymbol{\mu}; \quad \boldsymbol{\chi}(0) = \mathbf{0} \tag{6-66}$$

and

$$-\dot{\boldsymbol{\pi}} = H_{\mathbf{xx}}\boldsymbol{\chi} + H_{\mathbf{xu}}\boldsymbol{\mu} + H_{\mathbf{xp}}\boldsymbol{\pi}; \quad \boldsymbol{\pi}(t_f) = G_{\mathbf{x}_f\mathbf{x}_f}\boldsymbol{\chi}_f \tag{6-67}$$

where $\boldsymbol{\chi}_f = \boldsymbol{\chi}(t_f)$. Finally, an alternate and more useful expression for V_2 can be obtained as follows. Equation (6-66) first can be used to eliminate the third term appearing in (6-65) so that

$$V_2 = \tfrac{1}{2} \int_0^{t_f} \{\boldsymbol{\mu}' H_{\mathbf{uu}}\boldsymbol{\mu} + \boldsymbol{\mu}' H_{\mathbf{ux}}\boldsymbol{\chi} - \boldsymbol{\chi}' H_{\mathbf{xp}}\boldsymbol{\pi} + \boldsymbol{\pi}'\dot{\boldsymbol{\chi}}\}\, dt$$

is obtained. The last term appearing in the previous integrand then is integrated by parts and $\dot{\boldsymbol{\pi}}$ is eliminated using (6-67). These steps yield

$$V_2 = \tfrac{1}{2} \int_0^{t_f} \{\boldsymbol{\mu}' H_{\mathbf{uu}}\boldsymbol{\mu} + 2\boldsymbol{\mu}' H_{\mathbf{ux}}\boldsymbol{\chi} + \boldsymbol{\chi}' H_{\mathbf{xx}}\boldsymbol{\chi}\}\, dt + \tfrac{1}{2}\boldsymbol{\chi}_f' G_{\mathbf{x}_f\mathbf{x}_f}\boldsymbol{\chi}_f. \tag{6-68}$$

All relationships involving the first and second variations are of course evaluated with $\mathbf{x}(t)$ and $\mathbf{p}(t)$ corresponding to some $\mathbf{u}(t) \in C^m [0, t_f]$.

The first necessary condition for a weak relative minimum with respect to continuous forcing functions on a finite time interval is stated as

Theorem 6-1 *Suppose performance functional J is defined by* (6-52) *and* (6-53) *and is twice continuously differentiable. Then J has a weak relative minimum at* $\mathbf{u}(t) \in C^m [0, t_f]$ *only if*

$$\mathbf{0} = H_{\mathbf{u}} \tag{6-69}$$

holds on $[0, t_f]$.

Proof By virtue of Lemma 6-1, a weak relative minimum is achieved in this problem only if $V_1 = 0$ for all $\boldsymbol{\mu}(t) \in C^m [0, t_f]$. Gradient $H_{\mathbf{u}}$ belongs to $C^m [0, t_f]$ so let $\boldsymbol{\mu} = H_{\mathbf{u}}$. Evaluation of the first variation then yields

$$V_1 = \int_0^{t_f} \{H_{\mathbf{u}}' H_{\mathbf{u}}\} \, dt,$$

and hence $V_1 > 0$ occurs for this choice of $\boldsymbol{\mu}$ whenever $H_{\mathbf{u}}$ is not identically zero.

Conditions corresponding to stationarity at $\mathbf{u}(t) \in C^m [0, t_f]$ are summarized by (6-56), (6-57), and (6-69), and these three equations are called the *canonical equations* of the minimization problem defined by (6-52) and (6-53). Solutions to the canonical equations are called *extremals.*

The second necessary condition for a weak relative minimum with respect to continuous forcing functions on a finite time interval is stated as

Theorem 6-2 *Suppose performance functional J is defined by* (6-52) *and* (6-53), *is twice continuously differentiable, and is stationary at some* $\mathbf{u}(t) \in C^m [0, t_f]$. *Then J has a weak relative minimum at* $\mathbf{u}$ *only if* $H_{\mathbf{uu}}$ *is positive semi-definite on* $[0, t_f]$.

Proof By virtue of Lemma 6-1, a weak relative minimum is achieved in this problem only if $V_2 \geqslant 0$ for all $\boldsymbol{\mu}(t) \in C^m [0, t_f]$. Now suppose that $H_{\mathbf{uu}}$ is not positive semi-definite and hence has a negative eigenvalue at $t = t_0$. Then there exist a constant vector $\mathbf{v}$ and corresponding $\delta > 0$ for which $\mathbf{v}' H_{\mathbf{uu}} \mathbf{v} = -2\delta$ at $t = t_0$. Moreover, there exist points $t_1, t_2 \in [0, t_f]$ such that $t_2 > t_1$, $t_0 \in [t_1, t_2]$, and $\mathbf{v}' H_{\mathbf{uu}} \mathbf{v} < -\delta$ holds on $[t_1, t_2]$ because $\mathbf{v}' \mathbf{H}_{\mathbf{uu}} \mathbf{v}$ is continuous. Let

$$\epsilon = \frac{t_2 - t_1}{2} \quad \text{and} \quad \tau = \frac{t_2 + t_1}{2}.$$

Also let $\boldsymbol{\mu}(t) = \mathbf{v}\dot{f}(t)$ where

$$f(t) = \begin{Bmatrix} \dfrac{1}{\pi} \sin^2\left[\dfrac{\pi(t-\tau)}{\epsilon}\right] & \text{on} & [\tau-\epsilon, \tau+\epsilon] \\ 0 & & \text{otherwise} \end{Bmatrix}$$

and hence

$$\dot{f}(t) = \begin{Bmatrix} \dfrac{1}{\epsilon} \sin\left[\dfrac{2\pi(t-\tau)}{\epsilon}\right] & \text{on} & [\tau-\epsilon, \tau+\epsilon] \\ 0 & & \text{otherwise} \end{Bmatrix}.$$

Finally, a functional $I = I[\boldsymbol{\mu}]$ is defined by

$$I = \tfrac{1}{2} \int_0^{t_f} \{2\boldsymbol{\mu}' H_{\mathbf{ux}} \boldsymbol{\chi} + \boldsymbol{\chi}' H_{\mathbf{xx}} \boldsymbol{\chi}\}\, \mathrm{d}t + \tfrac{1}{2} \boldsymbol{\chi}_f' G_{\mathbf{x}_f \mathbf{x}_f} \boldsymbol{\chi}_f$$

so that function $g(\epsilon) = I[\mathbf{v}\dot{f}]$ is continuous on $[0, \infty]$ if $g(0)$ exists [3]. The second variation then becomes

$$V_2[\mathbf{v}\dot{f}] = g(\epsilon) + \tfrac{1}{2} \int_{\tau-\epsilon}^{\tau+\epsilon} \{\mathbf{v}' H_{\mathbf{uu}} \mathbf{v} \dot{f}^2(t)\}\, \mathrm{d}t < g(\epsilon) - \frac{\delta}{2} \int_{\tau-\epsilon}^{\tau+\epsilon} \dot{f}^2(t)\, \mathrm{d}t$$

and hence

$$V_2[\mathbf{v}\dot{f}] < g(\epsilon) - \frac{\delta}{2\epsilon}.$$

Because all elements of $H_{\mathbf{pu}}$ are continuous, use of (6-66) yields

$$\boldsymbol{\chi}(t) \to \begin{Bmatrix} H_{\mathbf{pu}} \mathbf{v} f(t) & \text{on} & [\tau-\epsilon, \tau+\epsilon] \\ \mathbf{0} & & \text{otherwise} \end{Bmatrix} \quad \text{as} \quad \epsilon \to 0$$

where $H_{\mathbf{pu}}$ is evaluated at $t = \tau$. Furthermore, because all elements of $H_{\mathbf{ux}}$ and $H_{\mathbf{xx}}$ are continuous, evaluation of $I[\mathbf{v}\dot{f}]$ as $\epsilon \to 0$ now yields

$$g(0) = \frac{1}{\pi^2} (\mathbf{v}' H_{\mathbf{ux}} H_{\mathbf{pu}} \mathbf{v}) \bigg|_{t=\tau}.$$

Therefore, there exists an $\epsilon > 0$ for which $V_2 < 0$ whenever $H_{\mathbf{uu}}$ has a negative eigenvalue at some point in $[0, t_f]$.

The result of Theorem 6-2 is called the Legendre necessary condition [2]. Extremals for which $H_{\mathbf{uu}}$ is singular in $[0, t_f]$ are called *singular*. All extremals considered in this chapter are assumed to be nonsingular and hence $H_{\mathbf{uu}}$ is positive definite on $[0, t_f]$ for all weak relative minima under discussion.

A third necessary condition for weak relative minima at nonsingular extremals is established from consideration of the quadratic functional $V = V(\beta)$ which is defined as

$$V = I_1 + \beta I_2 \tag{6-70}$$

where

$$I_1 = \frac{1}{2} \int_0^{t_f} \{(\boldsymbol{\mu} + H_{\mathbf{uu}}^{-1} H_{\mathbf{ux}} \boldsymbol{\chi})' H_{\mathbf{uu}} (\boldsymbol{\mu} + H_{\mathbf{uu}}^{-1} H_{\mathbf{ux}} \boldsymbol{\chi})\} \, dt, \tag{6-71}$$

$$I_2 = \frac{1}{2} \int_0^{t_f} \{\boldsymbol{\chi}' (H_{\mathbf{xx}} - H_{\mathbf{xu}} H_{\mathbf{uu}}^{-1} H_{\mathbf{ux}}) \boldsymbol{\chi}\} \, dt + \tfrac{1}{2} \boldsymbol{\chi}_f' G_{\mathbf{x}_f \mathbf{x}_f} \boldsymbol{\chi}_f, \tag{6-72}$$

and $\boldsymbol{\chi}$ is the solution of (6-66). This quadratic functional has the property $V(1) = V_2$. In addition, inequality $V(0) > 0$ holds, given any $\boldsymbol{\mu}(t) \in C^m [0, t_f]$ for which $\|\boldsymbol{\mu}(t)\| \neq 0$, when $H_{\mathbf{uu}}$ is positive definite on $[0, t_f]$. Now suppose that $\boldsymbol{\mu}(t)$ is defined by

$$-\dot{\boldsymbol{\pi}} = [\beta H_{\mathbf{xx}} + (1 - \beta) H_{\mathbf{xu}} H_{\mathbf{uu}}^{-1} H_{\mathbf{ux}}] \boldsymbol{\chi} + H_{\mathbf{xu}} \boldsymbol{\mu} + H_{\mathbf{xp}} \boldsymbol{\pi}; \quad \boldsymbol{\pi}(t_f) = \beta G_{\mathbf{x}_f \mathbf{x}_f} \boldsymbol{\chi}_f \tag{6-73}$$

and

$$\mathbf{0} = H_{\mathbf{ux}} \boldsymbol{\chi} + H_{\mathbf{uu}} \boldsymbol{\mu} + H_{\mathbf{up}} \boldsymbol{\pi} \tag{6-74}$$

so that $I_1 = I_1(\beta)$ and $I_2 = I_2(\beta)$. Equations (6-66), (6-73), and (6-74) are analogous to the canonical equations for $V(\beta)$ and are referred as the *accessory equations* of the minimization problem defined by (6-52) and (6-53) when $\beta = 1$.

Functional $V(\beta)$ can be evaluated in a convenient fashion when $\boldsymbol{\mu}(t)$ is specified by (6-73) and (6-74). Specifically, equation (6-74) is used to eliminate $\boldsymbol{\mu}$ from (6-66) and (6-73), thereby obtaining

$$-\begin{bmatrix} \dot{\boldsymbol{\chi}} \\ \dot{\boldsymbol{\pi}} \end{bmatrix} = \boldsymbol{\Delta}(\beta) \begin{bmatrix} \boldsymbol{\chi} \\ \boldsymbol{\pi} \end{bmatrix}; \quad \boldsymbol{\chi}(0) = \mathbf{0} \quad \text{and} \quad \boldsymbol{\pi}(t_f) = \beta G_{\mathbf{x}_f \mathbf{x}_f} \boldsymbol{\chi}_f \tag{6-75}$$

where

$$\boldsymbol{\Delta}(\beta) = \begin{bmatrix} -(H_{\mathbf{px}} - H_{\mathbf{pu}} H_{\mathbf{uu}}^{-1} H_{\mathbf{ux}}) & H_{\mathbf{pu}} H_{\mathbf{uu}}^{-1} H_{\mathbf{up}} \\ \beta (H_{\mathbf{xx}} - H_{\mathbf{xu}} H_{\mathbf{uu}}^{-1} H_{\mathbf{ux}}) & (H_{\mathbf{px}} - H_{\mathbf{pu}} H_{\mathbf{uu}}^{-1} H_{\mathbf{ux}})' \end{bmatrix}. \tag{6-76}$$

In addition, matrices $\boldsymbol{\Phi}$ and $\boldsymbol{\Psi}$ are defined by

$$-\begin{bmatrix}\dot{\boldsymbol{\Phi}}\\ \dot{\boldsymbol{\Psi}}\end{bmatrix} = \boldsymbol{\Delta}(\beta)\begin{bmatrix}\boldsymbol{\Phi}\\ \boldsymbol{\Psi}\end{bmatrix}; \quad \begin{bmatrix}\boldsymbol{\Phi}(t_f)\\ \boldsymbol{\Psi}(t_f)\end{bmatrix} = \begin{bmatrix}\mathbf{I}\\ \beta G_{\mathbf{x}_f\mathbf{x}_f}\end{bmatrix}. \tag{6-77}$$

Postmultiplication by $\boldsymbol{\chi}_f$ and comparison with (6-75) then yield

$$\boldsymbol{\chi}(t) = \boldsymbol{\Phi}(t)\boldsymbol{\chi}_f \quad \text{and} \quad \boldsymbol{\pi}(t) = \boldsymbol{\Psi}(t)\boldsymbol{\chi}_f \tag{6-78}$$

assuming $\boldsymbol{\chi}_f$ is selected so that $\boldsymbol{\chi}(0) = \mathbf{0}$. Furthermore, elimination of $\boldsymbol{\mu}$ and $\boldsymbol{\chi}$ from (6-71) and (6-72) yields

$$I_1 = \tfrac{1}{2}\boldsymbol{\chi}_f'\left[\int_0^{t_f} \{\boldsymbol{\Psi}' H_{\mathbf{pu}} H_{\mathbf{uu}}^{-1} H_{\mathbf{up}} \boldsymbol{\Psi}\}\, \mathrm{d}t\right]\boldsymbol{\chi}_f \tag{6-79}$$

and

$$I_2 = \tfrac{1}{2}\boldsymbol{\chi}_f'\left[\int_0^{t_f} \{\boldsymbol{\Phi}'(H_{\mathbf{xx}} - H_{\mathbf{xu}} H_{\mathbf{uu}}^{-1} H_{\mathbf{ux}})\boldsymbol{\Phi}\}\, \mathrm{d}t + G_{\mathbf{x}_f\mathbf{x}_f}\right]\boldsymbol{\chi}_f. \tag{6-80}$$

Finally, differentiation yields

$$-\frac{\mathrm{d}}{\mathrm{d}t}(\boldsymbol{\Phi}'\boldsymbol{\Psi}) = \boldsymbol{\Psi}' H_{\mathbf{pu}} H_{\mathbf{uu}}^{-1} H_{\mathbf{up}} \boldsymbol{\Psi} + \beta\boldsymbol{\Phi}'(H_{\mathbf{xx}} - H_{\mathbf{xu}} H_{\mathbf{uu}}^{-1} H_{\mathbf{ux}})\boldsymbol{\Phi}$$

and hence

$$V(\beta) = \tfrac{1}{2}[\boldsymbol{\Phi}(0)\boldsymbol{\chi}_f]'[\boldsymbol{\Psi}(0)\boldsymbol{\chi}_f] \tag{6-81}$$

then is obtained by integration and comparison with (6-79) and (6-80).

Previous relationships involving $\boldsymbol{\chi}_f$ are of course trivial unless $\boldsymbol{\Phi}(0)$ is singular. In fact, values of η for which $\boldsymbol{\Phi}(\eta)$ is singular are called *conjugate points* of $V(\beta)$ and have important significances in the study of quadratic functionals. For example, condition $V(\beta) = 0$ occurs for some nonzero $\boldsymbol{\mu}(t)$ if $V(\beta)$ has a conjugate point at $t = 0$ as can be seen from (6-81). Conjugate points, which also arise in Sturm–Liouville theory, possess a number of properties including the fact that conjugate points are isolated. Of particular significance to the study of weak relative minima is the result [4]

Proposition 6-1 *Let $V(\beta)$ be defined by* (6-66) *and* (6-70) *with continuous coefficient matrices. Then, corresponding to each conjugate point η of $V(\beta)$, there exists a continuous increasing function $\beta = \beta(\eta)$ having a range* $(0, 1]$.

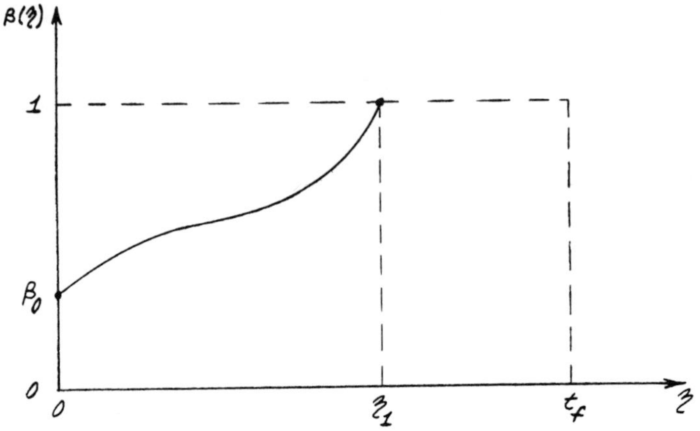

Figure 6-1 Typical example of parameter β as a function of conjugate point location.

A typical function $\beta(\eta)$ is depicted in Figure 6-1 assuming $V(1)$ has a conjugate point $\eta_1 \in (0, t_f)$ so that correspondingly $V(\beta_0)$ with $\beta_0 < 1$ must also have a conjugate point $\eta = 0$.

The first conclusion concerning conjugate points is called the Jacobi necessary condition and is stated as

Theorem 6-3 *Suppose performance functional J is defined by* (6-52) *and* (6-53), *is twice continuously differentiable, and is stationary at some* $\mathbf{u}(t) \in C^m[0, t_f]$ *for which* $H_{\mathbf{uu}}$ *is positive definite on* $[0, t_f]$. *Then J has a weak relative minimum at* $\mathbf{u}$ *only if there are no conjugate points of the second variation of J in* $(0, t_f)$.

Proof Assume V_2 has one or more conjugate points in $(0, t_f)$. Then, by Proposition 6-1, there exists a $\beta_0 \in (0, 1)$ for which $V(\beta_0)$ has a conjugate point at $t = 0$ and no conjugate point at any $t \in (0, t_f)$. Now let $\boldsymbol{\chi}_f$ be any nonzero vector for which $\boldsymbol{\Phi}(0)\boldsymbol{\chi}_f = \mathbf{0}$ where $\boldsymbol{\Phi}$ is the solution of (6-77) with $\beta = \beta_0$. Equality

$$I_1(\beta_0) + \beta_0 I_2(\beta_0) = 0$$

also holds in accordance with (6-70) and (6-81). Moreover, inequality $I_1(\beta_0) > 0$ holds because the integral appearing in (6-79) is positive semi-definite and the following contradiction occurs. Suppose $I_1(\beta_0) = 0$ holds so that

$$(H_{\mathbf{pu}} H_{\mathbf{uu}}^{-1} H_{\mathbf{up}})\pi = \mathbf{0}$$

must also hold on $[0, t_f]$. The first set of equations appearing in (6-75) then reduces to

$$\dot{\boldsymbol{\chi}} = (H_{\mathbf{px}} - H_{\mathbf{pu}} H_{\mathbf{uu}}^{-1} H_{\mathbf{ux}})\boldsymbol{\chi}; \quad \boldsymbol{\chi}(0) = \mathbf{0}$$

which contradicts the previous construction with $\boldsymbol{\chi}_f \neq \mathbf{0}$. Evaluation of the second variation using $\boldsymbol{\mu}(t)$ which corresponds to this nontrivial solution of (6-75) with $\beta = \beta_0$ now yields

$$V_2 = I_1(\beta_0) + I_2(\beta_0) = -\frac{1-\beta_0}{\beta_0} I_1(\beta_0) < 0.$$

Therefore, a weak relative minimum cannot exist at $\mathbf{u}(t) \in C^m[0, t_f]$ by virtue of Lemma 6-1.

Conjugate points of V_2, if there are any, usually must be located by solving (6-75) with $\beta = 1$ and evaluating $\det \boldsymbol{\Phi}(t)$.

The significance of the second conclusion concerning conjugate points is based on the following elementary result

Lemma 6-2 *Suppose performance functional J is twice continuously differentiable and is stationary at some* $\mathbf{u}(t) \in C^m[0, t_f]$. *Then J has a strict weak relative minimum if* $V_2 > 0$ *holds for all* $\boldsymbol{\mu}(t) \in C^m[0, t_f]$ *with* $\|\boldsymbol{\mu}(t)\| \neq 0$.

Proof The result follows from Theorem 2-3 noting that $f(\alpha) = J[\mathbf{u} + \alpha\boldsymbol{\mu}]$ is twice continuously differentiable and is stationary at $\alpha = 0$.

The Jacobi sufficiency condition can now be proved in the form

Theorem 6-4 *Suppose performance functional J is defined by* (6-52) *and* (6-53), *is twice continuously differentiable, and is stationary at some* $\mathbf{u}(t) \in C^m[0, t_f]$ *for which* $H_{\mathbf{uu}}$ *is positive definite on* $[0, t_f]$. *Then J has a strict weak relative minimum at* $\mathbf{u}$ *if there are no conjugate points of the second variation of J in* $[0, t_f)$.

Proof Let $\boldsymbol{\Phi}$ and $\boldsymbol{\Psi}$ be solutions of (6-77) with $\beta = 1$ and let

$$\boldsymbol{\Omega} = \boldsymbol{\Psi}\boldsymbol{\Phi}^{-1}. \tag{6-82}$$

Differentiation then yields

$$-\dot{\boldsymbol{\Omega}} = H_{\mathbf{xx}} + \boldsymbol{\Omega}H_{\mathbf{px}} + H'_{\mathbf{px}}\boldsymbol{\Omega} - \mathbf{K}'H_{\mathbf{uu}}\mathbf{K}; \quad \boldsymbol{\Omega}(t_f) = G_{\mathbf{x}_f\mathbf{x}_f} \tag{6-83}$$

where

$$\mathbf{K} = H_{\mathbf{uu}}^{-1}(H_{\mathbf{ux}} + H_{\mathbf{up}}\boldsymbol{\Omega}). \tag{6-84}$$

This set of differential equations implies that

$$\tfrac{1}{2}\int_0^{t_f} \{\boldsymbol{\chi}'[\dot{\boldsymbol{\Omega}} + H_{\mathbf{xx}} + \boldsymbol{\Omega}H_{\mathbf{px}} + H'_{\mathbf{px}}\boldsymbol{\Omega} - \mathbf{K}'H_{\mathbf{uu}}\mathbf{K}]\boldsymbol{\chi}\}\, dt = 0$$

holds for all $\boldsymbol{\chi}(t)$. If the previous integral is subtracted from V_2 as given in (6-68), then

$$V_2 = \tfrac{1}{2} \int_0^{t_f} \{(\boldsymbol{\mu} + \mathbf{K}\boldsymbol{\chi})' H_{\mathbf{uu}} (\boldsymbol{\mu} + \mathbf{K}\boldsymbol{\chi})\} \, \mathrm{d}t \tag{6-85}$$

is obtained for all $\boldsymbol{\mu}(t) \in C^m [0, t_f]$ using

$$\int_0^{t_f} \{\boldsymbol{\chi}' \dot{\boldsymbol{\Omega}} \boldsymbol{\chi}\} \, \mathrm{d}t = \boldsymbol{\chi}' \boldsymbol{\Omega} \boldsymbol{\chi} \Big|_{t=0}^{t=t_f} - 2 \int_0^{t_f} \{\boldsymbol{\chi}' \boldsymbol{\Omega} \dot{\boldsymbol{\chi}}\} \, \mathrm{d}t$$

and (6-66). Inequality $V_2 > 0$ therefore is obtained for all $\boldsymbol{\mu}(t) \in C^m [0, t_f]$ with $\|\boldsymbol{\mu}(t)\| \neq 0$.

In addition, Jacobi necessary and sufficiency conditions combine to yield

Corollary 6-1 *Suppose performance functional J is defined by (6-52) and (6-53), is twice continuously differentiable, and is stationary at some* $\mathbf{u}(t) \in C^m [-\epsilon, t_f]$ *for which $H_{\mathbf{uu}}$ is positive definite on $[-\epsilon, t_f]$ given some $\epsilon > 0$. Then J has a strict weak relative minimum at* $\mathbf{u}$ *when $G_{\mathbf{x}_f \mathbf{x}_f}$ is positive semi-definite and $(H_{\mathbf{xx}} - H_{\mathbf{xu}} H_{\mathbf{uu}}^{-1} H_{\mathbf{ux}})$ is positive semi-definite on $[-\epsilon, t_f]$.*

Proof Examination of (6-71) and (6-72) using a lower limit of integration of $-\epsilon$ instead of 0 reveals that $V \geqslant 0$ now holds for all $\boldsymbol{\mu}(t) \in C^m [-\epsilon, t_f]$ and hence V_2 has no conjugate points on $[0, t_f)$ by virtue of Theorem 6-3. The desired result then follows directly from Theorem 6-4.

This corollary identifies conditions that assure a strict weak relative minimum and can also be verified for some problems without the solution of differential equations. Moreover, the existence of a unique solution to the matrix Riccati equations given in (6-83) is guaranteed under these conditions.

6.4 OPTIMAL TIME-VARYING GAIN CONTROL

Optimal gain control which was introduced in Section 5.5 for stochastic design problems is revisited here by way of example of optimal forcing functions. This example is presented conveniently using the matrix form of canonical equations. Specifically, if matrices

$$\mathbf{F} = [\mathbf{f}_1 \quad \mathbf{f}_2 \quad \ldots], \quad \mathbf{U} = [\mathbf{u}_1 \quad \mathbf{u}_2 \quad \ldots]$$

and

$$\mathbf{X} = [\mathbf{x}_1 \quad \mathbf{x}_2 \quad \ldots]$$

are introduced in terms of vector partitions of **f**, **u**, and **x** respectively, then equations (6-52) and (6-53) are rewritten as

$$J = \int_0^{t_f} F(\mathbf{X}, \mathbf{U})\, dt + G(\mathbf{X}_f) \tag{6-86}$$

and

$$\dot{\mathbf{X}} = \mathbf{F}(\mathbf{X}, \mathbf{U}); \quad \mathbf{X}(0) = \mathbf{X}_0. \tag{6-87}$$

In addition, if co-state matrix

$$\mathbf{P} = [\mathbf{p}_1 \quad \mathbf{p}_2 \quad \ldots]$$

is introduced similarly, then the Hamiltonian given in (6-54) becomes

$$H = F(\mathbf{X}, \mathbf{U}) + \mathrm{tr}\,\{\mathbf{P}'\mathbf{F}(\mathbf{X}, \mathbf{U})\}. \tag{6-88}$$

Finally, canonical equations corresponding to this class of finite-time problems then become

$$\dot{\mathbf{X}} = H_{\mathbf{P}}; \quad \mathbf{X}(0) = \mathbf{X}_0, \tag{6-89}$$

$$-\dot{\mathbf{P}} = H_{\mathbf{X}}; \quad \mathbf{P}(t_f) = G_{\mathbf{X}_f}, \tag{6-90}$$

and

$$\mathbf{0} = H_{\mathbf{U}} \tag{6-91}$$

using the matrix form of (6-56), (6-57), and (6-69) respectively.

Optimal gain control for stochastic finite-time design problems is based on the performance functional formed with

$$F = E\{\mathbf{e}'(t)\mathbf{e}(t)\} \quad \text{and} \quad G = E\{\mathbf{x}'(t_f)\mathbf{T}\mathbf{x}(t_f)\}. \tag{6-92}$$

Error vector $\mathbf{e}(t)$ is defined in terms of state vector $\mathbf{x}(t)$ by

$$\mathbf{e} = (\mathbf{D} - \mathbf{EUH})\mathbf{x} \tag{6-93}$$

and

$$d\mathbf{x} = (\mathbf{A}\mathbf{x})\, dt + \mathbf{B}\, d\mathbf{u} + \sum_{k=1}^{m} \mathbf{S}_k \mathbf{x}\, dw_k; \quad \mathbf{x}(0) = \mathbf{x}_0 \tag{6-94}$$

where

$$\mathbf{A} = \mathbf{F} - \mathbf{GUH}. \tag{6-95}$$

Matrix $\mathbf{U}(t)$ denotes time varying feedback gains that are sought here as optimal forcing functions. Covariance matrix $\mathbf{X}(t)$ is defined as

$$\mathbf{X}(t) = E\{\mathbf{x}(t)\mathbf{x}'(t)\}$$

assuming that stochastic variables appearing in (6-94) have the properties given in Section 5.5. Also, function D is again defined by (5-9). The performance functiona given in (6-86) can now be expressed in terms of

$$F = \text{tr}\{\mathbf{WX}\} \quad \text{and} \quad G = \text{tr}\{\mathbf{TX}_f\} \tag{6-96}$$

where

$$\mathbf{W} = (\mathbf{D} - \mathbf{EUH})'(\mathbf{D} - \mathbf{EUH}) \tag{6-97}$$

and

$$\dot{\mathbf{X}} = \mathbf{XA}' + \mathbf{AX} + D(\mathbf{S}, \mathbf{X}) + \mathbf{BB}'; \quad \mathbf{X}(0) = \mathbf{X}_0. \tag{6-98}$$

Equation (6-98) is the stochastic equivalent of linear closed-loop state equations in matrix form.

Canonical equations given in (6-89), (6-90), and (6-91) can now be applied directly to optimal time-varying gain control on a finite time interval. The first canonical equation of course results in (6-98), and the second canonical equation results in

$$-\dot{\mathbf{P}} = \mathbf{PA} + \mathbf{A}'\mathbf{P} + D(\mathbf{S}', \mathbf{P}) + \mathbf{W}; \quad \mathbf{P}(t_f) = \mathbf{T}. \tag{6-99}$$

Furthermore, the third canonical equation yields

$$\mathbf{0} = 2[(\mathbf{E}'\mathbf{E})\mathbf{U}(\mathbf{HXH}') - (\mathbf{E}'\mathbf{D} + \mathbf{G}'\mathbf{P})(\mathbf{XH}')].$$

Optimal time varying gains then are given by

$$\mathbf{U} = (\mathbf{E}'\mathbf{E})^{-1}(\mathbf{E}'\mathbf{D} + \mathbf{G}'\mathbf{P})(\mathbf{XH}')(\mathbf{HXH}')^{-1} \tag{6-100}$$

assuming the extremal is nonsingular so that $(\mathbf{E}'\mathbf{E})$ and $(\mathbf{HXH}')$ are nonsingular.

Perhaps the most significant case of optimal gain control for stochastic design problems with finite time intervals occurs with $\mathbf{H} = \mathbf{I}$. This special case is called *linear optimal control* when optimal time-varying gains are given by (6-100). Linear optimal control thus is specified by

$$-\dot{\mathbf{P}} = (\mathbf{D}'\mathbf{D}) + \mathbf{PF} + \mathbf{F}'\mathbf{P} + D(\mathbf{S}', \mathbf{P}) - \mathbf{U}'(\mathbf{E}'\mathbf{E})\mathbf{U}; \quad \mathbf{P}(t_f) = \mathbf{T} \tag{6-101}$$

where

$$\mathbf{U} = (\mathbf{E}'\mathbf{E})^{-1}(\mathbf{E}'\mathbf{D} + \mathbf{G}'\mathbf{P}). \tag{6-102}$$

Co-state matrix $\mathbf{P}$ is seen to be the solution of constant-coefficient differential equations of the Riccati type for linear optimal control. Moreover, optimal gains are given by the constant matrix $\mathbf{U} = \mathbf{C}$ when $\mathbf{T} = \mathbf{\Lambda}$ is the positive semi-definite solution of (5-86). In other words, optimal time invariant gains found in Section 5.7 for linear optimal control correspond to an equilibrium solution of (6-101).

Under suitable circumstances, these differential equations can be used to obtain an approximation to **C** by selecting t_f to be suitably large when

$$\mathbf{\Lambda} = \lim_{t_f \to \infty} \mathbf{P}(0)$$

holds where $\mathbf{\Lambda}$ is the positive semi-definite solution of (5-86). This method of approximating optimal time-invariant gains involves considerably less computational complexity than the method discussed in Section 6.2.

6.5 PROPERTIES OF MATRIX RICCATI EQUATIONS

Matrix Riccati equations arise in a number of contexts associated with optimal forcing functions, and these equations have many interesting properties [5, 6]. Of particular interest here are asymptotic properties which are investigated in conjunction with approximations to optimal time-invariant gains. Specifically, solutions of

$$\dot{\mathbf{V}} = \mathbf{P} + \mathbf{VQ} + \mathbf{Q}'\mathbf{V} + D(\mathbf{S}', \mathbf{V}) - \mathbf{VRV}; \quad \mathbf{V}(0) = \mathbf{T} \tag{6-103}$$

are investigated. Coefficient matrices **P**, **Q**, **R**, and **S** are assumed to be constant unless explicitly stated otherwise, and matrices **P**, **R**, and **T** are assumed to be symmetric. Equilibrium solutions of (6-103) are denoted by $\mathbf{\Lambda}$ and hence satisfy (5-87). Matrix **A** is defined by (5-89), and matrix $\mathbf{\Delta}'$ denotes the coefficient matrix that corresponds to $[\mathbf{VA} + \mathbf{A}'\mathbf{V} + D(\mathbf{S}', \mathbf{V})]$ when equations and variables are ordered as in Section 5.7.

The first preliminary result concerns positive semi-definite solutions of (6-103) and is stated as

Lemma 6-3 *Suppose solution* $\mathbf{V}(t)$ *exists on* $[0, \tau]$ *for* $\mathbf{Q} = \mathbf{Q}(t)$, **T** *positive semi-definite, and* $\mathbf{P} = \mathbf{P}(t)$ *positive semi-definite on* $[0, \tau]$. *Then* $\mathbf{V}(t)$ *is positive semi-definite on* $[0, \tau]$.

Proof Let $\mathbf{v}(t)$ be defined on $[0, \tau]$ by

$$\dot{\mathbf{v}} = (\tfrac{1}{2}\mathbf{RV} - \mathbf{Q})\mathbf{v}; \quad \mathbf{v}(0) = \mathbf{v}_0$$

given any $\mathbf{v}_0 \in \mathscr{R}^n$. Differentiation then yields

$$\frac{\mathrm{d}}{\mathrm{d}t}(\mathbf{v}'\mathbf{V}\mathbf{v}) = \mathbf{v}'[\mathbf{P} + D(\mathbf{S}', \mathbf{V})]\mathbf{v}$$

so that

$$\mathbf{v}'(t)\mathbf{V}(t)\mathbf{v}(t) = \mathbf{v}_0'\mathbf{T}\mathbf{v}_0 + \int_0^t \mathbf{v}'(\xi)\{\mathbf{P}(\xi) + D[\mathbf{S}', \mathbf{V}(\xi)]\}\mathbf{v}(\xi)\,\mathrm{d}\xi$$

is obtained. Inequality

$$\mathbf{v}'(t)\mathbf{V}(t)\mathbf{v}(t) \geqslant 0$$

thus holds for all $\mathbf{v}(t)$, and vector $\mathbf{v}(t)$ can be selected to be any vector in $\mathscr{R}^n$ because $\mathbf{v}(t) = \mathbf{\Phi}(t)\mathbf{v}_0$ and state transition matrix $\mathbf{\Phi}(t)$ is nonsingular.

The second preliminary result concerns a partial ordering of solutions to (6-103) and is stated as

Lemma 6-4 *Suppose solutions* $\mathbf{V}_i(t)$ *to*

$$\dot{\mathbf{V}}_i = \mathbf{P}_i + \mathbf{V}_i\mathbf{Q} + \mathbf{Q}'\mathbf{V}_i + D(\mathbf{S}', \mathbf{V}_i) - \mathbf{V}_i\mathbf{R}_i\mathbf{V}_i; \quad \mathbf{V}_i(0) = \mathbf{T}_i$$

exist on $[0, \tau]$ *for* $i = 1, 2$. *Then matrix* $[\mathbf{V}_2(t) - \mathbf{V}_1(t)]$ *is positive semi-definite on* $[0, \tau]$ *when* $(\mathbf{T}_2 - \mathbf{T}_1)$ *and* $(\mathbf{R}_1 - \mathbf{R}_2)$ *are positive semi-definite and* $[\mathbf{P}_2(t) - \mathbf{P}_1(t)]$ *is positive semi-definite on* $[0, \tau]$.

Proof Let $\mathbf{W} = \mathbf{V}_2 - \mathbf{V}_1$ so that differentiation and regrouping of terms yield

$$\dot{\mathbf{W}} = [(\mathbf{P}_2 - \mathbf{P}_1) + \mathbf{V}_1(\mathbf{R}_1 - \mathbf{R}_2)\mathbf{V}_1] + \mathbf{W}(\mathbf{Q} - \mathbf{R}_2\mathbf{V}_1) + (\mathbf{Q} - \mathbf{R}_2\mathbf{V}_1)'\mathbf{W}$$
$$+ D(\mathbf{S}', \mathbf{W}) - \mathbf{W}\mathbf{R}_2\mathbf{W}; \quad \mathbf{W}(0) = \mathbf{T}_2 - \mathbf{T}_1.$$

The desired result now follows directly from Lemma 6-3.

Properties given in the previous two lemmas are used frequently in the sequel.

The third preliminary result concerns an overbound to solutions of (6-103), and this result is stated as

Lemma 6-5 *Suppose* $\mathbf{R}$ *is positive semi-definite and* $\mathbf{\Lambda}$ *is a positive semi-definite solution to (5-87) for which the corresponding matrix* $\mathbf{\Delta}$ *is asymptotically stable. Then solution* $\mathbf{V}(t)$ *exists on* $[0, \infty)$ *and* $\lim_{t \to \infty} \mathbf{V}(t) = \mathbf{\Lambda}$ *holds when* $(\mathbf{T} - \mathbf{\Lambda})$ *is positive semi-definite.*

Proof By Lemma 6-4, matrix $[\mathbf{V}(t) - \mathbf{\Lambda}]$ is positive semi-definite on any interval $[0, \tau]$ on which $\mathbf{V}(t)$ exists. In addition, the solution to

$$\dot{\mathbf{U}} = \mathbf{P} + \mathbf{\Lambda}\mathbf{R}\mathbf{\Lambda} + \mathbf{U}\mathbf{A} + \mathbf{A}'\mathbf{U} + D(\mathbf{S}', \mathbf{U}); \quad \mathbf{U}(0) = \mathbf{T}$$

exists on $[0, \infty)$ and $\lim_{t \to \infty} \mathbf{U}(t) = \mathbf{\Lambda}$ holds in accordance with (5-88) because $\mathbf{\Delta}$ is asymptotically stable. Now let $\mathbf{W} = \mathbf{U} - \mathbf{V}$ so that differentiation and regrouping of terms yield

$$\dot{\mathbf{W}} = (\mathbf{V} - \mathbf{\Lambda})\mathbf{R}(\mathbf{V} - \mathbf{\Lambda}) + \mathbf{W}\mathbf{A} + \mathbf{A}'\mathbf{W} + D(\mathbf{S}', \mathbf{W}); \quad \mathbf{W}(0) = \mathbf{0}.$$

Matrix $\mathbf{W}(t)$ therefore exists and is positive semi-definite on any interval $[0, \tau]$ on which $\mathbf{V}(t)$ exists. Moreover, matrix $\mathbf{V}(t)$ is continuously differentiable at τ when

$\mathbf{V}(\tau)$ exists and hence $\mathbf{V}(t)$ also exists on $(\tau, \tau + \epsilon]$ for some $\epsilon > 0$ by the continuation theorem of ordinary differential equations [1]. Matrix $\mathbf{V}(t)$ thus must exist on $[0, \infty)$, and condition $\lim_{t \to \infty} \mathbf{V}(t) = \mathbf{\Lambda}$ holds because $[\mathbf{\Lambda} - \mathbf{V}(\infty)]$ and $[\mathbf{V}(\infty) - \mathbf{\Lambda}]$ are positive semi-definite.

The last preliminary result concerns an underbound to solutions of (6-103), and this result is stated as

Lemma 6-6 *Suppose matrices* **P** *and* **R** *are positive semi-definite and (5-87) has a unique positive semi-definite solution* $\mathbf{\Lambda}$. *Then solution* $\mathbf{V}(t)$ *exists as a positive semi-definite matrix on* $[0, \infty)$ *and* $\lim_{t \to \infty} \mathbf{V}(t) = \mathbf{\Lambda}$ *holds when* $\mathbf{T} = \mathbf{0}$.

Proof By Lemma 6-4, matrix $[\mathbf{\Lambda} - \mathbf{V}(t)]$ is positive semi-definite on any interval $[0, \tau]$ on which $\mathbf{V}(t)$ exists. Let $\mathbf{X} = \mathbf{X}(t)$ be an arbitrary symmetric matrix with continuous elements so that the solution to

$$\dot{\mathbf{U}} = \mathbf{P} + \mathbf{U}(\mathbf{Q} - \mathbf{R}\mathbf{X}) + (\mathbf{Q} - \mathbf{R}\mathbf{X})'\mathbf{U} + D(\mathbf{S}', \mathbf{U}) + \mathbf{X}\mathbf{R}\mathbf{X}; \quad \mathbf{U}(0) = \mathbf{0}$$

also exists on any interval $[0, \tau]$ on which $\mathbf{X}$ exists. Moreover, by Lemma 6-4, matrix $[\mathbf{U}(t) - \mathbf{V}(t)]$ is positive semi-definite for all such $\mathbf{X}$ because regrouping of terms yields

$$\dot{\mathbf{U}} = \mathbf{P} + (\mathbf{X} - \mathbf{U})\mathbf{R}(\mathbf{X} - \mathbf{U}) + \mathbf{U}\mathbf{Q} + \mathbf{Q}'\mathbf{U} - \mathbf{U}\mathbf{R}\mathbf{U}; \quad \mathbf{U}(0) = \mathbf{0}.$$

In addition, let $\mathbf{\Phi}(t)$ be the state transition matrix defined by

$$\dot{\mathbf{\Phi}} = (\mathbf{Q} - \mathbf{R}\mathbf{X})'\mathbf{\Phi}; \quad \mathbf{\Phi}(0) = \mathbf{I}$$

and let

$$\mathbf{\Phi}(t, \xi) = \mathbf{\Phi}(t)\mathbf{\Phi}^{-1}(\xi).$$

Similarly, let $\mathbf{\Psi}(t)$ be the state transition matrix defined by

$$\dot{\mathbf{\Psi}} = (\mathbf{Q} - \mathbf{R}\mathbf{V})'\mathbf{\Psi}; \quad \mathbf{\Psi}(0) = \mathbf{I}$$

and let

$$\mathbf{\Psi}(t, \xi) = \mathbf{\Psi}(t)\mathbf{\Psi}^{-1}(\xi).$$

Matrix $\mathbf{V}(t)$ can then be expressed in the form

$$\mathbf{V}(t) = \int_0^t \mathbf{\Psi}(t, \xi)\{\mathbf{P} + D[\mathbf{S}', \mathbf{V}(\xi)] + \mathbf{V}(\xi)\mathbf{R}\mathbf{V}(\xi)\}\mathbf{\Psi}'(t, \xi)\, d\xi.$$

Now let $\mathbf{X}(t) = \mathbf{V}(t + \beta)$ for some constant $\beta \in [0, \tau - t]$. Relationships

$$\mathbf{\Phi}(t) = \mathbf{\Psi}(t + \beta, \beta) \quad \text{and} \quad \mathbf{\Phi}(t, \xi) = \mathbf{\Psi}(t + \beta, \xi + \beta)$$

can then be shown to hold. In addition, matrix $\mathbf{U}(t-\beta)$ can be expressed in the form

$$\mathbf{U}(t-\beta) = \int_{\beta}^{t} \mathbf{\Psi}(t,\xi)\{\mathbf{P} + D[\mathbf{S}', \mathbf{U}(\xi-\beta)] + \mathbf{V}(\xi)\mathbf{R}\mathbf{V}(\xi)\}\mathbf{\Psi}'(t,\xi)\,\mathrm{d}\xi$$

for $t \geqslant \beta$. If $\mathbf{W}(t)$ is defined as

$$\mathbf{W}(t) = \mathbf{V}(t) - \mathbf{U}(t-\beta)$$

and $\mathbf{U}(t)$ is taken to be $\mathbf{0}$ on $[-\beta, 0)$, then

$$\mathbf{W}(t) = \int_{0}^{t} \mathbf{\Psi}(t,\xi) D[\mathbf{S}', \mathbf{W}(\xi)]\mathbf{\Psi}'(t,\xi)\,\mathrm{d}\xi$$

$$+ \int_{0}^{\beta} \mathbf{\Psi}(t,\xi)[\mathbf{P} + \mathbf{V}(\xi)\mathbf{R}\mathbf{V}(\xi)]\mathbf{\Psi}'(t,\xi)\,\mathrm{d}\xi$$

follows directly for $t \geqslant \beta$. Moreover, matrix $\mathbf{W}(t)$ is positive semi-definite on $[0, \beta]$ by Lemma 6-3 and hence $\mathbf{W}(t)$ is seen to be positive semi-definite on $[\beta, \tau]$. Because $[\mathbf{V}(t) - \mathbf{U}(t-\beta)]$ and $[\mathbf{U}(t-\beta) - \mathbf{V}(t-\beta)]$ are positive semi-definite, matrix $[\mathbf{V}(t) - \mathbf{V}(t-\beta)]$ must also be positive semi-definite for all $t \in [0, \tau - \beta]$. Solution $\mathbf{V}(t)$ therefore must exist as a positive semi-definite matrix on $[0, \infty)$ and monotone convergence $\lim_{t\to\infty} \mathbf{V}(t) = \mathbf{\Lambda}$ is established in accordance with Proposition 4-4.

The solution described in Lemma 6-6 has the property that $t_2 \geqslant t_1 \geqslant 0$ implies $[\mathbf{V}(t_2) - \mathbf{V}(t_1)]$ is positive semi-definite, and hence equilibrium solution $\mathbf{\Lambda}$ is approached monotonely from below by $\mathbf{V}(t)$.

The major result of this section concerns asymptotic properties of solutions to (6-103) and can now be stated as

Theorem 6-5 *Suppose matrices* **P** *and* **R** *are positive semi-definite and* (5-87) *has a unique positive semi-definite solution* $\mathbf{\Lambda}$ *for which the corresponding matrix* $\mathbf{\Delta}$ *is asymptotically stable. Then solution* $\mathbf{V}(t)$ *exists as a positive semi-definite matrix on* $[0, \infty)$ *and* $\lim_{t\to\infty} \mathbf{V}(t) = \mathbf{\Lambda}$ *holds given any positive semi-definite* **T**.

Proof Let $\mathbf{U}(t)$ be the solution to (6-103) corresponding to some $\mathbf{V}(0)$ for which $[\mathbf{V}(0) - \mathbf{T}]$ and $[\mathbf{V}(0) - \mathbf{\Lambda}]$ are positive semi-definite. Also let $\mathbf{W}(t)$ be the solution to (6-103) corresponding to $\mathbf{V}(0) = \mathbf{0}$. Solutions $\mathbf{U}(t)$ and $\mathbf{W}(t)$ exist as positive semi-definite matrices on $[0, \infty)$ and

$$\lim_{t\to\infty} \mathbf{U}(t) = \lim_{t\to\infty} \mathbf{W}(t) = \mathbf{\Lambda}$$

holds by Lemmas 6-5 and 6-6. Moreover, matrices $[\mathbf{U}(\tau) - \mathbf{V}(\tau)]$ and $[\mathbf{V}(\tau) - \mathbf{W}(\tau)]$ are positive semi-definite when $\mathbf{V}(t)$ exists on $[0, \tau]$. These properties thus hold for all $\tau \geqslant 0$ and the proof is complete.

The conclusions of the previous theorem of course apply to stochastic gain-control problems with $\mathbf{S}_k \neq \mathbf{0}$ for which the preconditions of Theorem 5-6 hold. Moreover, these conclusions also apply to deterministic gain-control problems, stochastic gain control problems with $\mathbf{S}_k = \mathbf{0}$, and stochastic filter problems for which the preconditions of Theorem 4-13 hold.

As suggested previously, equation (6-103) can be solved for the purpose of obtaining an approximation to optimal time-invariant gains which correspond to equilibrium solution $\mathbf{\Lambda}$. This procedure is used, however, only when gains resulting in an asymptotically stable closed-loop system cannot be found by other more convenient means. In order to exhibit explicit transient solutions to (6-103) so that required length of solution might be estimated, the case where Theorem 6-5 applies with $\mathbf{T} = \mathbf{0}$, $\mathbf{S}_k = \mathbf{0}$, and $\mathbf{\Lambda}$ nonsingular is considered here. Solution $\mathbf{V}(t)$ can be given explicitly in terms of $e^{\mathbf{A}t}$ for this case. First let $\mathbf{W}(t)$ be defined by

$$\mathbf{V}(t) = \mathbf{\Lambda} - \mathbf{W}(t)$$

so that differentiation and regrouping of terms yields

$$\dot{\mathbf{W}} = \mathbf{WA} + \mathbf{A}'\mathbf{W} + \mathbf{WRW}; \quad \mathbf{W}(0) = \mathbf{\Lambda}$$

where $\mathbf{A}$ is defined by (5-89). Matrix $\mathbf{W}(t)$ of course exists on $[0, \infty)$ and is also found to be nonsingular on this interval because

$$\mathbf{w}'(t)\mathbf{W}(t)\mathbf{w}(t) = \mathbf{w}_0'\mathbf{\Lambda}\mathbf{w}_0$$

holds where

$$\dot{\mathbf{w}} = -(\tfrac{1}{2}\mathbf{RW} + \mathbf{A})\mathbf{w}; \quad \mathbf{w}(0) = \mathbf{w}_0.$$

Now let $\mathbf{U}(t)$ be defined on $[0, \infty)$ by

$$\mathbf{W}(t) = \mathbf{U}^{-1}(t)$$

so that differentiation yields

$$-\dot{\mathbf{U}} = \mathbf{UA}' + \mathbf{AU} + \mathbf{R}; \quad \mathbf{U}(0) = \mathbf{\Lambda}^{-1}.$$

The solution to this differential equation has the well-known form

$$\mathbf{U}(t) = e^{-\mathbf{A}t}\left[\mathbf{\Lambda}^{-1} - \int_0^t e^{\mathbf{A}\xi}\mathbf{R}\, e^{\mathbf{A}'\xi}\, d\xi\right] e^{-\mathbf{A}'t},$$

and hence (6-103) has the solution

$$\mathbf{V}(t) = \mathbf{\Lambda} - e^{\mathbf{A}'t}\,[\mathbf{\Lambda}^{-1} - \int_0^t e^{\mathbf{A}\xi}\mathbf{R}\,e^{\mathbf{A}'\xi}\,d\xi\,]^{-1}\,e^{\mathbf{A}t} \tag{6-104}$$

in this special case. Duration of the transient behavior of $\mathbf{V}(t)$ is seen to be comparable with the integrals given in (6-49) and (6-50) which determine optimal time-invariant gains on a finite time interval. In other words, solving (6-103) by numerical means constitutes a viable method of obtaining optimal time-invariant gains.

Conditions appearing in the statement of Theorem 6-5 are sufficient but of course not necessary for the existence of a positive semi-definite equilibrium solution to matrix Riccati equations. This point is illustrated by the example optimal-gain problem corresponding to $\mathbf{S}_k = \mathbf{0}$ and

$$\mathbf{P} = \begin{bmatrix} 0 & 0 \\ 0 & 0 \end{bmatrix}, \quad \mathbf{Q} = \begin{bmatrix} 0 & a \\ 1 & 0 \end{bmatrix}, \quad \mathbf{R} = \begin{bmatrix} 1 & 0 \\ 0 & 0 \end{bmatrix}, \quad \text{and} \quad \mathbf{T} = \begin{bmatrix} 0 & 0 \\ 0 & 1 \end{bmatrix}.$$

Matrix Riccati equations corresponding to these problem matrices with $a = 1$ have two positive semi-definite equilibrium solutions; namely

$$\mathbf{\Lambda} = \begin{bmatrix} 0 & 0 \\ 0 & 0 \end{bmatrix} \quad \text{and} \quad \mathbf{\Lambda} = \begin{bmatrix} 2 & 2 \\ 2 & 2 \end{bmatrix}.$$

These equilibrium solutions yield closed-loop Jacobians

$$\mathbf{A} = \begin{bmatrix} 0 & 1 \\ 1 & 0 \end{bmatrix} \quad \text{and} \quad \mathbf{A} = \begin{bmatrix} -2 & -1 \\ 1 & 0 \end{bmatrix}$$

respectively. However, the transient solution to these equations for $a = 1$ is found to be

$$\mathbf{V}(t) = \frac{1}{1 - \frac{1}{2}t + \frac{1}{4}\sinh 2t} \begin{bmatrix} \sinh^2 t & \frac{1}{2}\sinh 2t \\ \frac{1}{2}\sinh 2t & \cosh^2 t \end{bmatrix}.$$

On the other hand, these equations have only one positive semi-definite equilibrium solution for $a = -1$, and the transient solution to these equations for $a = -1$ is found to be

$$\mathbf{V}(t) = \frac{1}{1 + \frac{1}{2}t - \frac{1}{4}\sin 2t} \begin{bmatrix} \sin^2 t & \frac{1}{2}\sin 2t \\ \frac{1}{2}\sin 2t & \cos^2 t \end{bmatrix}.$$

In other words, the nonzero equilibrium solution is obtained when $a = 1$ but the zero equilibrium solution is obtained for $a = -1$. Correspondingly, the closed-loop

Jacobian is asymptotically stable for $a = 1$ whereas the closed-loop Jacobian is not asymptotically stable for $a = -1$. Open-loop Jacobian $\mathbf{Q}$ is not asymptotically stable for both $a = 1$ and $a = -1$ and hence $\{\mathbf{P}, \mathbf{Q}, \mathbf{R}\}$ is not regular for both of these cases.

6.6 METHODS OF COMPUTING OPTIMAL FORCING FUNCTIONS

Canonical equations for optimal forcing functions in general cannot be solved explicitly by analytical means. Furthermore, these equations generally are very difficult to solve by computational means and constitute the so-called two-point boundary-value problem which arises with optimal forcing functions. Two-point boundary-value problems involve both partial initial-state and partial final-state conditions, namely $\mathbf{x}(0) = \mathbf{x}_0$ and $\mathbf{p}(t_f) = G_{\mathbf{x}_f}$ respectively, which must be satisfied.

If forcing functions $\mathbf{u}(t)$ are given *a priori*, then the two-point boundary-value problem posed by (6-56) and (6-57) decouples into a sequence of two one-point boundary-value problems using the computational procedure outlined previously for finite-interval design problems with adjustable parameters. However, the condition given in (6-69) for the optimality of forcing functions couples state and co-state equations so that the equations now must be solved simultaneously by either numerical integration from $t = 0$ to $t = t_f$ or numerical integration from $t = t_f$ to $t = 0$. Numerical integration from $t = 0$ to $t = t_f$ requires a knowledge of $\mathbf{p}(0)$, whereas numerical integration from $t = t_f$ to $t = 0$ requires a knowledge of $\mathbf{x}(t_f)$. Of course, neither $\mathbf{p}(0)$ nor $\mathbf{x}(t_f)$ are known *a priori*, and hence the canonical equations for optimal forcing functions cannot be solved by direct computational means.

Perhaps the most obvious method for the computation of optimal forcing functions is called boundary-condition iteration and is based on successive approximations to $\mathbf{p}(0)$ or $\mathbf{x}(t_f)$. By way of example, the Newton method of successive substitutions to $\mathbf{x}(t_f)$ is investigated here. Specifically, state and co-state equations are rewritten as

$$\dot{\mathbf{x}} = H_{\mathbf{p}}; \quad \mathbf{x}(t_f) = \mathbf{c} \tag{6-105}$$

and

$$-\dot{\mathbf{p}} = H_{\mathbf{x}}; \quad \mathbf{p}(t_f) = G_{\mathbf{c}} \tag{6-106}$$

respectively where parameter vector $\mathbf{c}$ is to be selected by successive substitutions. State, co-state, and forcing function vectors now become functions of both t and $\mathbf{c}$ which is expressed notationally as

$$\mathbf{x}(t) = \mathbf{x}(t, \mathbf{c}), \quad \mathbf{p}(t) = \mathbf{p}(t, \mathbf{c}), \quad \text{and} \quad \mathbf{u}(t) = \mathbf{u}(t, \mathbf{c}).$$

In addition, these functions have the differentiability and continuity properties discussed in Section 6.1 for optimal parameters. For convenience, notation

$$\mathbf{\Phi}(t) = \mathbf{x_c}(t, \mathbf{c}), \quad \mathbf{\Psi}(t) = \mathbf{p_c}(t, \mathbf{c}), \quad \text{and} \quad \mathbf{M}(t) = \mathbf{u_c}(t, \mathbf{c})$$

is introduced for partial derivatives with respect to elements of $\mathbf{c}$ so that

$$\dot{\mathbf{\Phi}}(t) = [\mathbf{x}_t(t, \mathbf{c})]_{\mathbf{c}} \quad \text{and} \quad \dot{\mathbf{\Psi}}(t) = [\mathbf{p}_t(t, \mathbf{c})]_{\mathbf{c}}$$

hold. Differentiation of (6-105), (6-106), and (6-69) with respect to $\mathbf{c}$ then yields

$$\dot{\mathbf{\Phi}} = H_{\mathbf{px}}\mathbf{\Phi} + H_{\mathbf{pu}}\mathbf{M}; \quad \mathbf{\Phi}(t_f) = \mathbf{I},$$

$$-\dot{\mathbf{\Psi}} = H_{\mathbf{xx}}\mathbf{\Phi} + H_{\mathbf{xu}}\mathbf{M} + H_{\mathbf{xp}}\mathbf{\Psi}; \quad \mathbf{\Psi}(t_f) = G_{\mathbf{cc}},$$

and

$$\mathbf{0} = H_{\mathbf{ux}}\mathbf{\Phi} + H_{\mathbf{uu}}\mathbf{M} + H_{\mathbf{up}}\mathbf{\Psi}$$

respectively. If $H_{\mathbf{uu}}$ is assumed to be nonsingular, matrix $\mathbf{M}$ can be eliminated using the third of these equations so that the first two equations reduce to (6-77) with $\beta = 1$ and $\mathbf{x}_f = \mathbf{c}$.

For each selection of final state $\mathbf{c}$, solution of canonical equations from $t = t_f$ to $t = 0$ results in the extremal which corresponds to $\mathbf{x}(0, \mathbf{c})$. Moreover, if this extremal is nonsingular, then the corresponding accessory equations can also be solved from $t = t_f$ to $t = 0$. In general, condition $\mathbf{x}(0)$ to $\mathbf{x}_0$ does not hold so that a new value of $\mathbf{c}$ must be selected. Differentiation with respect to $\mathbf{c}$ yields

$$[\mathbf{x}_0 - \mathbf{x}(0)]_{\mathbf{c}} = -\mathbf{\Phi}(0)$$

so that, if the additional assumption is made that $\eta = 0$ is not a conjugate point of V_2, then the Newton method of successive substitutions is formed with direction vector

$$\mathbf{d} = \mathbf{\Phi}^{-1}(0)[\mathbf{x}_0 - \mathbf{x}(0)]. \tag{6-107}$$

That is, each new value of $\mathbf{c}$ is selected to be $(\mathbf{c} + \mathbf{d})$. This method converges quadratically to $\mathbf{c} = \mathbf{x}_f$ where $\mathbf{x}_f$ corresponds to the extremal with $\mathbf{x}(0) = \mathbf{x}_0$ when the method is initialized with sufficient accuracy. Moreover, this extremal is a strict weak relative minimum when $H_{\mathbf{uu}}$ is positive definite on $[0, t_f]$ and interval $[0, t_f)$ contains no conjugate points of V_2.

The boundary-condition iteration method corresponding to (6-107) requires the solution of (6-69) for $\mathbf{u}$ and a total of $2n(n + 1)$ first order differential equations from $t = t_f$ to $t = 0$ on each iteration where $n = \dim \mathbf{x}$. The primary advantages of this method are computer program simplicity, minimal computer memory requirements, and quadratic convergence in a suitably small neighborhood of a strict weak relative minimum. The primary disadvantages of this method are shared by all boundary-condition iteration methods. In particular, coupled state and co-state equations as well as the accessory equations are very unstable. Instability not only gives rise to the propagation of numerical integration errors but also gives rise to difficulties in the selection of a sufficiently accurate initial value of **c**. Difficulties caused by instability become increasingly severe as t_f increases. In fact, boundary-condition iteration methods are not practicable for most control system design problems of interest.

Another type of method for solving two-point boundary-value problems is based on decoupling the canonical equations by selecting forcing functions iteratively. That is, the state equation given in (6-56) can be solved from $t = 0$ to $t = t_f$ and then the co-state equation given in (6-57) can be solved from $t = t_f$ to $t = 0$ when $\mathbf{u}(t)$ is specified as a set of independent time functions. Iterative selection of forcing functions is called forcing-function iteration and can be performed so that the corresponding values of performance functional J form a decreasing sequence until (6-69) is satisfied on $[0, t_f]$. In addition, forcing-function iteration can be performed so that quadratic convergence to the optimal forcing-function vector can be achieved under some conditions [7].

Direction vectors for forcing-function iteration methods are also functions of time and hence cannot always be identified directly. Instead, efficient direction vectors for forcing functions can be formulated indirectly in terms of an auxiliary minimization problem formed with a quadratic performance functional and linear state equations [8]. The performance functional adopted here for this purpose is

$$V = V_1 + I_2 \tag{6-108}$$

where V_1 is the first variation of J given in (6-63) and I_2 is a model of the second variation of J given in (6-85). Specifically, functional I_2 is taken to be

$$I_2 = \frac{1}{2} \int_0^{t_f} \{(\boldsymbol{\mu} + \mathbf{K}\boldsymbol{\chi})'\mathbf{R}(\boldsymbol{\mu} + \mathbf{K}\boldsymbol{\chi})\}\, \mathrm{d}t. \tag{6-109}$$

where **K** and **R** are arbitrary except that their elements exist as continuous functions of t and **R** is positive definite on $[0, t_f]$. In addition, state vector $\boldsymbol{\chi}$ for this auxiliary minimization problem is the solution of (6-66). Direction vector $\boldsymbol{\mu}(t)$ is selected so that $V = V[\boldsymbol{\mu}]$ is minimized for the purpose of forcing function iteration.

Minimization of V with respect to $\boldsymbol{\mu}(t)$ is formulated conveniently in terms of a Hamiltonian $\mathscr{H}$ which accounts for the time dependence of $H_{\mathbf{u}}$, **K**, and **R**. State and co-state vectors are taken to be

$$\mathbf{y} = \begin{bmatrix} \boldsymbol{\chi} \\ t \end{bmatrix} \quad \text{and} \quad \mathbf{q} = \begin{bmatrix} \boldsymbol{\pi} \\ h \end{bmatrix}$$

respectively so that the corresponding Hamiltonian becomes

$$\mathscr{H} = \boldsymbol{\mu}' H_{\mathbf{u}} + \tfrac{1}{2}(\boldsymbol{\mu} + \mathbf{K}\boldsymbol{\chi})'\mathbf{R}(\boldsymbol{\mu} + \mathbf{K}\boldsymbol{\chi}) + \boldsymbol{\pi}'(H_{\mathbf{px}}\boldsymbol{\chi} + H_{\mathbf{pu}}\boldsymbol{\mu}) + h.$$

Co-state equations corresponding to $\boldsymbol{\chi}$ then are found to be

$$-\dot{\boldsymbol{\pi}} = \mathbf{K}'\mathbf{R}(\boldsymbol{\mu} + \mathbf{K}\boldsymbol{\chi}) + H_{\mathbf{xp}}\boldsymbol{\pi}; \quad \boldsymbol{\pi}(t_f) = \mathbf{0},$$

and the optimality equation which is analogous to (6-69) is found to be

$$\mathbf{0} = H_{\mathbf{u}} + \mathbf{R}(\boldsymbol{\mu} + \mathbf{K}\boldsymbol{\chi}) + H_{\mathbf{up}}\boldsymbol{\pi}.$$

These two equations can be written more conveniently as

$$-\dot{\boldsymbol{\pi}} = -\mathbf{K}'H_{\mathbf{u}} + \mathbf{A}'\boldsymbol{\pi}; \quad \boldsymbol{\pi}(t_f) = \mathbf{0} \tag{6-110}$$

and

$$\boldsymbol{\mu} = \boldsymbol{\delta} - \mathbf{K}\boldsymbol{\chi}, \tag{6-111}$$

where

$$\mathbf{A} = H_{\mathbf{px}} - H_{\mathbf{pu}}\mathbf{K} \tag{6-112}$$

and

$$\boldsymbol{\delta} = -\mathbf{R}^{-1}(H_{\mathbf{u}} + H_{\mathbf{up}}\boldsymbol{\pi}). \tag{6-113}$$

Direction vector $\boldsymbol{\mu}$ is thus formed by the linear feedback control equation given in (6-111) and the linearized state equation given in (6-66). The corresponding closed-loop system is therefore defined by

$$\dot{\boldsymbol{\chi}} = \mathbf{A}\boldsymbol{\chi} + H_{\mathbf{pu}}\boldsymbol{\delta}; \quad \boldsymbol{\chi}(0) = \mathbf{0}. \tag{6-114}$$

Matrix $\mathbf{A}$ is seen to be the Jacobian of the homogenous form of this closed-loop system, and vector $\boldsymbol{\delta}$ is seen to be the forcing function of this closed-loop system.

A convenient expression for the first variation of J can be found using

$$\int_0^{t_f} \{\boldsymbol{\chi}'(\dot{\boldsymbol{\pi}} - \mathbf{K}'H_{\mathbf{u}} + \mathbf{A}'\boldsymbol{\pi})\}\, \mathrm{d}t = 0$$

where $\boldsymbol{\pi}$ is the solution to (6-110). Integration by parts, substitution of (6-66), and subtraction of the resulting integral from (6-63) then results in

$$V_1 = -\int_0^{t_f} \{(\boldsymbol{\mu} + \mathbf{K}\boldsymbol{\chi})'\mathbf{R}\boldsymbol{\delta}\}\, \mathrm{d}t.$$

This form of the first variation of J reveals that $V_1[\boldsymbol{\delta} - \mathbf{K}\boldsymbol{\chi}] \leqslant 0$ and the equality holds if and only if $\|H_{\mathbf{u}}\| = 0$ because

$$V_1[\boldsymbol{\delta} - \mathbf{K}\boldsymbol{\chi}] = -\int_0^{t_f} \{\boldsymbol{\delta}'\mathbf{R}\boldsymbol{\delta}\}\, dt \tag{6-115}$$

and $\|\boldsymbol{\delta}(t)\| = 0$ holds if and only if $\|H_{\mathbf{u}}\| = 0$.

Forcing-function iteration methods can now be identified directly in terms of the notation introduced in Section 6.3. Specifically, direction vector $\boldsymbol{\mu}(t)$ was defined as

$$\boldsymbol{\mu}(t) = \frac{1}{\alpha}[\mathbf{u}(t, \alpha) - \mathbf{u}(t)]$$

in accordance with (6-58). In addition, the corresponding state vector $\boldsymbol{\chi}(t)$ is approximated as

$$\boldsymbol{\chi}(t) \cong \frac{1}{\alpha}[\mathbf{x}(t, \alpha) - \mathbf{x}(t)]$$

for nonlinear state equations in accordance with (6-64). Correspondingly, the linear feedback control equation given in (6-111) then results in the approximation

$$\mathbf{u}(t, \alpha) \cong \alpha\boldsymbol{\delta}(t) + \mathbf{k}(t) - \mathbf{K}(t)\mathbf{x}(t, \alpha)$$

where $\mathbf{k}(t)$ is given by

$$\mathbf{k} = \mathbf{u} + \mathbf{K}\mathbf{x} \tag{6-116}$$

when evaluated with $\mathbf{u}(t)$ and $\mathbf{x}(t)$. If $\mathbf{u}(t, \alpha)$ is identified with a new forcing function and $\mathbf{x}(t, \alpha)$ is identified with the resulting new state vector, then the forcing-function iteration method corresponds to

$$\mathbf{u} = \alpha\boldsymbol{\delta} + \mathbf{k} - \mathbf{K}\mathbf{x}. \tag{6-117}$$

The control equation given in (6-117) results in the function

$$f(\alpha) = J[\alpha\boldsymbol{\delta} + \mathbf{k} - \mathbf{K}\mathbf{x}]$$

on each iteration. Moreover, the first variation of J on each iteration, namely f_α evaluated at $\alpha = 0$, is given by (6-115) and hence is negative. Each iteration of forcing functions can therefore be performed so that a relative minimum of $f(\alpha)$ is located with $\alpha \geqslant 0$.

The simplest iteration method for forcing functions is formed with $\mathbf{R} = \mathbf{I}$ and $\mathbf{K} = \mathbf{0}$ which result in $\boldsymbol{\pi} = \mathbf{0}$ and hence $\boldsymbol{\delta} = -H_{\mathbf{u}}$. This method is analogous to steepest-descent and requires the solution of a total of $2n$ first order differential equations on each iteration where $n = \dim \mathbf{x}$. Each additional trial required to locate a relative minimum of $f(\alpha)$ requires the solution of n first order differential equations as well as evaluation of performance functional J. Furthermore, computer memory must be allocated so that a total of $(n + 3m)$ time functions can be stored in tabulated form where $m = \dim \mathbf{u}$.

A somewhat more complicated iteration method for forcing functions is formed with constant $\mathbf{R}$ and $\mathbf{K}$ and any nonzero $\mathbf{K}$. Then an additional set of n first order differential equations must be solved on each iteration because $\boldsymbol{\pi} \neq \mathbf{0}$. However, judicious selections of $\mathbf{K}$ generally result in significantly improved convergence rates as well as less trials per iteration in either locating a relative minimum of $f(\alpha)$ or merely establishing a monotone decreasing sequence of values of J.

In order to improve convergence rates and decrease the number of trials required per iteration, a method is needed for specifying judicious values of $\mathbf{R}$ and $\mathbf{K}$. For this purpose, model I_2 is taken to be of the same structural form as V_2 and hence is written as

$$I_2 = \tfrac{1}{2} \int_0^{t_f} \{\boldsymbol{\mu}'\mathbf{R}\boldsymbol{\mu} + 2\boldsymbol{\mu}'\mathbf{Q}\boldsymbol{\chi} + \boldsymbol{\chi}'\mathbf{P}\boldsymbol{\chi}\}\, dt + \tfrac{1}{2}\boldsymbol{\chi}_f'\mathbf{T}\boldsymbol{\chi}_f. \tag{6-118}$$

If $\mathbf{R}$ is positive definite on $[0, t_f]$ and I_2 given in (6-118) is assumed not to have conjugate points in $[0, t_f)$, then this model reduces to (6-109) using feedback gain matrix

$$\mathbf{K} = \mathbf{R}^{-1}(\mathbf{Q} + H_{\mathbf{up}}\boldsymbol{\Omega}) \tag{6-119}$$

where

$$-\dot{\boldsymbol{\Omega}} = \mathbf{P} + \boldsymbol{\Omega} H_{\mathbf{px}} + H_{\mathbf{px}}'\boldsymbol{\Omega} - \mathbf{K}'\mathbf{R}\mathbf{K}; \quad \boldsymbol{\Omega}(t_f) = \mathbf{T}. \tag{6-120}$$

Typical performance functionals are selected so that matrix $F_{\mathbf{uu}}$ is positive definite and matrices $(F_{\mathbf{xx}} - F_{\mathbf{xu}}F_{\mathbf{uu}}^{-1}F_{\mathbf{ux}})$ and $G_{\mathbf{x}_f\mathbf{x}_f}$ are positive semi-definite for all values of $\mathbf{x}$ and $\mathbf{u}$. Under these conditions, coefficient matrices appearing in (6-118) can be selected as

$$\mathbf{P} = \mathscr{H}_{\mathbf{xx}}, \quad \mathbf{Q} = \mathscr{H}_{\mathbf{ux}}, \quad \mathbf{R} = \mathscr{H}_{\mathbf{uu}}, \quad \text{and} \quad \mathbf{T} = G_{\mathbf{x}_f\mathbf{x}_f} \tag{6-121}$$

where

$$\mathscr{H} = F + \beta \mathbf{p}'\mathbf{f}. \tag{6-122}$$

Then matrix $\mathbf{R}$ is positive definite on $[0, t_f]$ and model I_2 does not have conjugate points in $[0, t_f)$ for some $\beta \in [0, 1]$ even when state equations are nonlinear.

The most important case of time dependent $\mathbf{R}$ and $\mathbf{K}$ given here of course occurs with $\beta = 1$. This case results in $I_2 = V_2$, and hence direction vector $\boldsymbol{\mu}(t)$ given by (6-111) minimizes the sum of the first and second variations of J. The iteration method given by (6-117) for this case with $\alpha = 1$ therefore results in one-step convergence for quadratic performance functionals and linear state equations. Moreover, use of $\alpha = \beta = 1$ in a suitably small neighborhood of a strict weak relative minimum, for which $H_{\mathbf{uu}}$ is nonsingular on $[0, t_f]$ and V_2 does not have a conjugate point at $\eta = 0$, results in quadratic convergence. This iteration method with $\alpha = \beta = 1$ thus is the Newton–Raphson method of successive substitutions for finding optimal forcing functions. On the other hand, the case with $\beta = 1$ and inaccurate initial forcing functions results in inadmissible time dependent **R** and **K** for many nonlinear state equations. The model I_2 given in (6-118) thus is often constructed with *convexity constraint* $\beta = 0$ for inaccurate initializations and nonlinear state equations. The convexity constraint is also often used to help reduce the propagation of numerical integration errors. The principal disadvantages of all iteration methods based on (6-118) are the additional $n(n + 1)/2$ first order differential equations that must be solved on each iteration and the additional computer memory required to store mn time dependent feedback gains in tabulated form.

A simple example serves to illustrate convergence rates attained with various models of the second variation of J, and the simplicity of this example allows the construction of these models with constant **R** and **K** from elementary analysis. The performance functional for the example is defined by

$$F = x_1^2 + x_2^2 + u_1^2, \quad G = 0, \quad \text{and} \quad t_f = 10.$$

Also, the state equations for the example are defined by

$$\mathbf{f} = \begin{bmatrix} (1 - x_2^2)x_1 - x_2 + u_1 \\ x_1 \end{bmatrix}, \quad \text{and} \quad \mathbf{x}_0 = \begin{bmatrix} 0 \\ 3 \end{bmatrix}.$$

All data presented are based on a crude method of selecting α which is based on the assumption that Newton–Raphson characteristics are achieved in a suitably small neighborhood of the extremal. In particular, the selection of α is performed by initializing with $\alpha = 1$, repeated halving of α until J decreases monotonely, and then doubling α if $\alpha \leqslant \frac{1}{2}$ for use in the next iteration. The total number of unsuccessful selections of α is denoted by N, and the total computer running time is denoted by T. Finally, all iterations are initialized by generating **u** from (6-117) with

$$\delta_1 = 0, \quad k_1 = 0, \quad k_{11} = 1 + 2^{3/4}, \quad \text{and} \quad k_{12} = 2^{1/2} - 1.$$

This value of **K** corresponds to the equilibrium solution of (6-119) and (6-120) with $\mathbf{x} = \mathbf{0}$ and $\beta = 1$.

The construction of model I_2 with constant **R** and **K** is based on the following elementary analysis. The value $r_{11} = 2$ is taken from (6-121) and (6-122) for all cases with constant **R** and **K.** Also, the constant matrix

$$\mathbf{A} = \begin{bmatrix} (1 - k_{11}) & -(1 + k_{12}) \\ 1 & 0 \end{bmatrix}$$

is found by evaluating (6-112) at $\mathbf{x} = \mathbf{0}$ and is used to estimate the damping ratio ζ and natural frequency ω of (6-114). These characteristics are given by

$$\zeta = \frac{k_{11} - 1}{2(1 + k_{12})^{1/2}} \quad \text{and} \quad \omega = (1 + k_{12})^{1/2}$$

for $\mathbf{x} = \mathbf{0}$. Various cases of constant **k** and **K** are defined in Table 6-1, and these are selected primarily in accordance with damping ratio. Case 1 corresponds to steepest-descent, and Case 3 corresponds to a constant-coefficient approximation to the Newton–Raphson method.

Table 6-1—Definition of cases with constant **k** and **K**

Case	k_1	k_{11}	k_{12}	ζ	ω
1	0.0	0.00	0.000	−0.500	1.00
2	0.0	1.00	0.414	0.000	1.19
3	0.0	2.68	0.414	0.707	1.19
4	0.0	5.76	0.414	2.000	1.19

Table 6-2—Values of performance functional for Cases 1 through 4

Iteration	Case 1	Case 2	Case 3	Case 4
0	25.37539	25.37539	25.37539	25.37539
1	25.32569	24.71869	22.04748	25.22020
2	25.32133	23.88076	21.55241	24.92939
3	25.23766	22.82092	21.45488	24.41867
4	25.19793	22.43837	21.42831	23.62971
5	25.19357	22.25134	21.42048	22.67345
6	25.09221	22.11994	21.41806	22.20423
7	25.05667	22.05400	21.41731	21.92705

Convergence characteristics for these four cases with constant **k** and **K** are summarized in Tables 6-2 and 6-3. Steepest-descent is seen to result in decidedly inferior convergence rate as well as a large number of unsuccessful trials. Convergence rate of the steepest-descent method could of course be improved by selecting α so that a relative minimum of $f(\alpha)$ is located on each iteration. Additional unsuccessful trials would be required, however, and total computer

Table 6-3—Running time and unsuccessful trials for Cases 1 through 4

Case	T	N
1	15.81	14
2	13.87	5
3	10.77	0
4	10.77	0

running time would not be improved appreciably. On the other hand, the constant-gain approximation to the Newton–Raphson method is seen to result in relatively rapid convergence, albeit not quadratic, and this convergence rate is achieved by successive substitutions with $\alpha = 1$. Cases 2 and 4 also indicate that reasonable convergence rates are obtained with relatively inaccurate constant-gain models of the second variation of J and that unsuccessful trials are reduced by employing increased damping which corresponds to a constant-coefficient approximation of $\mathbf{A}$.

Computer results are given in Tables 6-4 and 6-5 for Cases 5 and 6 which are based on (6-119) through (6-122) with $\beta = 1$ and $\beta = 0$ respectively. These data illustrate quadratic convergence characteristics for Case 5 once a suitably small neighborhood of the extremal is achieved. The time-dependent feedback-gain matrix for this case, however, results in a somewhat underdamped closed-loop

Table 6-4—Values of performance functional for Cases 5 and 6

Iteration	Case 5	Case 6
0	25.37539	25.37539
1	22.63782	21.70071
2	21.77779	21.43896
3	21.42465	21.41994
4	21.41697	21.41738
5	21.41695	21.41700
6	–	21.41696
7	–	21.41695

Table 6-5—Running time and unsuccessful trials for Cases 5 and 6

Case	T	N
5	12.76	2
6	15.88	0

system for gross approximations and hence results in a rather poor initial convergence rate as well as some unsuccessful trials. On the other hand, the time-dependent feedback-gain matrix for Case 6 does not exhibit this undesirable tendency. Quadratic convergence, however, is not achieved when convexity constraint $\beta = 0$ is imposed with nonlinear state equations.

Reference signal $\mathbf{k}(t)$ and feedback gains $\mathbf{K}(t)$ are plotted in Figure 6-2 for Case 5 at the extremal where $\boldsymbol{\delta} = \mathbf{0}$ on $[0, t_f]$. Similar time functions are plotted in Figure 6-3 for Case 6. These time functions can be compared with the constant approximations given in Table 6-1. Constant $\mathbf{k}$ and $\mathbf{K}$ for Case 3 are seen to be reasonably accurate approximations on the interval $[4.5, 6.5]$ but otherwise differ rather markedly from time dependent $\mathbf{k}$ and $\mathbf{K}$. Figures 6-2 and 6-3 also illustrate that the convexity constraint results in significantly increased position feedback gain k_{12} on the interval $[0, 2]$ where nonlinearities of the state equations are significant as determined from the optimal trajectory plotted in Figure 6-4.

6.7 LINEARIZED OPTIMAL CONTROL

In the case of linear optimal control such as discussed in Section 6.4, optimal forcing functions can be implemented in feedback control equation form for all initial states. However, implementation of optimal forcing functions in feedback control equation form generally is impossible for nonquadratic performance functionals, nonlinear state equations, or even linear state equations where access to the entire state vector is not permitted. Properties of the linear feedback control equation appearing in (6-117), however, suggest that optimal forcing functions can be approximated in a suitably small neighborhood of an extremal when access to the entire state vector is permitted. Feedback control equations of this type are discussed in this section from the point of view of approximating optimal forcing functions for initial states in a neighborhood of some $\mathbf{x}_0$. Throughout this section, the extremal corresponding to $\mathbf{x}_0$ is assumed to be a strict weak relative minimum which satisfies all of the conditions of Theorem 6-4.

The *family* of extremals corresponding to initial states in a neighborhood of $\mathbf{x}_0$ satisfies canonical equations

$$\dot{\mathbf{x}} = H_{\mathbf{p}}; \quad \mathbf{x}(0) = \mathbf{x}_0 + \alpha\mathbf{d}, \tag{6-123}$$

(6-57), and (6-69). Solutions to these equations are denoted by $\mathbf{x}(t, \alpha)$, $\mathbf{p}(t, \alpha)$, and $\mathbf{u}(t, \alpha)$ assuming direction vector $\mathbf{d}$ is fixed but arbitrary. In addition, notation introduced in Section 6.3 is employed again except for redefinitions

$$\mathbf{u}(t) = \mathbf{u}(t, \alpha)|_{\alpha=0} \quad \text{and} \quad \boldsymbol{\mu}(t) = \mathbf{u}_\alpha(t, \alpha)|_{\alpha=0}.$$

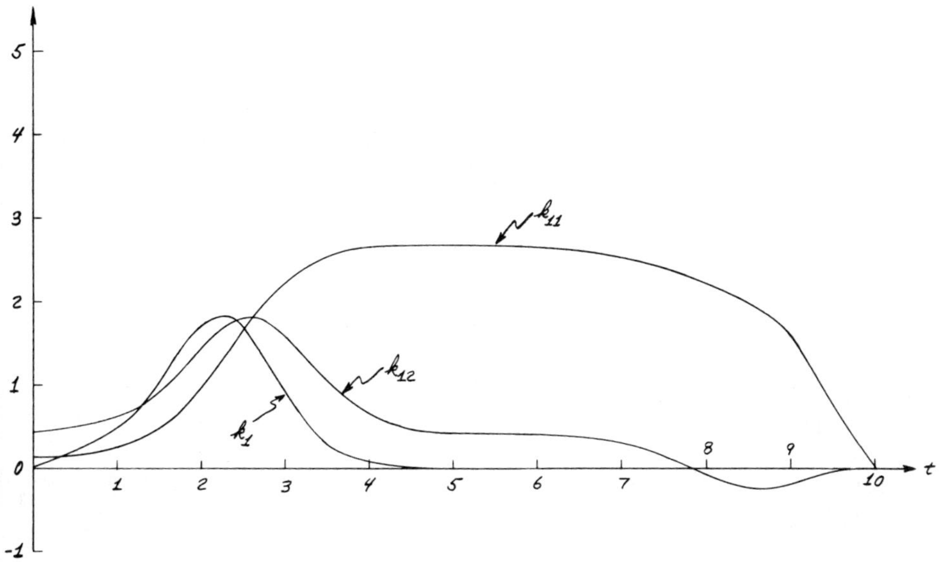

Figure 6-2 Reference signal and feedback gains for Case 5 ($\beta = 1$).

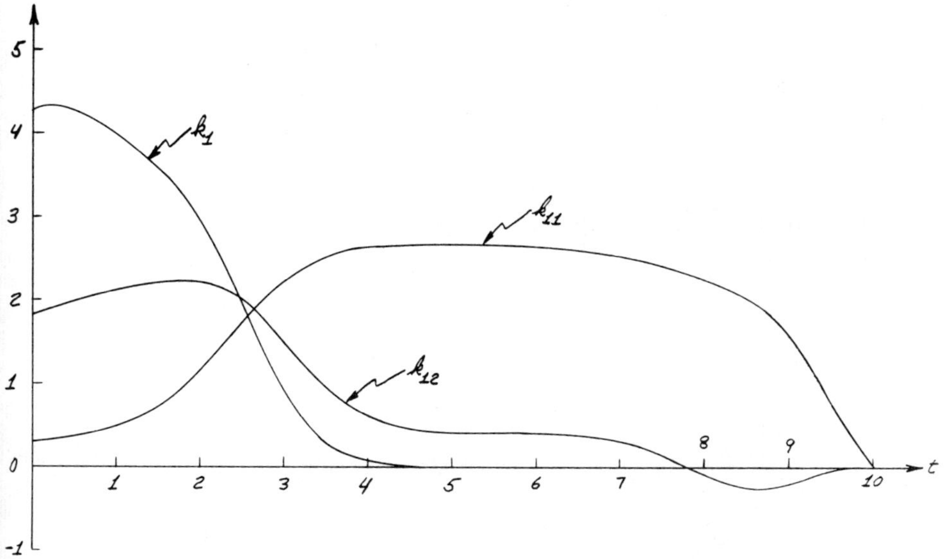

Figure 6-3 Reference signal and feedback gains for Case 6 ($\beta = 0$).

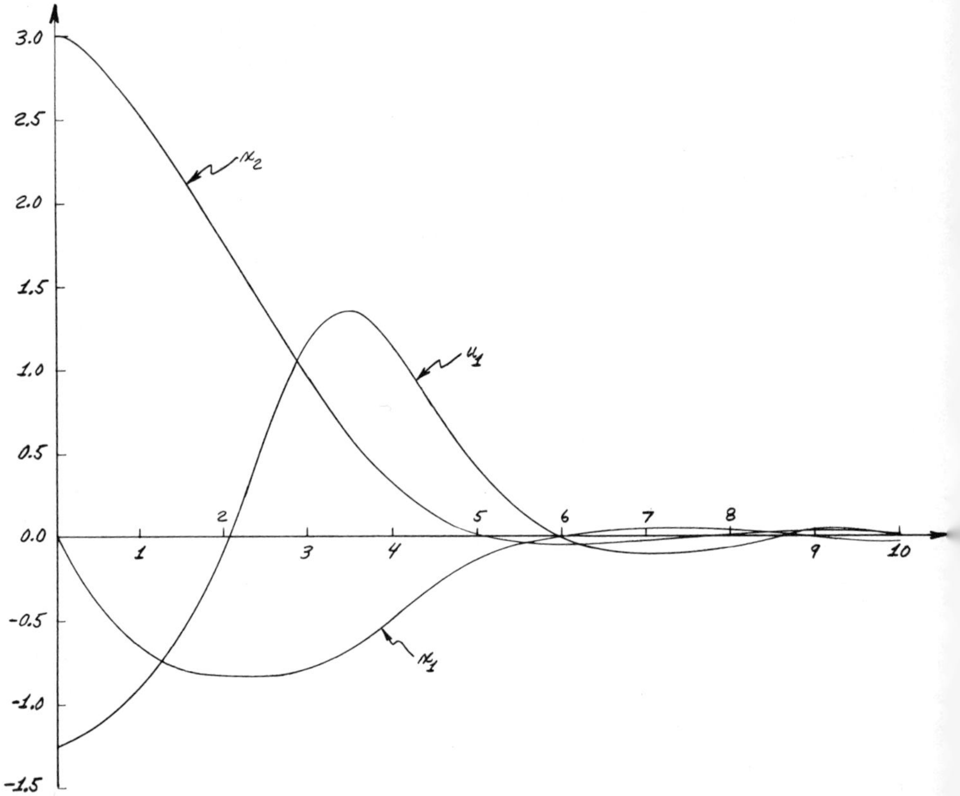

Figure 6-4 Optimal trajectory.

Differentiation of (6-123), (6-57), and (6-69) with respect to α and evaluation of the results at $\alpha = 0$ now yield

$$\dot{\boldsymbol{\chi}} = H_{\mathbf{px}}\boldsymbol{\chi} + H_{\mathbf{pu}}\boldsymbol{\mu}; \quad \boldsymbol{\chi}(0) = \mathbf{d}, \tag{6-124}$$

(6-67), and (6-74) respectively. Elimination of $\boldsymbol{\mu}$ using (6-74) further reduces (6-124) and (6-67) to the form

$$-\begin{bmatrix} \dot{\boldsymbol{\chi}} \\ \dot{\boldsymbol{\pi}} \end{bmatrix} = \boldsymbol{\Delta}(1)\begin{bmatrix} \boldsymbol{\chi} \\ \boldsymbol{\pi} \end{bmatrix}; \quad \boldsymbol{\chi}(0) = \mathbf{d}, \quad \boldsymbol{\pi}(t_f) = G_{\mathbf{x}_f\mathbf{x}_f}\boldsymbol{\chi}_f \tag{6-125}$$

where $\boldsymbol{\Delta}(1)$ is specified by (6-76). A comparison of (6-125) and (6-77) with $\beta = 1$ reveals that equalities

$$\boldsymbol{\chi}(t) = \boldsymbol{\Phi}(t)\boldsymbol{\Phi}^{-1}(0)\mathbf{d} \quad \text{and} \quad \boldsymbol{\pi}(t) = \boldsymbol{\Psi}(t)\boldsymbol{\Phi}^{-1}(0)\mathbf{d} \tag{6-126}$$

hold and hence

$$\boldsymbol{\pi}(t) = \boldsymbol{\Omega}(t)\boldsymbol{\chi}(t) \tag{6-127}$$

is obtained where $\boldsymbol{\Omega}$ is defined by (6-82) and satisfies (6-83). Substitution of this relationship for $\boldsymbol{\pi}(t)$ into (6-74) now yields

$$\boldsymbol{\mu} = -\mathbf{K}\boldsymbol{\chi} \tag{6-128}$$

where $\mathbf{K}$ is defined by (6-84). Finally, substitution of (6-128) into (6-124) results in

$$\dot{\boldsymbol{\chi}} = \mathbf{A}\boldsymbol{\chi}; \quad \boldsymbol{\chi}(0) = \mathbf{d} \tag{6-129}$$

where $\mathbf{A}$ is defined by (6-112).

Direction vector $\boldsymbol{\chi}(t)$ of the family of extremals thus is the solution of a linear state equation. Moreover, direction vector $\boldsymbol{\mu}(t)$ of the corresponding optimal forcing functions is given by a linear feedback control equation. The feedback gain matrix which appears in (6-128) is independent of direction vector $\mathbf{d}$, and (6-128) hence can be approximated in the same sense to obtain (6-117). That is, the definition of a partial derivative is used to approximate (6-128) as

$$\mathbf{u}(t, \alpha) \cong \mathbf{k}(t) - \mathbf{K}(t)\mathbf{x}(t, \alpha)$$

where $\mathbf{k}(t)$ is given by (6-116) when evaluated with $\mathbf{u}(t)$ and $\mathbf{x}(t)$. Therefore, *linearized optimal control* is defined by

$$\mathbf{u} = \mathbf{k} - \mathbf{K}\mathbf{x} \tag{6-130}$$

and is used to approximate optimal forcing functions for all state vectors that belong to a suitably small neighborhood of the extremal corresponding to initial state $\mathbf{x}_0$.

Reference signals and feedback gains which appear in (6-130) for linearized optimal control are identical to those appearing in (6-117) at an extremal when (6-119) through (6-122) are used with $\beta = 1$ to compute $\mathbf{K}$. In other words, linearized optimal control can be implemented directly from the Newton–Raphson method of successive approximations. If relatively accurate approximations to optimal forcing functions are required, then the neighborhood of the extremal corresponding to initial state $\mathbf{x}_0$ must be very small for some problems. Moreover, very undesirable behavior may be experienced outside of this neighborhood with unstable nonlinear state equations. As a matter of practical expediency, reference signals and feedback gains appearing in (6-130) are often computed with convexity constraint $\beta = 0$ in order to increase the neighborhood in which an acceptable approximation to optimal forcing functions is sought. This convexity constraint also acts as a stability constraint with nonlinear state equations.

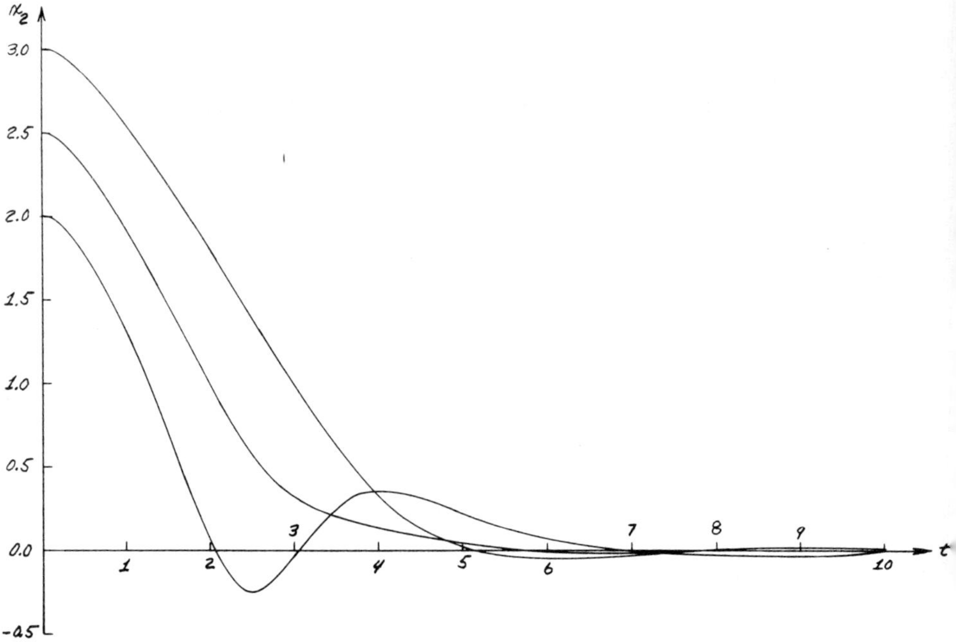

Figure 6-5 Position responses for linearized optimal control ($\beta = 1$).

An elementary example of linear control based on (6-130) is provided by the example introduced in Section 6.6. Linearized optimal control is computed from (6-117) through (6-122) with $\beta = 1$ and hence is constructed with the reference signal and feedback gains shown in Figure 6-2. Typical responses of the corresponding closed-loop system are shown in Figure 6-5. The response for $x_1(0) = 0.0$ and $x_2(0) = 2.5$ is a reasonable approximation to an extremal, whereas the response for $x_1(0) = 0.0$ and $x_2(0) = 2.0$ exhibits a tendency for instability and hence is unacceptable. Linear control with convexity constraint $\beta = 0$ is constructed with the reference signal and feedback gains shown in Figure 6-3. Typical responses are shown in Figure 6-6 for the corresponding closed-loop system, and

Table 6-6—Values of performance functional with $x_1(0) = 0.0$

$x_2(0)$	Optimal control	Linear control with $\beta = 1$	Linear control with $\beta = 0$
3.0	21.42	21.42	21.42
2.5	12.70	13.72	14.00
2.0	7.75	14.44	10.91

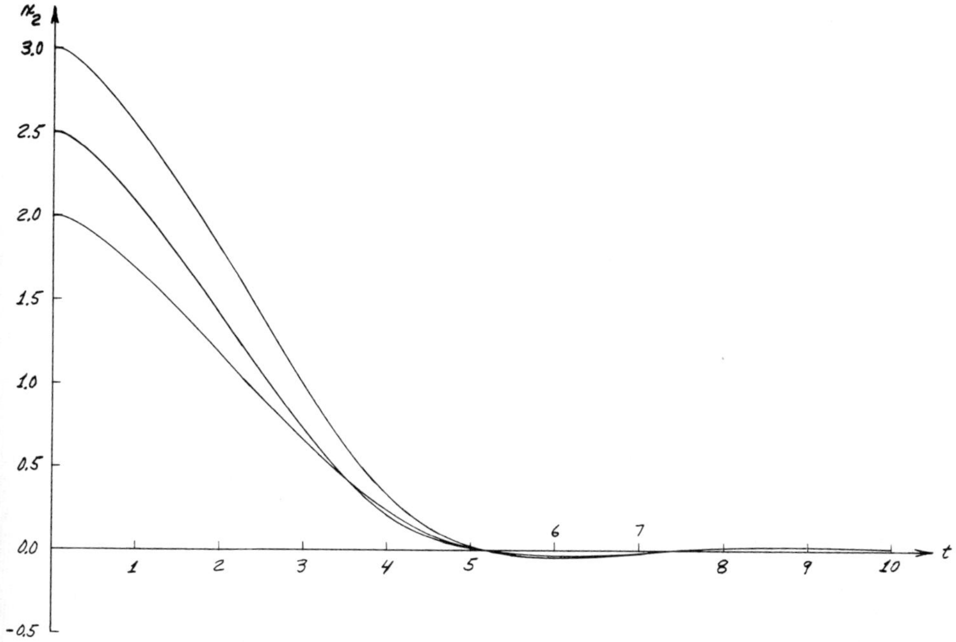

Figure 6-6 Position responses for linear control with $\beta = 0$.

all responses are considered to be acceptable approximations to extremals. Values of performance functional J are summarized in Table 6-6 for optimal nonlinear control, linearized optimal control, and linear control with convexity constraint $\beta = 0$.

6.8 OPTIMAL FORCING FUNCTIONS WITH OPTIMAL PARAMETERS AND TERMINAL EQUALITY CONSTRAINTS

Many design problems involving optimal forcing functions cannot be formulated in the simple form given in Section 6.3. For example, many such problems involving optimal forcing functions also involve optimal parameters and terminal equality constraints. This specific class of problems is presented here not only as a useful extension of previous results but also as a summary of previous concepts.

Finite-time design problems with optimal forcing functions and parameters as well as terminal equality constraints are introduced in this section as follows. Nonlinear state equations are taken to be

$$\dot{\mathbf{x}} = \mathbf{f}(\mathbf{x}, \mathbf{u}); \quad \mathbf{x}(0) = \boldsymbol{\gamma} + \boldsymbol{\Gamma}\mathbf{c}, \quad \mathbf{0} = \mathbf{g}(\mathbf{x}_f) \tag{6-131}$$

where $\mathbf{u}$ denotes adjustable forcing functions and $\mathbf{c}$ denotes adjustable parameters. In addition, the performance functional is taken to be

$$J = E(\mathbf{c}) + \int_0^{t_f} F(\mathbf{x}, \mathbf{u})\, \mathrm{d}t + G(\mathbf{x}_f) \tag{6-132}$$

where final time t_f is fixed and $\mathbf{x}_f = \mathbf{x}(t_f)$. Performance functional J is to be minimized with respect to both time functions $\mathbf{u}$ and parameters $\mathbf{c}$.

For the purpose of stating necessary and sufficient conditions for a relative minimum in a convenient form, the Hamiltonian given in (6-54) is introduced as well as the Lagrangian given in (6-4). Correspondingly, Lagrangian $\mathscr{L} = \mathscr{L}[\mathbf{u}; \mathbf{c}, \boldsymbol{\lambda}]$ is introduced so that the performance functional can be written as

$$\mathscr{L} = E + \int_0^{t_f} \{H - \mathbf{p}'\dot{\mathbf{x}}\}\, \mathrm{d}t + L. \tag{6-133}$$

State equations are also rewritten in terms of the Hamiltonian as

$$\dot{\mathbf{x}} = H_{\mathbf{p}}; \quad \mathbf{x}(0) = \boldsymbol{\gamma} + \boldsymbol{\Gamma}\mathbf{c}, \tag{6-134}$$

and the co-state vector $\mathbf{p}$ is defined by

$$-\dot{\mathbf{p}} = H_{\mathbf{x}}; \quad \mathbf{p}(t_f) = L_{\mathbf{x}_f} \tag{6-135}$$

under suitable assumptions.

Conditions presented in this section for optimal forcing functions and parameters are simplified by a number of assumptions. First, only forcing functions belonging to $C^m[0, t_f]$ are investigated. Second, Lagrangian $\mathscr{L}$ is assumed to be *twice continuously differentiable* on $C^m[0, t_f]$ in accordance with conditions

i) J is twice continuously differentiable on $C^m[0, t_f]$ for all $\mathbf{c} \in \mathscr{R}^k$,
ii) E is twice continuously differentiable on $\mathscr{R}^k$,

and

iii) L is twice continuously differentiable with respect to $\mathbf{x}_f \in \mathscr{R}^n$ given any $\boldsymbol{\lambda} \in \mathscr{R}^l$.

Third, constraint functions $\mathbf{g}$ are assumed to have a manifold defined on $\mathscr{R}^n$. Fourth, only forcing functions which result in a completely reachable terminal tangent space are investigated.

The fourth assumption imposed here is stated more precisely as

Definition 6-3 Let $\boldsymbol{\chi}$ be defined by

$$\dot{\boldsymbol{\chi}} = H_{\mathbf{px}}\boldsymbol{\chi} + H_{\mathbf{pu}}\boldsymbol{\mu}; \quad \boldsymbol{\chi}(0) = \boldsymbol{\chi}_0 \tag{6-136}$$

where $H_{\mathbf{px}}$ and $H_{\mathbf{pu}}$ exist and are continuous on $[0, t_f]$. Then the tangent space of $\mathbf{g}$ is *completely reachable* if and only if, given any $\boldsymbol{\chi}_0 \in \mathscr{R}^n$, there exists a $\boldsymbol{\mu}(t) \in C^m [0, t_f]$ such that

$$\mathbf{g}_{\mathbf{x}_f}\boldsymbol{\chi}(t_f) = \mathbf{0}.$$

A necessary and sufficient condition for complete reachability of the terminal tangent space is stated in terms of the integral

$$\mathbf{X} = \int_0^{t_f} \{\mathbf{V}' H_{\mathbf{pu}} H_{\mathbf{up}} \mathbf{V}\}\, \mathrm{d}t \tag{6-137}$$

where

$$-\dot{\mathbf{V}} = H_{\mathbf{xp}}\mathbf{V}; \quad \mathbf{V}(t_f) = \mathbf{g}'_{\mathbf{x}_f}. \tag{6-138}$$

This condition is based on the relationship

$$\mathbf{g}_{\mathbf{x}_f}\boldsymbol{\chi}(t_f) = \mathbf{V}'(0)\boldsymbol{\chi}_0 + \int_0^{t_f} \{\mathbf{V}' H_{\mathbf{pu}} \boldsymbol{\mu}\}\, \mathrm{d}t \tag{6-139}$$

and is stated as

Lemma 6-7 *The tangent space of* $\mathbf{g}$ *is completely reachable if and only if* $\mathbf{X}$ *is nonsingular.*

Proof Assume the tangent space of $\mathbf{g}$ is completely reachable but $\mathbf{X}$ is singular. Then there exists a nonzero constant vector $\mathbf{v} \in \mathscr{R}^l$ such that $\mathbf{v}'\mathbf{X}\mathbf{v} = 0$. Therefore, condition

$$\mathbf{v}'\mathbf{V}' H_{\mathbf{pu}} = \mathbf{0}'$$

must hold on $[0, t_f]$ and hence

$$\mathbf{v}'\mathbf{g}_{\mathbf{x}_f}\boldsymbol{\chi}(t_f) = \mathbf{v}'\mathbf{V}'(0)\boldsymbol{\chi}_0$$

holds for all $\boldsymbol{\mu}(t) \in C^m [0, t_f]$. Now let $\boldsymbol{\mu}$ be selected so that

$$\mathbf{0} = \mathbf{g}_{\mathbf{x}_f}\boldsymbol{\chi}(t_f)$$

holds given each $\boldsymbol{\chi}_0 \in \mathscr{R}^n$. The previous relationship then is seen to require that $\mathbf{V}(0)\mathbf{v} = \mathbf{0}$ for some nonzero $\mathbf{v}$. However, if $\mathbf{v} \neq \mathbf{0}$, then condition

$$\mathbf{g}'_{\mathbf{x}_f}\mathbf{v} \neq \mathbf{0}$$

must hold because rank $\mathbf{g}_{\mathbf{x}_f}$ = dim $\mathbf{g}$ by definition of a tangent space. Thus, by uniqueness of solutions to ordinary differential equations, a contradiction has been established and hence $\mathbf{X}$ must be nonsingular when the tangent space of $\mathbf{g}$ is completely reachable. Now assume $\mathbf{X}$ is nonsingular and let

$$\boldsymbol{\mu} = H_{\mathbf{up}}\mathbf{V}\mathbf{v}$$

where $\mathbf{v}$ is the constant vector

$$\mathbf{v} = -\mathbf{X}^{-1}\mathbf{V}'(0)\boldsymbol{\chi}_0.$$

Then evaluation of (6-139) yields

$$\mathbf{g}_{\mathbf{x}_f}\boldsymbol{\chi}(t_f) = \mathbf{0}$$

and hence the tangent space of $\mathbf{g}$ is completely reachable.

Complete reachability also yields the following elementary result

Lemma 6-8 *Let* $\mathbf{W}$ *be defined as*

$$\mathbf{W} = \int_0^{t_f} \{\mathbf{U}'H_{\mathbf{pu}}H_{\mathbf{uu}}^{-1}H_{\mathbf{up}}\mathbf{U}\}\,\mathrm{d}t$$

where

$$-\dot{\mathbf{U}} = (H_{\mathbf{px}} - H_{\mathbf{pu}}\mathbf{K})'\mathbf{U}; \quad \mathbf{U}(t_f) = \mathbf{g}'_{\mathbf{x}_f}$$

assuming that $\mathbf{K}$ *and* $H_{\mathbf{uu}}$ *exist and are continuous on* $[0, t_f]$. *Then matrix* $\mathbf{W}$ *is nonsingular for all* $\mathbf{K}$ *when the tangent space of* $\mathbf{g}$ *is completely reachable and* $H_{\mathbf{uu}}$ *is positive definite on* $[0, t_f]$.

Proof Equation (6-136) is rewritten as

$$\dot{\boldsymbol{\chi}} = (H_{\mathbf{px}} - H_{\mathbf{pu}}\mathbf{K})\boldsymbol{\chi} + H_{\mathbf{pu}}\mathbf{m}; \quad \boldsymbol{\chi}(0) = \boldsymbol{\chi}_0.$$

where

$$\mathbf{m} = (\boldsymbol{\mu} + H_{\mathbf{pu}}\mathbf{K}\boldsymbol{\chi}).$$

Then, by direct construction, there must exist an $\mathbf{m}(t) \in C^m[0, t_f]$ for which

$$\mathbf{g}_{\mathbf{x}_f}\boldsymbol{\chi}(t_f) = \mathbf{0}$$

holds given any $\boldsymbol{\chi}_0 \in \mathscr{R}^n$. The tangent space of $\mathbf{g}$ must still be completely reachable, and hence

$$\overline{\mathbf{W}} = \int_0^{t_f} \{\mathbf{U}'H_{\mathbf{pu}}H_{\mathbf{up}}\mathbf{U}\}\,\mathrm{d}t$$

must be nonsingular by Lemma 6-7. Moreover, matrix $\mathbf{W}$ is singular only if there exists a nonzero constant $\mathbf{v}$ such that

$$\mathbf{v}'\mathbf{U}'H_{\mathbf{pu}} = \mathbf{0}'$$

holds $[0, t_f]$. No such vector exists, however, because $\overline{\mathbf{W}}$ is nonsingular.

Complete reachability is introduced in order to assure that the Lagrange multiplier rule applies to terminal equality constraints.

The first and second variations of Lagrangian $\mathscr{L}$ are expressed in terms of forcing functions defined in (6-58) as well as

$$\mathbf{c}(\alpha) = \mathbf{c} + \alpha\mathbf{d} \quad \text{and} \quad \boldsymbol{\lambda}(\alpha) = \boldsymbol{\lambda} + \alpha l. \tag{6-140}$$

These directional derivatives are denoted by

$$\mathscr{V}_1 = \mathscr{L}_\alpha[\mathbf{u} + \alpha\boldsymbol{\mu};\ \mathbf{c} + \alpha\mathbf{d}, \boldsymbol{\lambda} + \alpha l]\,|_{\alpha=0} \tag{6-141}$$

and

$$\mathscr{V}_2 = \tfrac{1}{2}\mathscr{L}_{\alpha\alpha}[\mathbf{u} + \alpha\boldsymbol{\mu};\ \mathbf{c} + \alpha\mathbf{d}, \boldsymbol{\lambda} + \alpha l]\,|_{\alpha=0} \tag{6-142}$$

respectively. The first variation of $\mathscr{L}$ is derived by the same steps used to derive (6-63) and is found to be

$$\mathscr{V}_1 = \mathbf{d}'[E_\mathbf{c} + \boldsymbol{\Gamma}'\mathbf{p}(0)] + \int_0^{t_f} \{\boldsymbol{\mu}'H_\mathbf{u}\}\,dt + l'L_\lambda \tag{6-143}$$

using (6-134), (6-135), and $\mathbf{x}_\alpha(0, \alpha) = \boldsymbol{\Gamma}\mathbf{d}$. In addition, the second variation of $\mathscr{L}$ is derived by the same steps used to derive (6-68) and is found to be

$$\mathscr{V}_2 = \tfrac{1}{2}\mathbf{d}'E_{\mathbf{cc}}\mathbf{d} + \tfrac{1}{2}\int_0^{t_f} \{\boldsymbol{\mu}'H_{\mathbf{uu}}\boldsymbol{\mu} + 2\boldsymbol{\mu}'H_{\mathbf{ux}}\boldsymbol{\chi} + \boldsymbol{\chi}'H_{\mathbf{xx}}\boldsymbol{\chi}\}\,dt$$

$$+ \tfrac{1}{2}\boldsymbol{\chi}_f'L_{\mathbf{x}_f\mathbf{x}_f}\boldsymbol{\chi}_f + \boldsymbol{\chi}_f'L_{\mathbf{x}_f\lambda}\mathbf{l} \tag{6-144}$$

using

$$\dot{\boldsymbol{\chi}} = H_{\mathbf{px}}\boldsymbol{\chi} + H_{\mathbf{pu}}\boldsymbol{\mu};\quad \boldsymbol{\chi}(0) = \boldsymbol{\Gamma}\mathbf{d} \tag{6-145}$$

and

$$-\dot{\boldsymbol{\pi}} = H_{\mathbf{xx}}\boldsymbol{\chi} + H_{\mathbf{xu}}\boldsymbol{\mu} + H_{\mathbf{xp}}\boldsymbol{\pi};\quad \boldsymbol{\pi}(t_f) = L_{\mathbf{x}_f\mathbf{x}_f}\boldsymbol{\chi}_f + L_{\mathbf{x}_f\lambda}\mathbf{l}. \tag{6-146}$$

First and second variations $\mathscr{V}_1$ and $\mathscr{V}_2$ are noted to reduce to the first and second variations of performance functional J when admissibility conditions

$$\mathbf{0} = L_{\boldsymbol{\lambda}} \tag{6-147}$$

and

$$\mathbf{0} = L_{\boldsymbol{\lambda}\mathbf{x}_f}\boldsymbol{\chi}_f \tag{6-148}$$

are imposed.

The first necessary condition for a constrained relative minimum can now be stated as

Theorem 6-6 *Suppose performance functional J is defined by (6-131) and (6-132), and the corresponding Lagrangian is twice continuously differentiable on* $C^m [0, t_f]$. *Also let* $\mathbf{u}(t) \in C^m [0, t_f]$ *be an admissible forcing function for which the tangent space of* **g** *is completely reachable. Then J has a weak relative minimum at* **u** *only if (6-69) holds on* $[0, t_f]$ *and*

$$\mathbf{0} = E_{\mathbf{c}} + \boldsymbol{\Gamma}'\mathbf{p}(0) \tag{6-149}$$

holds for some real $\boldsymbol{\lambda}$.

Proof The first variation of performance functional J must vanish for all admissible **d** and $\boldsymbol{\mu}$ at a weak relative minimum. First let

$$\mathbf{d} = [E_{\mathbf{c}} + \boldsymbol{\Gamma}'\mathbf{p}(0)] \quad \text{and} \quad \boldsymbol{\mu} = H_{\mathbf{u}}.$$

Then, if this selection of **d** and $\boldsymbol{\mu}$ is admissible for some $\boldsymbol{\lambda} \in \mathscr{R}^l$, the first variation of J becomes

$$\mathscr{V}_1 = [E_{\mathbf{c}} + \boldsymbol{\Gamma}'\mathbf{p}(0)]'[E_{\mathbf{c}} + \boldsymbol{\Gamma}'\mathbf{p}(0)] + \int_0^{t_f} \{H_{\mathbf{u}}'H_{\mathbf{u}}\}\, dt > 0$$

unless (6-69) and (6-149) hold. Admissibility of this selection of **d** and $\boldsymbol{\mu}$ for some $\boldsymbol{\lambda} \in \mathscr{R}^l$ is established by introducing **r** and **V** in accordance with

$$-\dot{\mathbf{r}} = (F + \mathbf{r}'\mathbf{f})_{\mathbf{x}}; \quad \mathbf{r}(t_f) = G_{\mathbf{x}_f}$$

and (6-138). These variables result in

$$\mathbf{p} = \mathbf{r} + \mathbf{V}\boldsymbol{\lambda}$$

and

$$H_{\mathbf{u}} = (F + \mathbf{r}'\mathbf{f})_{\mathbf{u}} + H_{\mathbf{up}}\mathbf{V}\boldsymbol{\lambda}.$$

The relationship given in (6-139) then becomes

$$L_{\boldsymbol{\lambda}\mathbf{x}_f}\boldsymbol{\chi}_f = \mathbf{V}'(0)\boldsymbol{\Gamma}[E_\mathbf{c} + \boldsymbol{\Gamma}'\mathbf{r}(0)] + \int_0^{t_f} \{\mathbf{V}'H_{\mathbf{pu}}(F + \mathbf{r}'\mathbf{f})_\mathbf{u}\}\, dt$$
$$+ [\mathbf{X} + \mathbf{V}'(0)\boldsymbol{\Gamma}\boldsymbol{\Gamma}'\mathbf{V}(0)]\boldsymbol{\lambda},$$

and thus $\boldsymbol{\lambda}$ can be selected so that the admissibility condition given in (6-148) is satisfied.

Equations (6-134), (6-135), (6-69), (6-147), and (6-149) define extremals for (6-131) and (6-132) and hence are called the *canonical equations* for this class of design problems. Performance functional J is stationary at extremals.

The second necessary condition for an admissible weak relative minimum in $C^m[0, t_f]$ is again found to be $H_{\mathbf{uu}}$ positive semi-definite on $[0, t_f]$ under suitable assumptions. The proof of this result is identical to the proof given for Theorem 6-2. Only nonsingular admissible extremals are considered in the sequel so that $H_{\mathbf{uu}}$ is positive definite for all admissible weak relative minima under consideration.

A sufficient condition for strict weak relative minima which are admissible is based on the assumption of no conjugate points in $[0, t_f)$. Conjugate points are redefined for (6-131) and (6-132) in terms of the solution to

$$-\begin{bmatrix}\dot{\boldsymbol{\Phi}}\\ \dot{\boldsymbol{\Psi}}\end{bmatrix} = \boldsymbol{\Delta}(1)\begin{bmatrix}\boldsymbol{\Phi}\\ \boldsymbol{\Psi}\end{bmatrix}; \quad \begin{bmatrix}\boldsymbol{\Phi}(t_f)\\ \boldsymbol{\Psi}(t_f)\end{bmatrix} = \begin{bmatrix}\mathbf{I}\\ L_{\mathbf{x}_f\mathbf{x}_f}\end{bmatrix} \tag{6-150}$$

where $\boldsymbol{\Delta}(1)$ is defined by (6-76). Specifically, values of η for which $\boldsymbol{\Phi}(\eta)$ is singular are called conjugate points of $\mathscr{V}_2$. If interval $[0, t_f)$ contains no conjugate points, then matrix $\boldsymbol{\Omega}$ is again defined by (6-82) but now satisfies

$$-\dot{\boldsymbol{\Omega}} = H_{\mathbf{xx}} + \boldsymbol{\Omega}H_{\mathbf{px}} + H_{\mathbf{px}}'\boldsymbol{\Omega} - \mathbf{K}'H_{\mathbf{uu}}\mathbf{K}; \quad \boldsymbol{\Omega}(t_f) = L_{\mathbf{x}_f\mathbf{x}_f} \tag{6-151}$$

where $\mathbf{K}$ is given in (6-84). Moreover, the steps used to derive (6-85) now yield

$$\mathscr{V}_2 = \tfrac{1}{2}\mathbf{d}'[E_{\mathbf{cc}} + \boldsymbol{\Gamma}'\boldsymbol{\Omega}(0)\boldsymbol{\Gamma}]\mathbf{d} + \tfrac{1}{2}\int_0^{t_f} \{(\boldsymbol{\mu} + \mathbf{K}\boldsymbol{\chi})'H_{\mathbf{uu}}(\boldsymbol{\mu} + \mathbf{K}\boldsymbol{\chi})\}\, dt + \boldsymbol{\chi}_f'L_{\mathbf{x}_f\boldsymbol{\lambda}}\mathbf{l}. \tag{6-152}$$

This sufficiency condition also is stated in terms of the solution to

$$-\dot{\mathbf{U}} = \mathbf{A}'\mathbf{U}; \quad \mathbf{U}(t_f) = L_{\mathbf{x}_f\boldsymbol{\lambda}} \tag{6-153}$$

where $\mathbf{A}$ is given by (6-112) and in terms of the matrix

$$\mathbf{W} = \int_0^{t_f} \{\mathbf{M}'H_{\mathbf{uu}}\mathbf{M}\}\, dt \tag{6-154}$$

where

$$\mathbf{M} = -H_{\mathbf{uu}}^{-1} H_{\mathbf{up}} \mathbf{U}. \tag{6-155}$$

Specifically, this result is stated as

Theorem 6-7 *Suppose performance functional J is defined by* (*6-131*) *and* (*6-132*), *and the corresponding Lagrangian is twice continuously differentiable on* $C^m[0, t_f]$. *Also suppose* $\mathbf{u}(t) \in C^m[0, t_f]$ *is an admissible forcing function for which the tangent space of* **g** *is completely reachable. Then J has a strict weak relative minimum at* **u** *if J is stationary,* $H_{\mathbf{uu}}$ *is positive definite on* $[0, t_f]$, *no conjugate points of* $\mathscr{V}_2$ *occur in* $[0, t_f)$, *and matrix*

$$\mathbf{E} = E_{\mathbf{cc}} + \mathbf{\Gamma}'[\mathbf{\Omega}(0) + \mathbf{U}(0)\mathbf{W}^{-1}\mathbf{U}'(0)]\mathbf{\Gamma} \tag{6-156}$$

is positive definite.

Proof Let

$$\boldsymbol{\mu} + \mathbf{K}\boldsymbol{\chi} = \mathbf{v} + \mathbf{M}\mathbf{w}$$

where **w** is the constant vector

$$\mathbf{w} = \mathbf{W}^{-1}\mathbf{U}'(0)\mathbf{\Gamma}\mathbf{d}.$$

Then the admissibility condition given in (6-148) is found to require that

$$\mathbf{0} = \int_0^{t_f} \{\mathbf{M}' H_{\mathbf{uu}} \mathbf{v}\}\, dt$$

holds using a relationship that is analogous to (6-139). If **v** is assumed to be admissible in accordance with this requirement but otherwise an arbitrary forcing function belonging to $C^m[0, t_f]$, then the second variation becomes

$$\mathscr{V}_2 = \tfrac{1}{2}\mathbf{d}'\mathbf{E}\mathbf{d} + \tfrac{1}{2}\int_0^{t_f} \{\mathbf{v}' H_{\mathbf{uu}} \mathbf{v}\}\, dt$$

and hence $\mathscr{V}_2 > 0$ holds for all nonzero **d** and **v**.

The proof of this sufficiency theorem also reveals that matrix **E** must be positive semi-definite at a weak relative minimum when $H_{\mathbf{uu}}$ is nonsingular on $[0, t_f]$ and no conjugate points of $\mathscr{V}_2$ occur in $[0, t_f)$.

Forcing-function iteration methods discussed in Section 6.6 are easily extended to design problems which also include adjustable parameters and terminal equality constraints. A model of the second variation $\mathscr{V}_2$ can be introduced in the form of (6-152) for the purpose of extending these methods. In the interest of brevity, only an extension of the Newton–Raphson method is discussed here. The Newton–Raphson method for forcing function iteration is based on the solution of the canonical equations corresponding to

$$\mathscr{V} = \mathscr{V}_1 + \mathscr{V}_2 \tag{6-157}$$

and (6-145) where $\mathscr{V}_1$ and $\mathscr{V}_2$ are given by (6-143) and (6-152) respectively. Moreover, this method is based on the assumption that all of the conditions appearing in Theorem 6-7 hold except stationarity and admissibility.

If the co-state vector of this auxiliary minimization problem which corresponds to $\boldsymbol{\chi}$ is denoted by $\boldsymbol{\pi}$, then the required auxiliary canonical equations become (6-145),

$$-\dot{\boldsymbol{\pi}} = \mathbf{K}'H_{\mathbf{uu}}(\boldsymbol{\mu} + \mathbf{K}\boldsymbol{\chi}) + H_{\mathbf{xp}}\boldsymbol{\pi}; \quad \boldsymbol{\pi}(t_f) = L_{\mathbf{x}_f\boldsymbol{\lambda}}\mathbf{l},$$

$$\mathbf{0} = H_{\mathbf{u}} + H_{\mathbf{uu}}(\boldsymbol{\mu} + \mathbf{K}\boldsymbol{\chi}) + H_{\mathbf{up}}\boldsymbol{\pi},$$

$$\mathbf{0} = L_{\boldsymbol{\lambda}} + L_{\boldsymbol{\lambda}\mathbf{x}_f}\boldsymbol{\chi}_f,$$

and

$$\mathbf{0} = E_{\mathbf{c}} + \boldsymbol{\Gamma}'[\mathbf{p}(0) + \boldsymbol{\pi}(0)] + [E_{\mathbf{cc}} + \boldsymbol{\Gamma}'\boldsymbol{\Omega}(0)\boldsymbol{\Gamma}]\,\mathbf{d}.$$

These auxiliary canonical equations are reduced to a more convenient form by introducing vector $\boldsymbol{\rho}$ which is defined by

$$\boldsymbol{\pi} = \boldsymbol{\rho} + \mathbf{U}l$$

where $\mathbf{U}$ is the solution to (6-153). In addition, perturbation vector

$$\boldsymbol{\delta} = -H_{\mathbf{uu}}^{-1}(H_{\mathbf{u}} + H_{\mathbf{up}}\boldsymbol{\rho}) \tag{6-158}$$

and matrix $\mathbf{M}$ defined in (6-155) are introduced for convenience. The second auxiliary canonical equation then yields

$$-\dot{\boldsymbol{\rho}} = -\mathbf{K}'H_{\mathbf{u}} + \mathbf{A}'\boldsymbol{\rho}; \quad \boldsymbol{\rho}(t_f) = \mathbf{0} \tag{6-159}$$

where $\mathbf{K}$ and $\mathbf{A}$ are given by (6-84) and (6-112) respectively. The third auxiliary equation also yields

$$\boldsymbol{\mu} = \boldsymbol{\delta} + \mathbf{Ml} - \mathbf{K}\boldsymbol{\chi}. \tag{6-160}$$

If a relationship that is analogous to (6-139) is used, then the fourth auxiliary canonical equation yields

$$\mathbf{l} = \boldsymbol{\epsilon} + \mathbf{Nd} \tag{6-161}$$

where

$$\boldsymbol{\epsilon} = \mathbf{W}^{-1}\left(L_{\boldsymbol{\lambda}} - \int_0^{t_f} \{\mathbf{M}H_{\mathbf{uu}}\boldsymbol{\delta}\}\,\mathrm{d}t\right), \tag{6-162}$$

$$\mathbf{N} = \mathbf{W}^{-1}\mathbf{U}'(0)\boldsymbol{\Gamma}, \tag{6-163}$$

and $\mathbf{W}$ is given by (6-154). The final auxiliary canonical equation yields

$$\mathbf{d} = -\mathbf{E}^{-1}\{E_{\mathbf{c}} + \boldsymbol{\Gamma}'[\mathbf{p}(0) + \boldsymbol{\rho}(0) + \mathbf{U}(0)\boldsymbol{\epsilon}]\} \tag{6-164}$$

where $\mathbf{E}$ is given by (6-156). Direction vectors corresponding to forcing functions $\mathbf{u}$, Lagrange multipliers $\boldsymbol{\lambda}$, and parameters $\mathbf{c}$ are respectively given in (6-160), (6-161), and (6-164) for the Newton–Raphson method.

Successive approximations, however, generally are performed directly in terms of variables $\mathbf{u}, \boldsymbol{\lambda}$, and $\mathbf{c}$. The definition of a partial derivative is again used to reduce the method to this form. Actual computations therefore are performed by evaluating

$$\mathbf{m} = \mathbf{c}, \quad \mathbf{n} = \boldsymbol{\lambda} - \mathbf{N}\mathbf{c}, \quad \text{and} \quad \mathbf{k} = \mathbf{u} - \mathbf{M}\boldsymbol{\lambda} + \mathbf{K}\mathbf{x} \tag{6-165}$$

on each iteration. Then the next iteration is determined by

$$\mathbf{c} = \alpha\mathbf{d} + \mathbf{m} \tag{6-166}$$

$$\boldsymbol{\lambda} = \alpha\boldsymbol{\epsilon} + \mathbf{n} + \mathbf{N}\mathbf{c} \tag{6-167}$$

and

$$\mathbf{u} = \alpha\boldsymbol{\delta} + \mathbf{k} + \mathbf{M}\boldsymbol{\lambda} - \mathbf{K}\mathbf{x} \tag{6-168}$$

with $\alpha \geqslant 0$. The presence of terminal equality constraints causes the difficulty that function

$$f(\alpha) = \mathscr{L}[\mathbf{u} + \alpha\boldsymbol{\mu}; \mathbf{c} + \alpha\mathbf{d}, \boldsymbol{\lambda} + \alpha\mathbf{l}]$$

may not have an unconstrained relative minimum for $\alpha \geqslant 0$. Hence, successive approximations are performed with $\alpha = 1$ on each iteration when possible so that (6-166), (6-167), and (6-168) then specify the Newton–Raphson method of successive substitutions.

A simple example with terminal equality constraints but without adjustable parameters is introduced here for the purposes of illustrating optimal forcing functions and the practicability of the Newton–Raphson method of successive substitutions for such design problems. The performance functional for the example is defined by

$$F = \tfrac{1}{2}(x_1^2 + x_2^2 + u_1^2), \quad G = \tfrac{1}{2}\mathbf{x}_f'\mathbf{T}\mathbf{x}_f, \quad \mathbf{T} = \begin{bmatrix} (1 + 2^{3/4}) & (2^{1/2} - 1) \\ (2^{1/2} - 1) & (1 + 2^{5/4}) \end{bmatrix},$$

and $t_f = 2$.

Also, state equations for the example are defined by

$$\mathbf{f} = \begin{bmatrix} (1 - x_2^2)x_1 - x_2 + u_1 \\ x_1 \end{bmatrix}, \quad \mathbf{x}(0) = \begin{bmatrix} 0 \\ 1 \end{bmatrix}, \quad \text{and} \quad \mathbf{x}(2) = \begin{bmatrix} 0 \\ 0 \end{bmatrix}.$$

The fixed-point terminal-boundary condition is accounted for by introducing terminal constraint functions $\mathbf{g}(\mathbf{x}_f) = \mathbf{x}_f$. Successive substitutions are initialized with

$$\boldsymbol{\epsilon} = \mathbf{n} = \mathbf{0}, \quad \boldsymbol{\delta} = \mathbf{k} = \mathbf{0}, \quad \mathbf{M} = \mathbf{0}, \quad \text{and} \quad \mathbf{K} = [(1 + 2^{3/4}) \;\; (2^{1/2} - 1)]$$

so that the Newton–Raphson method is initialized with a constant feedback-gain approximation to optimal forcing functions without terminal equality constraints. Successive substitutions then result in the data summarized in Table 6-7. These data illustrate quadratic convergence and also illustrate that values of performance functional J may not form a nonincreasing sequence when constraints are imposed. The optimal forcing function and corresponding state variables are shown in Figure 6-7.

Table 6-7 Iterations for terminal equality constraints by the Newton–Raphson method

Iteration	J	$x_1(2)$	$x_2(2)$	λ_1	λ_2
0	1.4634	-3.27×10^{-1}	2.71×10^{-1}	0.000	0.000
1	1.7757	-1.78×10^{-2}	8.92×10^{-3}	-1.413	1.284
2	1.7752	-1.42×10^{-5}	2.92×10^{-6}	-1.359	1.285
3	1.7752	-1.29×10^{-7}	2.33×10^{-8}	-1.360	1.287

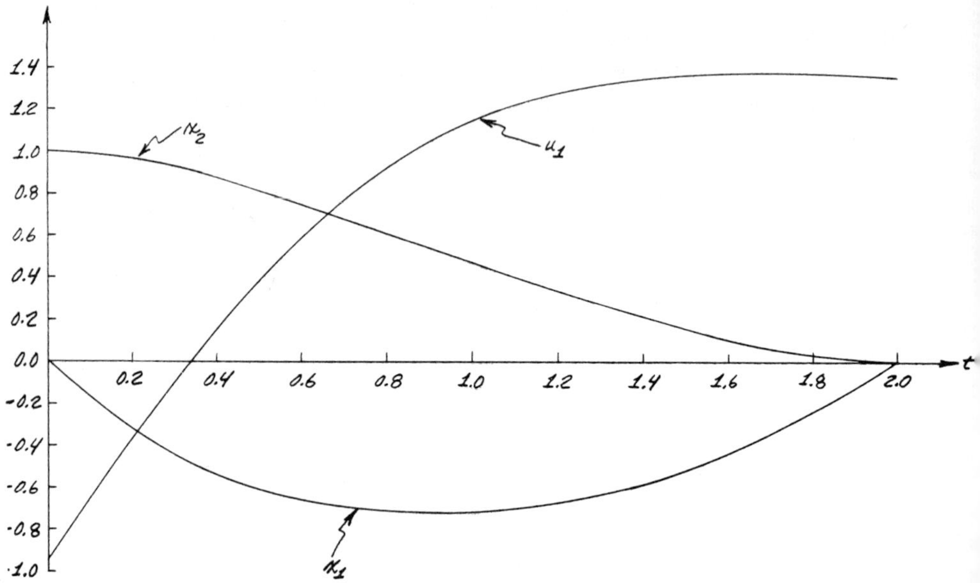

Figure 6-7 Optimal trajectory with fixed-point terminal-boundary condition.

6.9 OPTIMAL FORCING FUNCTIONS WITH MINIMAL FINAL TIME AND TERMINAL EQUALITY CONSTRAINTS

A number of design problems involve finite time intervals of variable length. One such class of design problems can be stated in terms of state equations

$$\dot{\mathbf{x}} = \mathbf{f}(\mathbf{x}, \mathbf{u}); \quad \mathbf{x}(0) = \mathbf{x}_0, \quad \mathbf{0} = \mathbf{g}(\mathbf{x}_f) \tag{6-169}$$

and performance functional

$$J = \int_0^{t_f} F(\mathbf{x}, \mathbf{u})\, \mathrm{d}t + G(\mathbf{x}_f). \tag{6-170}$$

In particular, performance functional J is to be minimized with respect to both forcing functions $\mathbf{u}$ and final time $t_f > 0$. This class of design problems is a special case of the class of design problems presented in Section 6.8 and is considered here as an example of previous results for which the conditions appearing in Theorem 6-7 are assumed to hold.

Variable final-time problems are formulated conveniently for computational purposes by the introduction of an alternate independent variable τ which is defined by

$$t = c\tau.$$

where the adjustable parameter c is defined by

$$c = \frac{t_f}{\tau_f}$$

for some fixed $\tau_f > 0$. If forcing-function and state vectors are redefined in terms of τ as

$$\mathbf{v}(\tau) = \mathbf{u}(t) \quad \text{and} \quad \mathbf{y}(\tau) = \mathbf{x}(t),$$

then state equations given in (6-169) become

$$\dot{\mathbf{y}} = c\mathbf{f}(\mathbf{y}, \mathbf{v}); \quad \mathbf{y}(0) = \mathbf{x}_0, \quad \mathbf{0} = \mathbf{g}(\mathbf{y}_f)$$

and the performance functional given in (6-170) becomes

$$J = \int_0^{\tau_f} cF(\mathbf{y}, \mathbf{v})\, d\tau + G(\mathbf{y}_f)$$

where $\mathbf{y}_f = \mathbf{y}(\tau_f)$. In other words, variable final-time problems are restated easily in terms of problems with an adjustable parameter and a fixed final time. This procedure is invariably employed when solutions are sought by computational methods because time functions of constant length can now be stored in tabulated form.

In order to reduce the design problem being discussed here to the form presented in Section 6.8, an additional state variable $z = z(\tau)$ is introduced and is defined by

$$\dot{z} = 0; \quad z(0) = c.$$

State and co-state vectors thus are taken to be

$$\begin{bmatrix} \mathbf{y} \\ z \end{bmatrix} \quad \text{and} \quad \begin{bmatrix} \mathbf{q} \\ r \end{bmatrix}$$

respectively where $\mathbf{q} = \mathbf{q}(\tau)$ and $r = r(\tau)$. The Hamiltonian and Lagrangian for this design problem then are expressed as

$$\mathcal{H} = zH \quad \text{where} \quad H = F(\mathbf{y}, \mathbf{v}) + \mathbf{q}'\mathbf{f}(\mathbf{y}, \mathbf{v})$$

and

$$L = G(\mathbf{y}_f) + \boldsymbol{\lambda}'\mathbf{g}(\mathbf{y}_f)$$

respectively. Canonical equations for this class of design problems thus become

$$\begin{bmatrix} \dot{\mathbf{y}} \\ \dot{z} \end{bmatrix} = \begin{bmatrix} zH_{\mathbf{q}} \\ 0 \end{bmatrix}; \quad \begin{bmatrix} \mathbf{y}(0) \\ z(0) \end{bmatrix} = \begin{bmatrix} \mathbf{x}_0 \\ c \end{bmatrix}, \tag{6-171}$$

$$-\begin{bmatrix} \dot{\mathbf{q}} \\ \dot{r} \end{bmatrix} = \begin{bmatrix} zH_{\mathbf{y}} \\ H \end{bmatrix}; \quad \begin{bmatrix} \mathbf{q}(\tau_f) \\ r(\tau_f) \end{bmatrix} = \begin{bmatrix} L_{\mathbf{y}_f} \\ 0 \end{bmatrix}, \tag{6-172}$$

and

$$0 = zH_{\mathbf{v}}, \quad 0 = L_{\boldsymbol{\lambda}}, \quad 0 = r(0) \tag{6-173}$$

in accordance with Section 6.8. Solution of (6-173) generally is achieved by selecting $\mathbf{u}, \boldsymbol{\lambda}$, and c by successive approximations in the manner also discussed in the previous section.

An alternate and more frequently employed form for these canonical equations is obtained by first noting

$$\frac{\mathrm{d}H}{\mathrm{d}\tau} = \dot{\mathbf{y}}'H_{\mathbf{y}} + \dot{\mathbf{q}}'H_{\mathbf{q}} + \dot{\mathbf{v}}'H_{\mathbf{v}} = 0$$

and hence H is a constant at an extremal. The last canonical equation then yields

$$0 = \tau_f H$$

and hence the Hamiltonian vanishes on $[0, \tau_f]$ at an extremal. Secondly, co-state vector

$$\mathbf{p}(t) = \mathbf{q}(\tau)$$

is introduced, and the Hamiltonian H and Lagrangian L are rewritten in the form of (6-54) and (6-4) respectively. Canonical equations corresponding to (6-169) and (6-170) then become

$$\dot{\mathbf{x}} = H_{\mathbf{p}}; \quad \mathbf{x}(0) = \mathbf{x}_0 \tag{6-174}$$

$$-\dot{\mathbf{p}} = H_{\mathbf{x}}; \quad \mathbf{p}(t_f) = L_{\mathbf{x}_f} \tag{6-175}$$

and

$$\mathbf{0} = H_{\mathbf{u}}, \quad \mathbf{0} = L_{\boldsymbol{\lambda}}, \quad 0 = H. \tag{6-176}$$

This form of the canonical equation is useful for analytical purposes.

An elementary example of optimal forcing functions with minimal final time and terminal equality constraints is stated as follows. Performance functional J

is defined by

$$F = 1 + \frac{1}{2a} u_1^2 \quad \text{and} \quad G = 0$$

where the constant term appearing in F is included so that J includes the term t_f. In addition, state equations are defined by

$$\mathbf{f} = \begin{bmatrix} 0 & 0 \\ 1 & 0 \end{bmatrix} \mathbf{x} + \begin{bmatrix} 1 \\ 0 \end{bmatrix} \mathbf{u}, \quad \mathbf{x}(0) = \begin{bmatrix} 0 \\ 1 \end{bmatrix}, \quad \text{and} \quad \mathbf{x}(t_f) = \begin{bmatrix} 0 \\ 0 \end{bmatrix},$$

and the fixed-point terminal-boundary condition is accounted for by constraint functions $\mathbf{g}(\mathbf{x}_f) = \mathbf{x}_f$ so that $\mathbf{p}(t_f) = \boldsymbol{\lambda}$. Solution of (6-175) then is expressed as

$$\mathbf{p}(t) = \begin{bmatrix} 1 & -t \\ 0 & 1 \end{bmatrix} \mathbf{p}(0),$$

and the optimal forcing function is given by

$$u_1(t) = -ap_1(0) + atp_2(0)$$

so that the Hamiltonian becomes

$$H = 1 - \frac{a}{2} [p_1(0) - tp_2(0)]^2 + p_2(0)x_1(t).$$

Furthermore, solution of (6-174) yields

$$\mathbf{x}(t) = \begin{bmatrix} -atp_1(0) + \dfrac{a}{2} t^2 p_2(0) \\ 1 - \dfrac{a}{2} t^2 p_1(0) + \dfrac{a}{6} t^3 p_2(0) \end{bmatrix}.$$

The remaining conditions to be satisfied are $x_1(t_f) = x_2(t_f) = 0$ and $H = 0$ at say $t = t_f$. These conditions yield

$$t_f = \left(\frac{18}{a}\right)^{1/4}$$

and

$$\mathbf{p}(0) = \begin{bmatrix} \dfrac{1}{3}\left(\dfrac{18}{a}\right)^{1/2} \\ \dfrac{2}{3}\left(\dfrac{18}{a}\right)^{1/4} \end{bmatrix}.$$

Finally, the optimal forcing function is found to be

$$u_1(t) = (2a)^{1/2}\left(-1 + 2\frac{t}{t_f}\right),$$

and the corresponding state vector becomes

$$\mathbf{x}(t) = \begin{bmatrix} (72a)^{1/4}\left(-1 + \dfrac{t}{t_f}\right)\left(\dfrac{t}{t_f}\right) \\ 1 + \left(-3 + 2\dfrac{t}{t_f}\right)\left(\dfrac{t}{t_f}\right)^2 \end{bmatrix}.$$

These time functions are plotted in Figure 6-8.

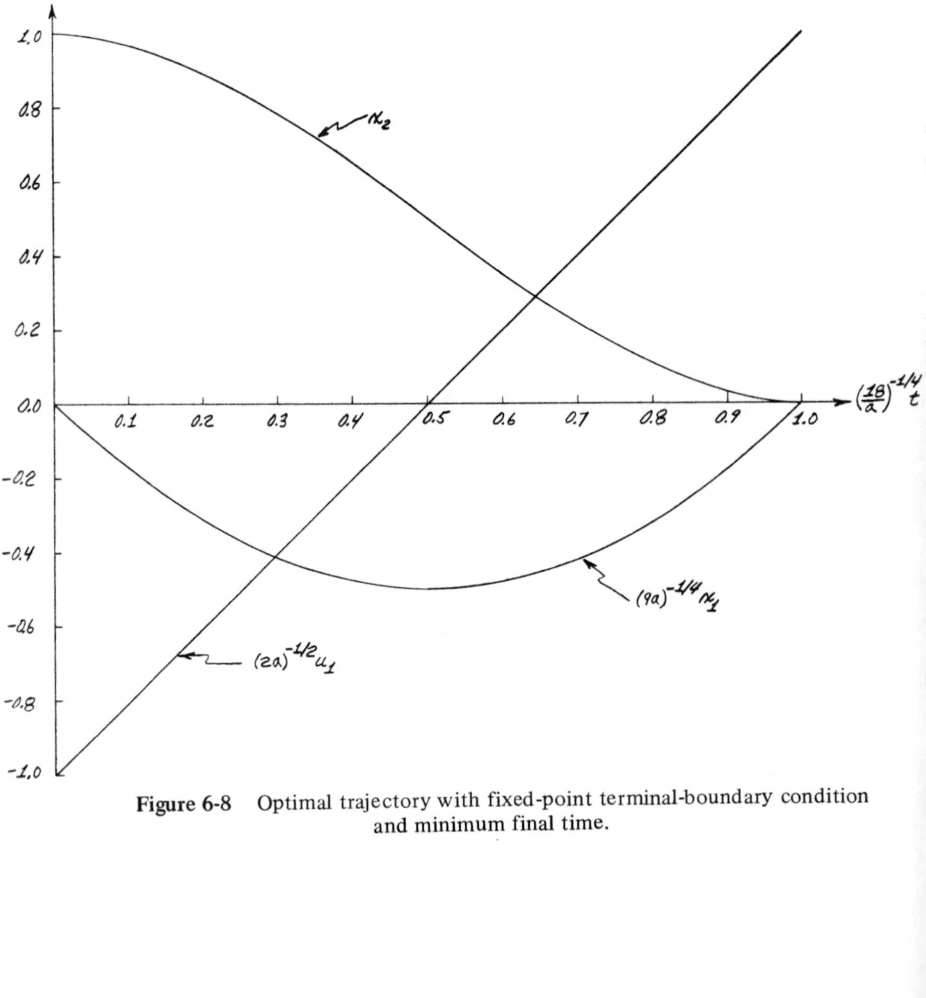

Figure 6-8 Optimal trajectory with fixed-point terminal-boundary condition and minimum final time.

6.10 SUMMARY

The formulation of both deterministic and stochastic design problems with finite time intervals is presented here in considerable detail. These problems involve both adjustable parameters and adjustable forcing functions as well as terminal equality constraints. Conditions for optimality are derived under a number of simplifying assumptions concerning continuity and differentiability. These assumptions permit the use of directional derivatives in the context of time functions as a device for deriving necessary and sufficient conditions for optimal parameters and forcing functions. Optimality conditions then follow directly from those derived in Chapter 2 for parameter optimization of functions and from properties of quadratic functionals. Optimality conditions given for adjustable forcing functions are the traditional necessary and sufficient conditions given for weak relative minima in the context of the calculus of variations [2]. Design problems with both optimal parameters and forcing functions are shown to have as a special case a class of minimal time problems that involves continuous forcing functions.

Dependence of optimal parameters on final time t_f is investigated in the context of optimal gain control. In particular, the behavior of these parameters as $t_f \to \infty$ is examined for the purpose of approximating optimal feedback gains which are discussed in Chapter 4. Gross approximations, which nevertheless result in asymptotically stable closed-loop systems, are sometimes found in this fashion for the purpose of initializing computational methods also discussed in Chapter 4. Similarly, dependence of optimal forcing functions on final time is investigated in the context of linear optimal control. The behavior of these forcing functions as $t_f \to \infty$ is examined for the purpose of approximating optimal feedback gains discussed in Chapters 4 and 5 so that asymptotically stable closed-loop systems are obtained. These forcing functions are determined from the transient solution of matrix Riccati equations, and hence properties of these differential equations are examined in some detail.

Computational methods for solving two-point boundary-value problems which arise in the context of optimal forcing functions are also discussed. Both boundary-condition iteration and forcing-function iteration methods are derived using the device of a directional derivative. Advantages and disadvantages of these methods are discussed in some detail. Moreover, conditions for which quadratic convergence is obtained are identified for both kinds of methods. Forcing-function iteration methods are easily extended to design problems that also include optimal parameters and terminal equality constraints. In other words, forcing-function iteration methods are easily constructed for use with variable final-time problems including a class of minimal time problems.

Linearized optimal control is discussed as a practical method of implementing approximations to optimal forcing functions in a neighborhood of a nominal extremal. Reference signals and feedback gains for linearized optimal control are computed and hence immediately available at the nominal extremal when the Newton–Raphson method of forcing-function iteration is used to solve the two-point boundary-value problem. These reference signals and feedback gains also yield optimal forcing functions for all initial states when state equations are linear and the performance functional is quadratic. Although not discussed, this method of approximating optimal forcing functions extends readily to design problems which also include optimal parameters and terminal equality constraints. Moreover, a similar implementation of linearized optimal control can be constructed for these more general design problems.

A number of design problems which arise are of course not amenable to the material presented in this book. For instance, many conditions given here do not apply to design problems which result in a conjugate point at $\eta = 0$ or in singular extremals. Furthermore, the sufficiency conditions given here do not apply to strong relative minima [9] where direction vector $\boldsymbol{\mu}(t)$ does not belong to $C^m[0, t_f]$. In fact, relative minima for which forcing functions $\mathbf{u}(t)$ do not belong to $C^m[0, t_f]$ generally require more advanced theory than put forth here such as Bellman's Principle of Optimality [10] or Pontryagin's Maximum Principle [11]. However, most realistic design problems that can be resolved in a practical manner by automated design techniques are subject to the theoretical considerations, computational methods, and system implementation techniques presented in this book.

BIBLIOGRAPHY

1. Coddington, E. A. and N. Levinson: *Theory of Ordinary Differential Equations,* McGraw-Hill Book Company, Inc., New York, 1955.
2. Gelfand, I. M. and S. V. Fomin: *Calculus of Variations*, Prentice-Hall, Inc., Englewood Cliffs, 1963.
3. Zemanian, A. H.: *Distribution Theory and Transform Analysis*, McGraw-Hill Book Company, Inc., New York, 1965.
4. Atkinson, F. V.: *Discrete and Continuous Boundary Problems*, Academic Press, New York, 1964.
5. Potter, J. E.: "A matrix equation arising in statistical filtering theory", Report No. RE-9, Experimental Astronomy Laboratory, M.I.T., Cambridge, 1965.
6. Wonham, W. M.: "On a matrix Riccati equation of stochastic control", *SIAM Journal on Control*, **6**, No. 4, 1968.
7. Merriam, C. W., III: *Optimization Theory and the Design of Feedback Control Systems*, McGraw-Hill Book Company, Inc., New York, 1964.
8. Merriam, C. W., III: "A computation method for feedback control optimization", *Information and Control*, **8**, No. 2, April 1965.

9. Hestenes, M. R.: *Calculus of Variations and Optimal Control Theory*, John Wiley and Sons, Inc., New York, 1966.
10. Bellman, R.: *Dynamic Programming*, Princeton University Press, Princeton, 1957.
11. Pontryagin, L. S. *et al.*: *The Mathematical Theory of Optimal Processes*, Interscience Publishers, Inc., New York, 1962.
12. Gill, S.: "A process for the step-by-step integration of differential equations in an automatic digital computing machine", *Proc. Cambridge Phil. Soc.*, **47**, Pt. 1, June 1950.

PROBLEMS

6-1 Demonstrate that the Hamiltonian introduced in Section 6.1 for parameter optimization is constant with respect to t.

6-2 Without the introduction of a Hamiltonian, find an expression for gradient $J_{\mathbf{c}}$ when the performance functional is given by (6-2) and state equations are given in (6-1). Also find an expression for $J_{\mathbf{cc}}$ without the introduction of a Hamiltonian if possible and explain your results.

6-3 Determine the canonical equations for minimizing performance functional

$$J = \int_0^{t_f} F(\mathbf{x}, t)\, \mathrm{d}t + G(\mathbf{x}_f)$$

with respect to $\mathbf{c}$ where

$$\dot{\mathbf{x}} = \mathbf{f}(\mathbf{x}, t); \quad \mathbf{x}(0) = \boldsymbol{\gamma} + \boldsymbol{\Gamma}\mathbf{c}.$$

6-4 Determine canonical equations for minimizing performance functional

$$J = \int_0^{t_f} F(\mathbf{x}, \mathbf{u})\, \mathrm{d}t + G(\mathbf{x}_f)$$

with respect to $\mathbf{c}$ where

$$\dot{\mathbf{x}} = \mathbf{f}(\mathbf{x}, \mathbf{u}); \quad \mathbf{x}(0) = \boldsymbol{\gamma} + \boldsymbol{\Gamma}\mathbf{c}$$

and

$$\mathbf{h}(\mathbf{x}, \mathbf{u}) = \mathbf{0}$$

hold on $[0, t_f]$. Assume $\mathbf{h_x}$ and $\mathbf{h_u}$ exist and $\mathbf{h_u}$ is nonsingular over $\mathbf{x}$ and $\mathbf{u}$.

6-5 Write a computer program for evaluating (6-10) and (6-11), and incorporate this program into the computer program written for Problem 3-28. For simplicity, use a fixed increment of integration and straight-line interpolation. A suitable algorithm for numerical integration is the Gill modification of the Runge–Kutta method [12].

6-6 Find the optimal constant-gain matrix **C** for Wiener–Kalman filters by minimizing

$$J = \mathrm{tr}\,\{\Sigma(t_f)\}$$

where

$$\dot{\Sigma} = \Sigma \mathbf{A}' + \mathbf{A}\Sigma + [\mathbf{G} - \mathbf{CJ}] \begin{bmatrix} \mathbf{I} & \mathbf{K}' \\ \mathbf{K} & \mathbf{I} \end{bmatrix} [\mathbf{G} \quad -\mathbf{CJ}]'; \quad \Sigma(0) = \Sigma_0$$

and

$$\mathbf{A} = \mathbf{F}. - \mathbf{CH}.$$

Assuming **A** is asymptotically stable, discuss the asymptotic dependence of optimal gains on t_f as $t_f \to \infty$.

6-7 Find the optimal constant-gain matrix for stochastic control by minimizing

$$J = \text{tr}\,\{\mathbf{WX}(t_f)\}.$$

where **X** is defined by (6-98). Matrices **A** and **W** are given in (4-105) and (4-106) respectively.

6-8 Assuming J is twice continuously differentiable and $\mathbf{u}(t) \in C^m[0, t_f]$, show that **u** is continuously differentiable with respect to t at a nonsingular extremal. Also show that the Hamiltonian is constant with respect to t at a nonsingular extremal.

6-9 The quadratic functional

$$J = \int_0^{t_f} \{\tfrac{1}{2}\mathbf{u}'\mathbf{u} - \tfrac{1}{2}\mathbf{x}'\mathbf{x} + \mathbf{x}'\mathbf{a}\}\, dt$$

is to be minimized with respect to forcing functions **u** where **a** is constant and

$$\dot{\mathbf{x}} = \mathbf{u}; \quad \mathbf{x}(0) = \mathbf{0}.$$

a) Solve the canonical equations in terms of an arbitrary t_f.
b) Locate all conjugate points of V_2 as a function of t_f.
c) Find a value of t_f for which the Jacobi necessary condition is satisfied but the Jacobi sufficient condition is not satisfied.

6-10 Let V be the quadratic function defined by (6-70), (6-71), and (6-72) when specialized to Problem 6-9. Find conjugate points of V as functions of β. Also construct time functions $\boldsymbol{\mu}(t) \in C^m[0, t_f]$ for which $V_2[\boldsymbol{\mu}] < 0$ when $t_f = \pi$.

6-11 Derive the canonical equations for the minimization of

$$J = \int_0^{t_f} F(\mathbf{x}, \mathbf{u}, t)\, dt + G(\mathbf{x}_f)$$

with respect to forcing functions **u** where

$$\dot{\mathbf{x}} = \mathbf{f}(\mathbf{x}, \mathbf{u}, t); \quad \mathbf{x}(0) = \mathbf{x}_0.$$

Identify the smallest set of equations that are required to determine extremals.

6-12 Give an expression for the time derivative of **u** at extremals determined in Problem 6-11, and state conditions which assure the existence of this derivative. Evaluate the total time derivative of the Hamiltonian under these conditions.

6-13 Let the performance functional and state equations given in Problem 6-11 be specialized to

$$F = \mathbf{x}'\mathbf{r} + \mathbf{u}'\mathbf{q} + \tfrac{1}{2}\mathbf{x}'\mathbf{P}\mathbf{x} + \mathbf{u}'\mathbf{Q}\mathbf{x} + \tfrac{1}{2}\mathbf{u}'\mathbf{R}\mathbf{u},$$

$$G = \mathbf{x}_f'\mathbf{t} + \tfrac{1}{2}\mathbf{x}_f'\mathbf{T}\mathbf{x}_f, \quad \text{and} \quad \mathbf{f} = \mathbf{F}\mathbf{x} + \mathbf{G}\mathbf{u}$$

where all coefficients of F and $\mathbf{f}$ may be functions of t. Assume $\mathbf{R}$ is positive definite on $[0, t_f]$, and assume there are no conjugate points of V_2 in $[0, t_f)$.
a) Give explicit expressions for the canonical equations that define the extremal.
b) Demonstrate that the co-state vector has the form

$$\mathbf{p}(t) = \omega(t) + \mathbf{\Omega}(t)\mathbf{x}(t)$$

where $\mathbf{\Omega}$ is symmetric.
c) Give explicit expressions for reference signals and feedback gains appearing in the feedback control equation that specifies optimal forcing functions.
d) Explain why this feedback control equation yields optimal forcing functions for all $\mathbf{x}(t) \in \mathscr{R}^n$ given any $t \in [0, t_f]$.
e) Find a set of linear equations that can be used to compute $\mathbf{\Omega}$.
f) Give explicit expressions for coefficients of $J = J(\mathbf{x}_0)$ when J is evaluated with optimal forcing functions.

6-14 Consider the design problem defined by Problem 6-13 where all coefficients are zero except

$$\mathbf{R} = [1], \quad \mathbf{T} = \begin{bmatrix} 0 & 0 \\ 0 & 1 \end{bmatrix}, \quad \mathbf{F} = \begin{bmatrix} 0 & 0 \\ 1 & 0 \end{bmatrix}, \quad \text{and} \quad \mathbf{G} = \begin{bmatrix} 1 \\ 0 \end{bmatrix}.$$

a) Compute reference signals $\mathbf{k}$ and feedback gains $\mathbf{K}$ as functions of $(t_f - t)$.
b) What values of $(t_f - t)$ result in $(\mathbf{F} - \mathbf{GK})$ with eigenvalues having negative real parts?

6-15 Consider the design problem defined by Problem 6-13 where all coefficients are zero except

$$\mathbf{P} = n\mathbf{I}, \quad \mathbf{R} = \mathbf{I}, \quad \mathbf{F} = -n\mathbf{I} + \mathbf{1}, \quad \text{and} \quad \mathbf{G} = \mathbf{I}$$

where $\mathbf{1}$ denotes an $n \times n$ matrix having unit elements only.
a) Is $\mathbf{F}$ asymptotically stable?
b) Is $\{\mathbf{F}, \mathbf{G}\}$ completely controllable?
c) Compute the feedback gain matrix $\mathbf{K}$ as a function of $\tau = (t_f - t)$ and let $\tau \to \infty$.
d) Is $(\mathbf{F} - \mathbf{GK})$ asymptotically stable for $\tau = \infty$?
e) Specify the null space of $\mathbf{\Omega}$ for $\tau = \infty$.

6-16 Consider the design problem defined by Problem 6-13 where all coefficients are zero except

$$\mathbf{P} = \begin{bmatrix} 5 & 0 \\ 0 & 4 \end{bmatrix}, \quad \mathbf{R} = [1], \quad \mathbf{T} = \begin{bmatrix} 3 & 2 \\ 2 & 6 \end{bmatrix}, \quad \mathbf{F} = \begin{bmatrix} 0 & 0 \\ 1 & 0 \end{bmatrix}, \quad \text{and} \quad \mathbf{G} = \begin{bmatrix} 1 \\ 0 \end{bmatrix}.$$

Suppose contributions to F and G due to x_2 are replaced by similar contributions due to $(x_2 - r)$ where r is the truncated Taylor series approximation

$$r(\xi) \cong \sum_{k=0}^{K} \frac{1}{k!} (\xi - t)^k \frac{d^k r(t)}{dt^k} \quad \text{for} \quad \xi \in [t, t_f].$$

a) Reformulate problem coefficient matrices so that this Taylor series approximation corresponds to an uncontrolled subsystem.
b) Give numerical values of all system gains assuming $K = 3$ and $t_f \to \infty$.
c) Sketch the control system which corresponds to part b).

6-17 The following is an over-simplified version of an aircraft landing problem obtained by approximating the transfer function between altitude and elevator deflection as

$$H(s) = \frac{k_0 V}{s^2} \Delta_e(s).$$

The desired altitude as a function of time is also simplified to

$$h_d(t) = \frac{h_0}{1 - e^{-4}} (e^{-t/5} - e^{-4}) \quad \text{for} \quad 0 \leqslant t \leqslant 20.$$

Let $x_1 = \dot{h}$, $x_2 = h$, $u_1 = \delta_e$ and assume $k_0 = -1$, $V = 256$, $h_0 = 100$. The control system is designed to minimize

$$J = \int_0^{20} \left\{ \frac{10^{-4}}{2} (h - h_d)^2 + \tfrac{1}{2} \delta_e^2 \right\} dt.$$

a) Compute feedback gain matrix $\mathbf{K}$ as a function of $\tau = 20 - t$, and specify these gains as $\tau \to \infty$.

b) Modify J so that the constant gains found in part a) persist over the entire interval [0, 20] and so that the terminal penalty function G is zero only if $x_1 = \dot{h}_d$ and $x_2 = h_d$ at $t = 20$.

6-18:

a) Using the problem formulation of Problem 6-17, part b), compute reference signals $\mathbf{k}$ which correspond to h_d.

b) Using the problem formulation of Problem 6-17, part b), compute additional gains corresponding to the approximation of desired altitude

$$h_d(\xi) \cong h_d(t) + (\xi - t)\dot{h}_d(t) \quad \text{for} \quad \xi \in [t, 20]$$

with the further approximation $\tau = \infty$.

c) Compare the results of parts a) and b) using $h_d(t)$ and $\dot{h}_d(t)$ given in Problem 6-17.

6-19 Determine optimal time-varying gains for the deterministic design problem discussed in Section 6.2. Also specialize these results to the case where $\mathbf{H} = \mathbf{I}$.

6-20 Give a condition on problem matrices appearing in Problem 6-19 which assures that co-state matrix $\mathbf{P}$ is identically zero.

6-21 Determine optimal time-varying gains for the Wiener–Kalman filters corresponding to Problem 6-6.

6-22 Prove that the linear differential equations which can be used to compute the solution to (6-103) are never asymptotically stable.

6-23 Give an explicit expression for the solution to (6-103) assuming $\mathbf{S}_k = \mathbf{0}$, $\mathbf{T} = \mathbf{0}$, $\{\mathbf{P}, \mathbf{Q}, \mathbf{R}\}$ is regular, and $\mathbf{\Lambda}$ is singular. The expression should be left in terms of a nonsingular transformation $\mathbf{N}$ having the property

$$\mathbf{N}'\mathbf{\Lambda}\mathbf{N} = \begin{bmatrix} \mathbf{I} & 0 \\ 0 & 0 \end{bmatrix}.$$

6-24 Verify the one-step convergence property of the Newton method for boundary condition iteration given in Section 6.6 for the special case

$$F = \tfrac{1}{2}\mathbf{x}'\mathbf{x} + \tfrac{1}{2}\mathbf{u}'\mathbf{u}, \quad G = 0, \quad \text{and} \quad \mathbf{f} = \mathbf{u}.$$

Use $\mathbf{x}_f = \mathbf{0}$ to initialize the iterations.

6-25 Does the boundary-condition iteration method given in Section 6.6 always result in nonincreasing sequences of values of J? Explain your results.

6-26 Derive the Newton method for boundary condition iteration which is based on iterating $\mathbf{p}(0)$ instead of $\mathbf{x}_f$. Compare this method with that given in Section 6.6.

6-27 Verify the one-step convergence property of the Newton–Raphson method for forcing-function iteration in the special case defined in Problem 6-24. Use $\mathbf{u}(t) = \mathbf{0}$ to initialize the iterations.

6-28 Find a form for partial derivative

$$f_\alpha(\alpha) = J_\alpha[\alpha\boldsymbol{\delta} + \mathbf{k} - \mathbf{Kx}]$$

which is relatively convenient computationally for use with forcing-function iteration methods in locating relative minima of $f(\alpha)$.

6-29 Repeat Problem 6-9 for the fixed-point terminal-boundary condition $\mathbf{x}(t_f) = \mathbf{0}$.

6-30 Verify the one-step convergence property of the Newton–Raphson method for forcing-function iteration in the special case defined in Problem 6-24 when subject to the fixed-point terminal-boundary condition $\mathbf{x}(t_f) = \mathbf{0}$. Use $\mathbf{u}(t) = \mathbf{0}$ and $\boldsymbol{\lambda} = \mathbf{0}$ to initialize the iterations.

6-31 Write a computer program for the Newton–Raphson method of successive substitutions which is derived in Section 6.8. For simplicity, use a fixed increment of integration, the Gill method of integration, and straight-line interpolation.

6-32 Determine the relationship between complete responsiveness on a finite time interval as defined in Problem 4-40 and complete reachability as defined in Section 6.8 with constant coefficient matrices.

6-33 The performance functional defined by

$$F = 1 + u_1^2, \quad G = 0, \quad f_1 = u_1, \quad x_1(0) = 1, \quad \text{and} \quad x_1(t_f) = 0$$

is to be minimized with respect to $u_1(t)$ and t_f. Determine t_f and sketch $u_1(t)$ and $\mathbf{x}(t)$.

6-34 Repeat Problem 6-33 for

$$F = 1 + \frac{1}{2a}u_1^2, \quad G = 0, \quad \mathbf{f} = \begin{bmatrix} 0 & 0 \\ 1 & 0 \end{bmatrix}\mathbf{x} + \begin{bmatrix} 1 \\ 0 \end{bmatrix}\mathbf{u}, \quad \mathbf{x}(0) = \begin{bmatrix} 0 \\ 1 \end{bmatrix}, \quad \text{and} \quad x_2(t_f) = 0.$$

6-35 Verify that the sufficiency condition given in Theorem 6-7 applies to the solution obtained for Problem 6-33.

6-36 Consider the design problem formed by (6-52), (6-53), and equality constraints

$$\mathbf{g}(\mathbf{x}, \mathbf{u}) = \mathbf{0}$$

which hold on $[0, t_f]$. In order to determine the optimal forcing functions, form Lagrangian

$$L = J + \int_0^{t_f} \{\boldsymbol{\lambda}'\mathbf{g}\}\, dt$$

and Hamiltonian

$$H = F + \mathbf{p}'\mathbf{f} + \boldsymbol{\lambda}'\mathbf{g}$$

in terms of Lagrange multipliers $\boldsymbol{\lambda} = \boldsymbol{\lambda}(t)$. Also assume Lagrangian L is twice continuously differentiable.

a) Determine the first and second variations of L in terms of H.
b) Identify an assumption which assures that the Lagrange multiplier rule holds on $[0, t_f]$.
c) Give the necessary condition for a weak relative minimum that is analogous to Theorem 6-1.
d) Give the necessary condition for a weak relative minimum that is analogous to Theorem 6-2.
e) Redefine a nonsingular extremal suitably for this design problem.
f) Give the necessary condition for a weak relative minimum that is analogous to Theorem 6-3.
g) Give the sufficient condition for a strict weak relative minimum that is analogous to Theorem 6-4.

6-37 Determine the Newton–Raphson method for forcing function iteration that can be used to solve design problems stated in Problem 6-36.

6-38 Consider the minimization problem defined in Problem 6-36 where

$$F = \tfrac{1}{2}\mathbf{x}'\mathbf{P}\mathbf{x} + \mathbf{u}'\mathbf{Q}\mathbf{x} + \tfrac{1}{2}\mathbf{u}'\mathbf{R}\mathbf{u}, \quad G = \tfrac{1}{2}\mathbf{x}_f'\mathbf{T}\mathbf{x}_f,$$

$$\mathbf{f} = \mathbf{F}\mathbf{x} + \mathbf{G}\mathbf{u}, \quad \text{and} \quad \mathbf{g} = \mathbf{E}(\mathbf{u} + \mathbf{D}\mathbf{x}).$$

a) Under the assumptions found for part g) of Problem 6-36, determine matrices **L** and **K** which appear in

$$\boldsymbol{\lambda} + \mathbf{L}\mathbf{x} = \mathbf{0} \quad \text{and} \quad \mathbf{u} + \mathbf{K}\mathbf{x} = \mathbf{0}$$

for optimal forcing functions and Lagrange multipliers.
b) Determine a matrix $\boldsymbol{\Delta}$ which is defined as the particular value of **D** that results in $\boldsymbol{\lambda} = \mathbf{0}$ independently of initial state and **E**.
c) Suppose **D** is selected in partitioned form as

$$\mathbf{D} = [\mathbf{0}_{11} \quad \boldsymbol{\Delta}_{12}].$$

Specialize results of part a) to this case and discuss the significance of these constraints when $\mathbf{E} = \mathbf{I}$.

6-39 Consider the design problem formed by (6-52), (6-53), and equality constraints $\mathbf{j} = \mathbf{a}$ where **a** is given and

$$\mathbf{j} = \int_0^{t_f} \mathbf{g}(\mathbf{x}, \mathbf{u})\, dt.$$

In order to determine optimal forcing functions, form Lagrangian

$$L = J + \boldsymbol{\lambda}'(\mathbf{j} - \mathbf{a})$$

and Hamiltonian

$$H = F + \mathbf{p}'\mathbf{f} + \boldsymbol{\lambda}'\mathbf{g}.$$

Also assume Lagrangian L is twice continuously differentiable.
a) Determine the first and second variations of L in terms of H.
b) Identify an assumption analogous to complete reachability which assures that the Lagrange multiplier rule holds.
c) Give the necessary condition for a weak relative minimum that is analogous to Theorem 6-6.
d) Give the sufficient condition for a strict weak relative minimum that is analogous to Theorem 6-7.

6-40 Determine the Newton–Raphson method for forcing function iteration that can be used to solve design problems stated in Problem 6-39.

6-41 Consider the design problem defined by

$$J = \int_0^{t_f} \{\tfrac{1}{2}\mathbf{x}'\mathbf{P}\mathbf{x} + \mathbf{u}'\mathbf{Q}\mathbf{x} + \tfrac{1}{2}\mathbf{u}'\mathbf{R}\mathbf{u}\}\, dt + \tfrac{1}{2}\mathbf{x}_f'\mathbf{T}\mathbf{x}_f$$

and

$$\dot{\mathbf{x}} = \mathbf{F}\mathbf{x} + \mathbf{G}\mathbf{u}; \quad \mathbf{x}(0) = \mathbf{x}_0, \quad \mathbf{x}(t_f) = \mathbf{x}_f$$

where $\mathbf{x}_0$ and $\mathbf{x}_f$ are given.

a) Formulate this problem in the manner specified in Problem 6-39.

b) Let $\mathbf{p} = \mathbf{\Omega}\mathbf{x} + \mathbf{U}\boldsymbol{\lambda}$. Derive differential equations and boundary conditions for $\mathbf{\Omega}$ and $\mathbf{U}$.

c) Let $\mathbf{A}$ denote the Jacobian of the optimal closed-loop system with $\boldsymbol{\lambda}$ fixed, and let $\mathbf{\Phi}(t)$ denote the corresponding state transition matrix. Determine $\boldsymbol{\lambda}$ in terms of $\mathbf{\Phi}$ and quantities given in the statement of the minimization problem.

d) Evaluate the minimum value of J in terms of $\mathbf{x}_0$ and $\boldsymbol{\lambda}$.

e) Discuss the general effects of $\boldsymbol{\lambda}$ on t_f and $(\mathbf{x}_f - \mathbf{x}_0)$.

6-42 Discuss the implementation of linearized optimal control for the class of design problems discussed in Section 6.8.

Appendix A

A library of twenty-two **APL/360** computer programs is described here in sufficient detail for student use. These computer programs divide into two basic types.

The first type is a main program which implements a computational method discussed in previous chapters. For example, program MIN1 implements both the steepest-descent and conjugate-gradient methods of successive approximations.

The second type is a subprogram which is either supplied by the library or the user. Subprograms supplied by the library are called by main programs and are of no particular interest to the user unless additional main programs are being defined by the user. For example, program INV is supplied by the library and is used for computing the inverse of a real matrix. Subprograms supplied by the user define specific design problems. For example, program MIN1S1 computes function $f(\mathbf{c})$ and gradient $f_{\mathbf{c}}$ for use with MIN1.

A.1 CONVENTIONS

All programs in the library are subject to standard **APL/360** programming conventions [1, 2], and these programs execute using the default options for workspace commands)DIGITS,)ORIGIN, and)WIDTH. Library program storage requires 29,632 bytes of memory which leaves an additional 2,236 bytes of memory from a nominal workspace that can be used for user supplied programs. Execution is typically performed in another workspace which contains only the subset of library programs being used so that sufficient memory is available during execution.

Additional conventions employed by this library of computer programs are summarized as

1) Names of user supplied subprograms end in S1. The remaining letters of these names are the same as the name of the main program which calls the subprogram. For example, MIN1S1 is called by MIN1.

2) Global variables with names beginning in D are derivatives. For example, DXV is the name of $\dot{\mathbf{x}}$ and DFV is the name of $f_{\mathbf{c}}$.

3) Global variables ending in V are column vectors. For example, CV is the name of **c**.

4) Global variables with names ending in M are matrices. For example, FM is the name of **F**.

5) Global variables with names ending in A are three-dimensional arrays. For example, BA is the name of the array formed by placing $\mathbf{B}_1$ in the top horizontal plane, $\mathbf{B}_2$ in the next lower horizontal plane, etc. This three-dimensional array is denoted by $\mathbf{B}_1/\mathbf{B}_2/\ldots$ which identifies subscript i corresponding to $\mathbf{B}_i$ as the first of three indices in accordance with APL/360 conventions.

6) Global variables with names ending in AA are four-dimensional arrays. For example, DBAA is the name of

$$(\mathbf{B}_1/\mathbf{B}_2/\ldots)_{c_1}/(\mathbf{B}_1/\mathbf{B}_2/\ldots)_{c_2}/\ldots$$

so that the first two indices of this array are i and j corresponding to c_i and $\mathbf{B}_j$ respectively.

7) Variables which are identically zero need not be programmed in user supplied subprograms. Moreover, only the upper triangular portion of symmetric matrices need be programmed in user supplied subprograms. For example

$$f_{cc} = \begin{bmatrix} 1 & 0 & 1 \\ 0 & 1 & 0 \\ 1 & 0 & 1 \end{bmatrix}$$

is defined by the statement

DFM[1; 1] ← DFM[2; 2] ← DFM [3; 3] ← DFM[1; 3] ← 1.

8) Input data corresponding to a nonscalar is entered as a row vector. These input data must be preceded by the shaping operator (dyadic ρ) unless a column vector with more than one element is being entered. For example, CV ← 1ρ1 and SV ← 0ρ1 respectively define $\mathbf{c} = [1]$ and $\mathbf{s}$ without elements.

9) The first step performed in main programs is the output listing of input data. Symbol ☐: signifies the end of the listing. Execution is continued by typing 0 and then depressing RETURN. On the other hand, execution is terminated by typing → and then depressing RETURN.

A.2 DESCRIPTIONS

Programs in the library are defined by material appearing in previous chapters and by the following descriptions.

1) DXM

Function: Compute

$$\dot{\mathbf{X}} = \mathbf{XA} + \mathbf{A}'\mathbf{X} + \sum_{k=1}^{m} \mathbf{B}_k'\mathbf{XB}_k + \mathbf{P} - \mathbf{XRX}$$

as an operator.

Arguments:

(l) FORMAT = 0 for upper triangular portions of $\mathbf{X}$ and $\dot{\mathbf{X}}$ in vector form or 1 for $\mathbf{X}$ and $\dot{\mathbf{X}}$ in matrix form

(r) XM = $\mathbf{X}$

Global Input:

AM = $\mathbf{A}$

BA = $\mathbf{B}_1/\mathbf{B}_2/\ldots$

PM = $\mathbf{P}$

RM = $\mathbf{R}$

Global Output:

None

Subroutines:

None

2) EVAL

Function: Compute

$$\mathbf{C}(s\mathbf{I} - \mathbf{A})^{-1}\mathbf{B}$$

and factor numerator and denominator polynomials.

Arguments:
None

Global Input:
MODE = 0 for factored form of polynomials or 1 for unfactored form only
AM = $\mathbf{A}$
BM = $\mathbf{B}$
CM = $\mathbf{C}$

Global Output:
None

Subroutines:
FACT

3) FACT

Function: Compute the real ($\mathbf{z}_1$) and imaginary ($\mathbf{z}_2$) parts of the zero locations of

$$\sum_{n=0}^{N} p_{N+1-n} s^n$$

as an operator.

Arguments:
(r) PV = $\mathbf{p}$

Global Input:
None

Global Output:
None

Subroutines:
None

4) FORM1

Function: Shape arrays for use of LOAD1 with MIN1S1 or MIN3S1.

Arguments:
None

Global Input:
NX = dim $\mathbf{x}$
NU = dim $\mathbf{u}$
NC = dim $\mathbf{c}$

Global Output:
AM = $\mathbf{A}$
CM = $\mathbf{C}$
DFM = $f_{\mathbf{X}}$
BA = $\mathbf{B}_1/\mathbf{B}_2/\ldots$
AA = $\mathbf{A}_{c_1}/\mathbf{A}_{c_2}/\ldots$
DCA = $\mathbf{C}_{c_1}/\mathbf{C}_{c_2}/\ldots$
DXA = $\mathbf{X}_{c_1}/\mathbf{X}_{c_2}/\ldots$

DBAA = $(\mathbf{B}_1/\mathbf{B}_2/\ldots)_{c_1}/(\mathbf{B}_1/\mathbf{B}_2/\ldots)C_2/\ldots$
Subroutines:
None

5) FORM2
Function: Compute

$$\mathbf{A} = \mathbf{F} - \mathbf{G}(\mathbf{E}'\mathbf{E})^{-1}\mathbf{E}'\mathbf{D},$$

$$\mathbf{P} = \mathbf{D}'[\mathbf{I} - \mathbf{E}(\mathbf{E}'\mathbf{E})^{-1}\mathbf{E}']\mathbf{D} + \mathbf{T},$$

and

$$\mathbf{R} = \mathbf{G}(\mathbf{E}'\mathbf{E})^{-1}\mathbf{G}'.$$

for use with QUAD.

Arguments:
None
Global Input:
FM = $\mathbf{F}$
GM = $\mathbf{G}$
DM = $\mathbf{D}$
EM = $\mathbf{E}$
TM = $\mathbf{T}$
Global Output:
AM = $\mathbf{A}$
PM = $\mathbf{P}$
RM = $\mathbf{R}$
Subroutines:
INV
SP

6) FORM3
Function: Compute

$$\mathbf{A} = [\mathbf{F} - \mathbf{G}(\mathbf{JK})'(\mathbf{JJ}')^{-1}\mathbf{H}]',$$

$$\mathbf{R} = \mathbf{H}'(\mathbf{JJ}')^{-1}\mathbf{H},$$

and

$$\mathbf{P} = \mathbf{G}[\mathbf{I} - (\mathbf{JK})'(\mathbf{JJ}')^{-1}(\mathbf{JK})]\mathbf{G}'$$

for use with QUAD.

Global Input:
FM = $\mathbf{F}$
GM = $\mathbf{G}$
HM = $\mathbf{H}$
JM = $\mathbf{J}$
KM = $\mathbf{K}$
Global Output:
AM = $\mathbf{A}$
PM = $\mathbf{P}$
RM = $\mathbf{R}$
Subroutines:
INV
SP

7) GAIN1

Function: Compute

$$\mathbf{C} = (\mathbf{E}'\mathbf{E})^{-1}(\mathbf{E}'\mathbf{D} + \mathbf{G}'\mathbf{X})$$

as an operator.

Arguments:
None

Global Input:
GM = $\mathbf{G}$
DM = $\mathbf{D}$
EM = $\mathbf{E}$
XM = $\mathbf{X}$

Global Output:
None

Subroutines:
INV
SP

8) GAIN2

Function: Compute

$$\mathbf{C} = [\mathbf{G}(\mathbf{J}\mathbf{K})' + \mathbf{X}\mathbf{H}'](\mathbf{J}\mathbf{J}')^{-1}$$

as an operator.

Arguments:
None

Global Input:
GM = $\mathbf{G}$
HM = $\mathbf{H}$
JM = $\mathbf{J}$
KM = $\mathbf{K}$
XM = $\mathbf{X}$

Global Output:
None

Subroutines:
INV
SP

9) INC

Function: Compute the inverse of the reduced equivalent coefficient matrix of

$$\mathbf{X}\mathbf{A} + \mathbf{A}'\mathbf{X} + \sum_{k=1}^{m} \mathbf{B}_k'\mathbf{X}\mathbf{B}_k$$

as an operator.

Arguments:
(l) AM = $\mathbf{A}$
(r) BA = $\mathbf{B}_1/\mathbf{B}_2/\ldots$

Global Input:
None

Global Output:
None

Subroutines:
INV

10) INT
Function: Integrate

$$\dot{\mathbf{x}} = \mathbf{f}(\mathbf{x}, t); \quad \mathbf{x}(t_0) = \mathbf{x}_0.$$

Arguments:
None
Global Input:
NI = number of integration increments
NP = number of integration increments omitted between output
T = t_0
DT = length of integration increments
XV = $\mathbf{x}_0$
Global Output:
T = t
XV = $\mathbf{x}$
Subroutines:
INTS1 for the computation of
DXV = $\mathbf{f}(\mathbf{x}, t)$

11) INV
Function: Compute the inverse of a real matrix **M** as an operator.

Arguments:
(r) MM = **M**
Global Input:
None
Global Output:
None
Subroutines:
None

12) LIN1
Function: Compute the solution to

$$\mathbf{XA} + \mathbf{BX} + \mathbf{C} = \mathbf{0}$$

as an operator.

Arguments:
(l) MODE = 0 for **X** not symmetric or 1 for **X** symmetric
(r) CM = **C**
Global Input:
AM = **A**
BM = **B**
Global Output:
None
Subroutines:
INV
SP

13) LIN2

Function: Compute the solution to

$$\mathbf{XA} + \mathbf{A}'\mathbf{X} + \sum_{k=1}^{m} \mathbf{B}_k'\mathbf{X}\mathbf{B}_k + \mathbf{C} = \mathbf{0}$$

as an operator.

Arguments:

(l) The inverse of the reduced equivalent coefficient matrix of

$$\mathbf{XA} + \mathbf{A}'\mathbf{X} + \sum_{k=1}^{m} \mathbf{B}_k'\mathbf{X}\mathbf{B}_k$$

(r) CM = $\mathbf{C}$

Global Input:

None

Global Output:

None

Subroutines:

None

14) LOAD1

Function: Compute the solution to

$$\mathbf{XA}' + \mathbf{AX} + \sum_{k=1}^{m} \mathbf{B}_k\mathbf{X}\mathbf{B}_k' + \mathbf{C} = \mathbf{0},$$

$$\mathbf{X}_{c_i}\mathbf{A}' + \mathbf{A}\mathbf{X}_{c_i} + \sum_{k=1}^{m} \mathbf{B}_k\mathbf{X}_{c_i}\mathbf{B}_k' + \{\mathbf{C}_{c_i} + \mathbf{X}\mathbf{A}_{c_i}' + \mathbf{A}_{c_i}\mathbf{X} + \sum_{k=1}^{m} [(\mathbf{B}_k)_{c_i}\mathbf{X}\mathbf{B}_k' + \mathbf{B}_k\mathbf{X}(\mathbf{B}_k')_{c_i}]\} = \mathbf{0},$$

and

$$f_i = \operatorname{tr}\{f_{\mathbf{X}}\mathbf{X}_{c_i}\}$$

for use with MIN1S1 or MIN3S1.

Arguments:

None

Global Input:

AM = $\mathbf{A}$

CM = $\mathbf{C}$

DFM = $f_{\mathbf{X}}$

BA = $\mathbf{B}_1/\mathbf{B}_2/\ldots$

DAA = $\mathbf{A}_{c_1}/\mathbf{A}_{c_2}/\ldots$

DCA = $\mathbf{C}_{c_1}/\mathbf{C}_{c_2}/\ldots$

DBAA = $(\mathbf{B}_1/\mathbf{B}_2/\ldots)_{c_1}/(\mathbf{B}_1/\mathbf{B}_2/\ldots)_{c_2}/\ldots$

Global Output:

XM = $\mathbf{X}$

DXA = $\mathbf{X}_{c_1}/\mathbf{X}_{c_2}/\ldots$

DFV = $\mathbf{f}$

Subroutines:
INC
INV
LIN2
MD

15) LOAD2
Function: Compute f and f_c for optimal gain control problems defined by

$$\mathbf{A} = \mathbf{F} - \mathbf{GCH}, \quad \mathbf{W} = (\mathbf{D} - \mathbf{ECH})'(\mathbf{D} - \mathbf{ECH}),$$

and

$$f = \text{tr}\{\mathbf{T}'\mathbf{XT}\}$$

for use with MIN1S1 or MIN3S1.

Arguments:
None
Global Input:
FM = **F**
GM = **G**
HM = **H**
DM = **D**
EM = **E**
TM = **T**
Global Output:
F = f
DFV = f_c
Subroutines:
INV
LIN1
SP

16) MD
Function: Compute the major diagonal of **AB** as an operator.

Arguments:
(l) AM = **A**
(r) BM = **B**
Global Input:
None
Global Output:
None
Subroutines:
None

17) MIN1
Function: Steepest-descent and conjugate gradient methods of successive approximations for minimizing functions without constraints.

Arguments:
None

Global Input:
NI = maximum number of iterations
NT = maximum number of trials per iteration
NR = number of iterations performed before restarting the conjugate gradient method (1 gives steepest-descent)
RATIO = minimum value of $|\dot{h}(\alpha)/\dot{h}(0)|$ requiring additional trials
DFMIN = minimum value of $|f_{\mathbf{c}}|$ requiring additional trials and iterations
FMIN = lower bound on $f(\mathbf{c})$
FMAX = upper bound on $f(\mathbf{c})$
ALMIN = minimum initial value of α
ALMAX = maximum initial value of α
RMIN = minimum fraction of $(\alpha - \beta)$ used in interval reduction (<1)
RMAX = maximum fraction of $(\alpha - \beta)$ used in extrapolation (>1)
CV = $\mathbf{c}$
Global Output:
CV = $\mathbf{c}$
Subroutines:
MIN1S1 for the computation of
F = $f(\mathbf{c})$
DFV = $f_{\mathbf{c}}$

18) MIN2
Function: Newton–Raphson method of successive substitutions for minimizing functions with equality and inequality constraints.
Arguments:
None
Global Input:
NI = maximum number of iterations
DFMIN = minimum value of $|L_{\mathbf{c}}|$ requiring additional iterations
CV = $\mathbf{c}$
LAMV = $\boldsymbol{\lambda}$
SV = $\mathbf{s}$
EICV = vector of effective (1) and ineffective (0) inequality constraints
MDCV = vector of maximum permissible magnitudes of increments in the elements of $\mathbf{c}$
Global Output:
CV = $\mathbf{c}$
LAMV = $\boldsymbol{\lambda}$
SV = $\mathbf{s}$
Subroutines:
INV
MIN2S1 for the computation of
F = $f(\mathbf{c})$
DFV = $f_{\mathbf{c}}$
DFM = $f_{\mathbf{cc}}$
GV = $\mathbf{g}(\mathbf{c})$
DGM = $\mathbf{g}_{\mathbf{c}}$
DGA = $(g_1)_{\mathbf{cc}}/(g_2)_{\mathbf{cc}}/\ \ldots$

19) MIN3

Function: Min (Fletcher–Powell)–max (Newton–Raphson) method of successive approximations for minimizing functions with equality and inequality constraints.

Arguments:

None

Global Input:

NI = maximum number of iterations

NT = maximum number of trials per iteration

NR = number of iterations performed before maximization steps

NEC = number of equality constraints

RATIO = minimum value of $|\dot{h}(\alpha)/\dot{h}(0)|$ requiring additional trials

DFMIN = minimum value of $|L_{\mathbf{c}}|$ requiring additional trials and iterations

GMAX = maximum error in satisfying constraints not requiring additional iterations

FMIN = lower bound on $f(\mathbf{c})$

FMAX = upper bound on $f(\mathbf{c})$

ALMIN = minimum initial value of α

ALMAX = maximum initial value of α

RMIN = minimum fraction of $(\alpha - \beta)$ used in interval reduction (<1)

RMAX = maximum fraction of $(\alpha - \beta)$ used in extrapolation (>1)

CV = $\mathbf{c}$

LAMV = $\boldsymbol{\lambda}$

DLIM = positive definite approximation to $L_{\mathbf{cc}}^{-1}$

Global Output:

CV = $\mathbf{c}$

LAMV = $\boldsymbol{\lambda}$

DLIM = positive definite approximation to $L_{\mathbf{cc}}^{-1}$

Subroutines:

INV

SP

MIN3S1 for the computation of

F = $f(\mathbf{c})$

DFV = $f_{\mathbf{c}}$

GV = $\mathbf{g}(\mathbf{c})$

DGM = $\mathbf{g}_{\mathbf{c}}$

20) QUAD

Function: Compute the solution to

$$\mathbf{XA} + \mathbf{A}'\mathbf{X} + \sum_{k=1}^{m} \mathbf{B}_k'\mathbf{X}\mathbf{B}_k + \mathbf{P} - \mathbf{XRX} = \mathbf{0}.$$

Arguments:

None

Global Input:

NI = maximum number of iterations

MODE = 0 for use with the initialization step to obtain $\dot{\mathbf{X}}$ negative semi-definite or 1 for use without the initialization step

DXMIN = minimum value of $\|\dot{\mathbf{X}}\|$ requiring additional iterations

AM = $\mathbf{A}$

BA = $\mathbf{B}_1/\mathbf{B}_2/\ldots$

PM = $\mathbf{P}$
RM = $\mathbf{R}$
XM = positive semi-definite $\mathbf{X}_0$ for which the equivalent coefficient matrix of

$$\mathbf{X}(\mathbf{A} - \mathbf{R}\mathbf{X}_0) + (\mathbf{A} - \mathbf{R}\mathbf{X}_0)'\mathbf{X} + \sum_{k=1}^{m} \mathbf{B}_k'\mathbf{X}\mathbf{B}_k$$

is asymptotically stable.
Global Output:
XM = $\mathbf{X}$
Subroutines:
DXM
INC
INV
LIN2

21) SP
Function: Compute the symmetric product **AB** as an operator.
Arguments:
(l) AM = $\mathbf{A}$
(r) BM = $\mathbf{B}$
Global Input:
None
Global Output:
None
Subroutines:
None

22) SYN
Function: Compute

$$\mathbf{D} = \sum_{k=0}^{K} \mathbf{P}_k\mathbf{H}\mathbf{F}^k,$$

$$\mathbf{E}_l = \left[\sum_{k=0}^{K-1-l} \mathbf{P}_{k+1+l}\mathbf{H}\mathbf{F}^k\right]\mathbf{G} \quad \text{for} \quad l = 0, 1, \ldots, (K-1),$$

$$\mathbf{K} = (\mathbf{E}_0'\mathbf{E}_0)^{-1}\mathbf{E}_0'\mathbf{D},$$

$$\mathbf{A} = \mathbf{F} - \mathbf{G}\mathbf{K},$$

and

$$\mathbf{B} = \mathbf{G}(\mathbf{E}_0'\mathbf{E}_0)^{-1}\mathbf{E}_0'.$$

Arguments:
None
Global Input:
FM = $\mathbf{F}$
GM = $\mathbf{G}$
HM = $\mathbf{H}$
PA = $\mathbf{P}_0/\mathbf{P}_1/\ldots$

Global Output:
  DM = **D**
  EM = $\mathbf{E}_0$
  AM = **A**
  BM = **B**

Subroutines:
  INV
  SP

A.3 EXAMPLE

An elementary example of the student use of library programs is summarized by the output listing given in Table A-1. Subprogram MIN1S1 first is defined by the user. This subprogram computes

$$f(\mathbf{c}) = 1 - \mathbf{c}'\mathbf{f} + \tfrac{1}{2}\mathbf{c}'\mathbf{F}\mathbf{c}$$

and

$$f_{\mathbf{c}} = -\mathbf{f} + \mathbf{F}\mathbf{c}.$$

where

$$\mathbf{f} = \begin{bmatrix} 0 \\ 1 \\ 2 \end{bmatrix} \quad \text{and} \quad \mathbf{F} = \begin{bmatrix} 1 & 1 & 1 \\ 1 & 2 & 2 \\ 1 & 2 & 3 \end{bmatrix}.$$

Input data for MIN1 then is entered by the user. The user next types MIN1 and depresses RETURN in order to start execution of MIN1 and obtain an output listing of the input data. Finally, the user types 0 and depresses RETURN so that the remaining output listing is obtained. These data illustrate three-step convergence of the conjugate-gradient method for a quadratic function of three variables.

Table A-1

```
      ∇MIN1S1;FV;FM
[1]   FV←0 1 2
[2]   FM←3 3⍴1 1 1 1 2 2 1 2 3
[3]   F←1+(-CV+.×FV)+0.5×CV+.×FM+.×CV
[4]   DFV←(-FV)+FM+.×CV
[5]   ∇
      NI←3
      NT←4
      NR←3
      RATIO←0.01
      DFMIN←1E¯6
      FMIN←0
      FMAX←5
      ALMIN←0.1
      ALMAX←10
      RMIN←0.5
      RMAX←10
      CV←1 0 0
      MIN1
```

Table A-1 (continued)

```
NI= 3    NT= 4    NR= 3
RATIO= 0.01    DFMIN= 1E¯6    FMIN= 0    FMAX= 5
ALMIN= 0.1    ALMAX= 10    RMIN= 0.5    RMAX= 10
CV
1  0  0
⎕:
      0

0
ITERATION= 0    TRIAL= 0    F= 1.5    ALPHA= 0    DF= ¯2    |DFV|=
      1.414213562

CV
1  0  0
ITERATION= 0    TRIAL= 1    F= 0.75    ALPHA= 1.5    DF= 1    |DFV|=
      2.692582404

CV
¯0.5  0  1.5
ITERATION= 1    TRIAL= 0    F= 0.5    ALPHA= 0    DF= ¯3    |DFV|=
      1.732050808

CV
0  0  1
ITERATION= 1    TRIAL= 1    F= 2.25    ALPHA= 1    DF= 6.5    |DFV|=
      3.774917218

CV
¯2.5  ¯1  1.5
ITERATION= 2    TRIAL= 0    F= 0.02631578947    ALPHA= 0    DF=
      ¯0.01662049861    |DFV|= 0.1289205128

CV
¯0.7894736842  ¯0.3157894737  1.157894737
ITERATION= 2    TRIAL= 1    F= 0.02132891294    ALPHA= 0.3157894737    DF=
      ¯0.01496305277    |DFV|= 0.1160641736

CV
¯0.8104679983  ¯0.2842980026  1.142149001
ITERATION= 2    TRIAL= 2    F= 0.0002473649075    ALPHA= 3.473684211    DF=
      0.001611405683    |DFV|= 0.01249921869

CV
¯1.020411139  0.03061670797  0.984691646
ITERATION= 3    TRIAL= 0    F= 0    ALPHA= 0    DF= ¯2.963158775E¯28    |DF
      = 1.721382809E¯14

CV
¯1  ¯1.01307851E¯15  1
ITERATION LIMIT
```

A.4 SUMMARY

The library of APL/360 programs described here is listed in Table A-2.

Table A-2

```
     ∇ Z←F DXM XM;D;L;I;J;YA;YM;K;NX;NV
[1]    NX←(ρAM)[1]
[2]    NV←(ρBA)[1]
[3]    →L1×ιF≠1
[4]    Z←(NX,NX)ρ0
[5]    →L5
[6]   L1:Z←ι0
[7]    D←XM
[8]    XM←(NX,NX)ρ0
[9]    L←0
[10]   J←0
[11]  L3:J←J+1
[12]   I←0
[13]  L4:I←I+1
[14]   L←L+1
[15]   XM[J;I]←XM[I;J]←D[L]
[16]   →L4×ιI<J
[17]   →L3×ιJ<NX
[18]  L5:YA←(NV,NX,NX)ρ0
[19]   YM←RM+.×XM
[20]   K←0
[21]  L6:→L7×ιK=NV
[22]   K←K+1
[23]   YA[K;;]←XM+.×BA[K;;]
[24]   →L6
[25]  L7:L←0
[26]   J←0
[27]  L8:J←J+1
[28]   I←0
[29]  L9:I←I+1
[30]   D←PM[I;J]
[31]   K←0
[32]  L10:→L11×ιK=NV
[33]   K←K+1
[34]   D←D+BA[K;;I]+.×YA[K;;J]
[35]   →L10
[36]  L11:D←D+(XM[I;]+.×AM[;J])+(AM[;I]+.×XM[;J])-XM[I;]+.×YM[;J]
[37]   →L12×ιF≠1
[38]   Z[J;I]←Z[I;J]←D
[39]   →L13
[40]  L12:Z←Z,D
[41]  L13:→L9×ιI<J
[42]   →L8×ιJ<NX
     ∇


     ∇ EVAL;J;D;ZM;L;ZA;YM;K;M;PV;N;NU;NX;NY
[1]    'MODE= ';MODE
[2]    'AM'
[3]    AM
[4]    'BM'
[5]    BM
```

Table A-2 (continued)

```
[6]   'CM'
[7]   CM
[8]   ⎕
[9]   NU←(ρBM)[2]
[10]  NX←(ρAM)[1]
[11]  NY←(ρCM)[1]
[12]  J←NX+1
[13]  ZA←(J,NY,NU)ρ0
[14]  PV←Jρ0
[15]  D←1
[16]  ZM←(NX,NX)ρ1,NXρ0
[17]  L←0
[18] L1:L←L+1
[19]  PV[L]←D
[20]  ZA[L;;]←CM+.×ZM+.×BM
[21]  YM←ZM+.×AM
[22]  D←-(+/ 1 1 ⍉YM)÷L
[23]  ZM←YM+(NX,NX)ρD,NXρ0
[24]  →L1×⍳L≤NX
[25]  'COEFFICIENTS OF DENOMINATOR POLYNOMIAL WHICH HAS DEGREE ';NX
[26]  PV
[27]  →L7×⍳MODE
[28]  'REAL AND IMAGINARY PARTS OF ZERO LOCATIONS'
[29]  FACT PV
[30] L7:L←0
[31] L2:L←L+1
[32]  K←0
[33] L3:K←K+1
[34]  D←|ZA[J;L;K]
[35]  M←0
[36] L4:M←M+1
[37]  →L5×⍳D<|ZA[M;L;K]
[38]  →L4×⍳M<NX
[39] L5:N←NX-M
[40]  'COEFFICIENTS OF NUMERATOR POLYNOMIAL ';L;',';K;' WHICH HAS DEG
      E ';N
[41]  ⎕←PV←(N+1)↑(M-1)↓ZA[;L;K]
[42]  →L6×⍳N≤0
[43]  →L6×⍳MODE
[44]  'REAL AND IMAGINARY PARTS OF ZERO LOCATIONS'
[45]  FACT PV
[46] L6:→L3×⍳K<NU
[47]  →L2×⍳L<NY
    ∇

    ∇ Z←FACT H;B;C;D;E;F;I;J;K;M;P;Q;R;CB;IR;NC;NP;QP;PP
[1]   IR←-1
[2]   NC←(ρH)[1]
[3]   Z←(2,NC-1)ρ0
[4]   B←C←0×⍳NC+1
[5]   P←Q←R←0
[6]  L0:→L1×⍳0≠H[NC]
[7]   NC←NC-1
```

```
[8]    Z[;NC]← 0 0
[9]    →L0
[10]  L1:→L2×ι0≠H[1]
[11]   NC←NC-1
[12]   H←1ΦH
[13]   →L1
[14]  L2:→0×ι1=NC
[15]   →L3×ι2≠NC
[16]   R←-H[1]÷H[2]
[17]   →L20
[18]  L3:→L4×ι3≠NC
[19]   P←H[2]÷H[3]
[20]   Q←H[1]÷H[3]
[21]   →L23
[22]  L4:→L8×ι(|H[2]÷H[1])≤|H[NC-1]÷H[NC]
[23]   IR←-IR
[24]   M←NC÷2
[25]   I←0
[26]  L5:I←I+1
[27]   J←NC+1-I
[28]   F←H[J]
[29]   H[J]←H[I]
[30]   H[I]←F
[31]   →L5×ιI<M
[32]   →L6×ι0≠Q
[33]   P←0
[34]   →L7
[35]  L6:P←P÷Q
[36]   Q←1÷Q
[37]  L7:→L8×ι0=R
[38]   R←1÷R
[39]  L8:E←1E¯7
[40]   B[NC]←C[NC]←H[NC]
[41]   B[NC+1]←C[NC+1]←0
[42]   NP←NC-1
[43]  L9:J←0
[44]  L10:J←J+1
[45]   I←0
[46]  L11:I←I+1
[47]   K←NC-I
[48]   B[K]←H[K]+R×B[K+1]
[49]   C[K]←B[K]+R×C[K+1]
[50]   →L11×ιI<NP
[51]   →L20×ιE≥|B[1]÷H[1]
[52]   →L12×ι0≠C[2]
[53]   R←R+1
[54]   →L13
[55]  L12:R←R-B[1]÷C[2]
[56]  L13:I←0
[57]  L14:I←I+1
[58]   K←NC-I
[59]   B[K]←H[K]-(P×B[K+1])+Q×B[K+2]
[60]   C[K]←B[K]-(P×C[K+1])+Q×C[K+2]
[61]   →L14×ιI<NP
```

Table A-2 (continued)

```
[62]   →L15×ι0≠H[2]
[63]   →L17×ιE<|B[2]÷H[1]
[64]   →L16
[65]  L15:→L17×ιE<|B[2]÷H[2]
[66]  L16:→L23×ιE≥|B[1]÷H[1]
[67]  L17:CB←C[2]-B[2]
[68]   D←(C[3]*2)-CB×C[4]
[69]   →L18×ι0≠D
[70]   P←P-2
[71]   Q←Q×Q+1
[72]   →L19
[73]  L18:P←P+((B[2]×C[3])-B[1]×C[4])÷D
[74]  L19:→L10×ι(J<15)
[75]   Q←Q+((B[1]×C[3])-B[2]×CB)÷D
[76]   E←E×10
[77]   →L9
[78]  L20:NC←NC-1
[79]   →L21×ι0≤IR
[80]   Z[1;NC]←1÷R
[81]   →L22
[82]  L21:Z[1;NC]←R
[83]  L22:H←1ΦB
[84]   →L2
[85]  L23:NC←NC-2
[86]   →L24×ι0≤IR
[87]   QP←1÷Q
[88]   PP←P÷Q×2
[89]   →L25
[90]  L24:QP←Q
[91]   PP←P÷2
[92]  L25:F←(PP*2)-QP
[93]   →L26×ι0≤F
[94]   Z[1;NC+1]←Z[1;NC]←-PP
[95]   Z[2;NC+1]←(-F)*0.5
[96]   Z[2;NC]←-Z[2;NC+1]
[97]   →L27
[98]  L26:Z[1;NC+1]←-(PP÷|PP)×(|PP)+F*0.5
[99]   Z[2;NC+1]←Z[2;NC]←0
[100] Z[1;NC]←QP÷Z[1;NC+1]
[101]L27:H←2ΦB
[102] →L2
    ∇


    ∇ FORM1
[1]    'NX= ';NX;'    NU= ';NU;'    NC= ';NC
[2]    ⎕
[3]    AM←CM←DFM←(NX,NX)ρ0
[4]    DAA←DCA←DXA←(NC,NX,NX)ρ0
[5]    BA←(NU,NX,NX)ρ0
[6]    DBAA←(NC,NU,NX,NX)ρ0
    ∇
```

```
      ∇ FORM2;D;E;F
[1]     'FM'
[2]     FM
[3]     'GM'
[4]     GM
[5]     'DM'
[6]     DM
[7]     'EM'
[8]     EM
[9]     'TM'
[10]    TM
[11]    ⎕
[12]    E←INV(⍉EM) SP EM
[13]    D←(⍉EM)+.×DM
[14]    F←GM+.×E
[15]    RM←F SP⍉GM
[16]    AM←FM-F+.×D
[17]    PM←TM+((⍉DM) SP DM)-(⍉D) SP E+.×D
      ∇

      ∇ FORM3;E;D;F
[1]     'FM'
[2]     FM
[3]     'GM'
[4]     GM
[5]     'HM'
[6]     HM
[7]     'JM'
[8]     JM
[9]     'KM'
[10]    KM
[11]    ⎕
[12]    E←INV JM SP⍉JM
[13]    D←JM+.×KM
[14]    F←GM+.×(⍉D)+.×E
[15]    RM←(⍉HM) SP E+.×HM
[16]    AM←⍉(FM-F+.×HM)
[17]    PM←(GM-F+.×D) SP⍉GM
      ∇

      ∇ Z←GAIN1
[1]     Z←(INV(⍉EM) SP EM)+.×((⍉EM)+.×DM)+(⍉GM)+.×XM
      ∇

      ∇ Z←GAIN2
[1]     Z←((GM+.×⍉JM+.×KM)+XM+.×⍉HM)+.×INV JM SP⍉JM
      ∇

      ∇ Z←AM INC BA;NX;NV;L;J;I;K;M;N;IM;P
[1]     NX←(⍴AM)[1]
[2]     NV←(⍴BA)[1]
[3]     IM←(NX,NX)⍴0
```

Table A-2 (continued)

```
[4]     L←0
[5]     J←0
[6]     L1:J←J+1
[7]     I←0
[8]     L2:I←I+1
[9]     L←L+1
[10]    IM[J;I]←IM[I;J]←L
[11]    →L2×ıI<J
[12]    →L1×ıJ<NX
[13]    Z←(L,L)ρ0
[14]    L←0
[15]    J←0
[16]    L3:J←J+1
[17]    I←0
[18]    L4:I←I+1
[19]    L←L+1
[20]    K←0
[21]    L6:K←K+1
[22]    M←IM[I;K]
[23]    N←IM[K;J]
[24]    Z[L;M]←Z[L;M]+AM[K;J]
[25]    Z[L;N]←Z[L;N]+AM[K;I]
[26]    M←0
[27]    L7:→L9×ıM=NV
[28]    M←M+1
[29]    N←0
[30]    L8:N←N+1
[31]    P←IM[K;N]
[32]    Z[L;P]←Z[L;P]+BA[M;K;I]×BA[M;N;J]
[33]    →L8×ıN<NX
[34]    →L7
[35]    L9:→L6×ıK<NX
[36]    →L10×ıI=J
[37]    Z[L;]←Z[L;]+Z[L;]
[38]    →L4
[39]    L10:→L3×ıJ<NX
[40]    Z←INV Z
        ∇


        ∇ INT;D;G1;G2;G3;G4;G5;G6;G7;G8;NIP;NIC;DV;DTH;NX
[1]     'NI= ';NI;'    NP= ';NP;'    T= ';T;'    DT= ';DT
[2]     'XV'
[3]     XV
[4]     ☐
[5]     NX←(ρXV)[1]
[6]     DXV←NXρ0
[7]     D←0.5*0.5
[8]     G1←DT×1-D
[9]     G2←DT×1+D
[10]    D←3×D
[11]    G6←D-2
[12]    G8←-(D+2)
```

```
[13]   G3←DT÷6
[14]   G4←-DT÷3
[15]   D←2*0.5
[16]   G5←2-D
[17]   G7←2+D
[18]   D←0
[19]   NIP←NP
[20]   DTH←0.5×DT
[21]   NIC←0
[22]  L1:INTS1
[23]   NIP←NIP+1
[24]   →L2×⍳NIP≤NP
[25]   'STEP= ';NIC;'    T= ';T;'    ||DXV||= ';⌈/|DXV
[26]   'XV'
[27]   XV
[28]   NIP←0
[29]   →0×⍳D
[30]  L2:DV←DXV
[31]   XV←XV+DTH×DXV
[32]   T←T+DTH
[33]   INTS1
[34]   XV←XV+G1×DXV-DV
[35]   DV←(G5×DXV)+G6×DV
[36]   INTS1
[37]   XV←XV+G2×DXV-DV
[38]   DV←(G7×DXV)+G8×DV
[39]   T←T+DTH
[40]   INTS1
[41]   XV←XV+(G3×DXV)+G4×DV
[42]   NIC←NIC+1
[43]   →L1×⍳NIC<NI
[44]   D←1
[45]   NIP←NP
[46]   →L1
     ∇
```

```
     ∇ Z←INV M;C;I;J;K;N;P;R;V
[1]    K←⍳0
[2]    V←⍳N←(⍴M)[2]
[3]    I←1
[4]   L1:R←|M[I;V]
[5]    J←V[R⍳⌈/R]
[6]    P←1÷M[I;J]
[7]    C←P×M[;J]
[8]    M←M-C∘.×R←M[I;]
[9]    M[;J]←C
[10]   M[I;]←-P×R
[11]   M[I;J]←P
[12]   V←(V≠J)/V
[13]   K←K,J
[14]   →L1×⍳N≥I←I+1
[15]   Z←M[⍋K;K]
     ∇
```

Table A-2 (continued)

```
     ∇ Z←MODE LIN1 CM;LM;MM;NM;L;D;XM;M;N
[1]    M←(ρCM)[1]
[2]    N←(ρCM)[2]
[3]    D←1
[4]    XM←(M,M)ρ1,Mρ0
[5]    LM←(N,N)ρ0
[6]    MM←(M,N)ρ0
[7]    NM←(M,M)ρ0
[8]    L←0
[9]   L1:L←L+1
[10]   LM←(-LM+.×AM)+(N,N)ρD,Nρ0
[11]   MM←(-MM+.×AM)+NM+.×CM
[12]   →L2×⍳L>M
[13]   NM←(NM+.×BM)+(M,M)ρD,Mρ0
[14]   XM←XM+.×BM
[15]   D←-(+/ 1 1 ⍉XM)÷L
[16]   XM←XM+(M,M)ρD,Mρ0
[17]   →L1
[18]  L2:→L3×⍳MODE
[19]   Z←MM+.×INV LM
[20]   →0
[21]  L3:Z←MM SP INV LM
     ∇
```

```
     ∇ Z←AM LIN2 CM;NX;YV;L;J;I
[1]    NX←(ρCM)[1]
[2]    Z←(NX,NX)ρ0
[3]    J←((NX+1)×NX)÷2
[4]    YV←Jρ0
[5]    L←0
[6]    J←0
[7]   L1:J←J+1
[8]    I←0
[9]   L2:I←I+1
[10]   L←L+1
[11]   YV[L]←CM[I;J]
[12]   →L3×⍳I=J
[13]   YV[L]←YV[L]+YV[L]
[14]   →L2
[15]  L3:→L1×⍳J<NX
[16]   YV←-AM+.×YV
[17]   L←0
[18]   J←0
[19]  L4:J←J+1
[20]   I←0
[21]  L5:I←I+1
[22]   L←L+1
[23]   Z[J;I]←Z[I;J]←YV[L]
[24]   →L5×⍳I<J
[25]   →L4×⍳J<NX
     ∇
```

```
    ∇ LOAD1;N;M;L;Y;I;J;U;K
      Y←⍉AM INC BA
      N←(⍴CV)[1]
      M←(⍴BA)[1]
      L←(⍴AM)[1]
      XM←Y LIN2 CM
      J←0
    L6:J←J+1
      I←0
    L7:I←I+1
      DFM[J;I]←DFM[I;J]
      →L7×⍳I<J
      →L6×⍳J<L
      K←0
    L1:K←K+1
      J←0
    L4:J←J+1
      I←0
    L5:I←I+1
      DCA[K;J;I]←DCA[K;I;J]
      →L5×⍳I<J
      →L4×⍳J<L
      DXA[K;;]←DCA[K;;]
      J←0
    L2:→L3×⍳J=M
      J←J+1
      U←DBAA[K;J;;]+.×XM+.×⍉BA[J;;]
      DXA[K;;]←DXA[K;;]+U+⍉U
      →L2
    L3:U←DAA[K;;]+.×XM
      DXA[K;;]←Y LIN2 DXA[K;;]+U+⍉U
      DFV[K]←+/DFM MD DXA[K;;]
      →L1×⍳I<N
    ∇
```

```
    ∇ LOAD2;C;M;N;D;AM;X;Y;BM
      M←(⍴GM)[2]
      N←(⍴HM)[1]
      C←((M,N)⍴CV)+.×HM
      AM←FM-GM+.×C
      BM←⍉AM
      D←DM-EM+.×C
      X←1 LIN1(⍉D) SP D
      BM←AM
      AM←⍉AM
      Y←(1 LIN1 TM SP⍉TM)+.×⍉HM
      F←+/(⍉TM) MD X+.×TM
      DFV←,2×((((⍉EM) SP EM)+.×C)-(((⍉EM)+.×DM)+(⍉GM)+.×X))+.×Y
    ∇
```

```
    ∇ Z←AM MD BM;M;L
      M←(⍴AM)[1]
      Z←⍳0
```

Table A-2 (continued)

```
[3]    L←0
[4]   L1:L←L+1
[5]    Z←Z,AM[L;]+.×BM[;L]
[6]    →L1×⍳L<M
     ∇

     ∇ MIN1;D1;D2;D3;NIC;NTC;NRC;A;B;C;FA;FB;DFA;DFB;DC;DF;SDF;SVP;DCV;
       SCV;NC
[1]    'NI= ';NI;'    NT= ';NT;'    NR= ';NR
[2]    'RATIO= ';RATIO;'    DFMIN= ';DFMIN;'    FMIN= ';FMIN;'    FMAX=
       FMAX
[3]    'ALMIN= ';ALMIN;'    ALMAX= ';ALMAX;'    RMIN= ';RMIN;'    RMAX=
       RMAX
[4]    'CV'
[5]    CV
[6]    ⎕
[7]    NC←(⍴CV)[1]
[8]    F←0
[9]    DFV←NC⍴0
[10]   NIC←NTC←C←0
[11]   NRC←NR
[12]   MIN1S1
[13]   SCV←CV
[14]   DCV←-DFV
[15]   SVP←-DF←DCV+.×DFV
[16]   SDF←RATIO×|DF
[17]   D3←SVP*0.5
[18]   'ITERATION= ';NIC;'    TRIAL= ';NTC;'    F= ';F;'    ALPHA= ';C;'
       DF= ';DF;'    |DFV|= ';D3
[19]   'CV'
[20]   CV
[21]   →L1×⍳FMAX>F
[22]   'RESET CV'
[23]   →0
[24]  L1:→L2×⍳DFMIN<D3
[25]   'DESIGN COMPLETE'
[26]   →0
[27]  L2:DC←2×(FMIN-F)÷DF
[28]  L3:DC←ALMIN⌈DC⌊ALMAX
[29]  L5:NTC←NTC+1
[30]   A←C
[31]   FA←F
[32]   DFA←DF
[33]   C←A+DC
[34]   CV←SCV+C×DCV
[35]   MIN1S1
[36]   DF←DCV+.×DFV
[37]   D1←DFV+.×DFV
[38]   D2←D1*0.5
[39]   →L18×⍳SDF≥|DF
[40]   'ITERATION= ';NIC;'    TRIAL= ';NTC;'    F= ';F;'    ALPHA= ';C;'
       DF= ';DF;'    |DFV|= ';D2
[41]   'CV'
```

```
[42]  CV
[43]  →L8×⍳FMAX>F
[44] L6:→L7×⍳NT>NTC
[45]  →L12
[46] L7:DC←RMIN×DC
[47]  C←A
[48]  F←FA
[49]  DF←DFA
[50]  →L5
[51] L8:→L9×⍳DFMIN<D2
[52]  'DESIGN COMPLETE'
[53]  →L20
[54] L9:→L13×⍳0<DF
[55] L10:→L12×⍳NT≤NTC
[56]  →L11×⍳((1-1÷RMAX)×DFA)≤DF
[57]  DC←DC×RMAX
[58]  →L5
[59] L11:DC←DC×DFA÷DFA-DF
[60]  →L5
[61] L12:'TRIAL LIMIT'
[62]  →L20
[63] L13:B←C
[64]  FB←F
[65]  DFB←DF
[66] L14:D1←DFA+DFB+3×(FA-FB)÷DC
[67]  D2←2×DFA+D1
[68]  D3←DFA+DFB+2×D1
[69]  →L15×⍳(1E¯5×|D2)≤|D3
[70]  C←B-(DFA+2×D1)×DC÷D2
[71]  →L16
[72] L15:D2←((D1*2)-DFA×DFB)*0.5
[73]  C←B-(DFB+D1-D2)×DC÷D3
[74] L16:CV←SCV+C×DCV
[75]  MIN1S1
[76]  DF←DCV+.×DFV
[77]  D1←DFV+.×DFV
[78]  D2←D1*0.5
[79]  →L18×⍳SDF≥|DF
[80]  NTC←NTC+1
[81]  'ITERATION= ';NIC;'   TRIAL= ';NTC;'   F= ';F;'   ALPHA= ';C;'
      DF= ';DF;'   |DFV|= ';D2
[82]  'CV'
[83]  CV
[84]  →L6×⍳FMAX≤F
[85]  →L12×⍳NT≤NTC
[86]  →L17×⍳0>DF
[87]  DC←C-A
[88]  →L13
[89] L17:A←C
[90]  FA←F
[91]  DFA←DF
[92]  DC←B-C
[93]  →L14
[94] L18:NIC←NIC+1
```

Table A-2 (continued)

```
[95]  DC←C
[96]  C←NTC←0
[97]  →L19×ιNRC>NIC
[98]  SVP←1E37
[99]  NRC←NRC+NR
[100]L19:B←D1÷SVP
[101] SVP←D1
[102] SCV←CV
[103] DCV←(-DFV)+B×DCV
[104] DF←DCV+.×DFV
[105] SDF←RATIO×|DF
[106] 'ITERATION= ';NIC;'   TRIAL= ';NTC;'   F= ';F;'   ALPHA= ';C;'
      DF= ';DF;'   |DFV|= ';D2
[107] 'CV'
[108] CV
[109] →L3×ιNI>NIC
[110] 'ITERATION LIMIT'
[111]L20:CV←SCV
    ∇

    ∇ MIN2;N1;N2;N3;N4;N5;N6;N7;NCC;ALPHA;I;J;L;LV;LM;L1M;L2M;HV;DM;D
      ;IV;NC;NIC;NEC
[1]   'NI= ';NI;'   DFMIN= ';DFMIN
[2]   'CV'
[3]   CV
[4]   'LAMV'
[5]   LAMV
[6]   'SV'
[7]   SV
[8]   'EICV'
[9]   EICV
[10]  'MDCV'
[11]  MDCV
[12]  ⎕
[13]  NC←(ρCV)[1]
[14]  NIC←(ρSV)[1]
[15]  N1←(ρLAMV)[1]
[16]  NEC←N1-NIC
[17]  N2←N1+NIC
[18]  N3←N2+NC
[19]  N4←NEC×NIC
[20]  N5←NIC×NIC
[21]  N6←NIC×NC
[22]  N7←N1×N1
[23]  IV←(NCρ1),EICV,(NECρ1),EICV
[24]  F←NCC←ALPHA←0
[25]  DFV←NCρ0
[26]  DFM←(NC,NC)ρ0
[27]  GV←N1ρ0
[28]  DGM←(N1,NC)ρ0
[29]  DGA←(N1,NC,NC)ρ0
[30]  LV←N3ρ0
[31]  DLM←(NC,NC)ρ0
```

```
[32] L1:CV←CV+ALPHA×NC↑LV
[33]  SV←SV+ALPHA×NIC↑NC↓LV
[34]  LAMV←LAMV+ALPHA×(-N1)↑LV
[35]  MIN2S1
[36]  HV←GV+(NECρ0),0.5×SV×SV
[37]  L←F+LAMV+.×HV
[38]  LV←(DFV+(⍉DGM)+.×LAMV),(SV×(-NIC)↑LAMV),HV
[39]  I←(LV+.×LV)*0.5
[40]  'ITERATION= ';NCC;'   F= ';F;'   L= ';L;'   ALPHA= ';ALPHA;'   |D
      LV|= ';I
[41]  'CV'
[42]  CV
[43]  →L6×ι0=N1
[44]  'GV'
[45]  GV
[46]  'LAMV'
[47]  LAMV
[48]  →L7×ιDFMIN<I
[49]  'DESIGN COMPLETE'
[50]  →0
[51] L7:→L6×ι0=NIC
[52]  'SV'
[53]  SV
[54] L6:NCC←NCC+1
[55]  →0×ιNCC>NI
[56]  I←0
[57] L2:I←I+1
[58]  J←I-1
[59] L3:J←J+1
[60]  DLM[J;I]←DLM[I;J]←DFM[I;J]+LAMV+.×DGA[;I;J]
[61]  →L3×ιNC>J
[62]  →L2×ιNC>I
[63]  DM←(N1,NIC)ρ(N4ρ0),(N5ρ1,NICρ0)\SV
[64]  L1M←(N3,NC)ρ(,DLM),(N6ρ0),,DGM
[65]  L2M←(N3,NIC)ρ(N6ρ0),((N5ρ1,NICρ0)\(-NIC)↑LAMV),,DM
[66]  LM←(N3,N1)ρ(,⍉DGM),(,⍉DM),N7ρ0
[67]  LM←(N3,N3)ρ(,⍉L1M),(,⍉L2M),,⍉LM
[68]  LM←IV/[1] IV/LM
[69]  LM←IV\IV\[1] INV LM
[70]  LV←-LM+.×LV
[71]  ALPHA←1
[72]  I←0
[73] L4:I←I+1
[74]  →L5×ι(ALPHA×|LV[I])≤MDCV[I]
[75]  ALPHA←MDCV[I]÷|LV[I]
[76] L5:→L4×ιI<NC
[77]  →L1
    ∇


    ∇ MIN3;D1;D2;D3;NCC;NTC;NRC;A;B;C;FA;FB;DFA;DFB;DC;DF;SDF;DCV;SCV;
      NC;N1;N2;N3;N4;N5;N6;L;DSV;DLV
[1]   'NI= ';NI;'   NT= ';NT;'   NR= ';NR;'   NEC= ';NEC;'   RATIO= ';
      RATIO
```

Table A-2 (continued)

```
[2]   'DFMIN= ';DFMIN;'    GMAX= ';GMAX;'    FMIN= ';FMIN;'    FMAX= ';
      FMAX
[3]   'ALMIN= ';ALMIN;'    ALMAX= ';ALMAX;'    RMIN= ';RMIN;'    RMAX= ';
      RMAX
[4]   'CV'
[5]   CV
[6]   'LAMV'
[7]   LAMV
[8]   'DLIM'
[9]   DLIM
[10]  ⎕
[11]  NC←(ρCV)[1]
[12]  N1←(ρLAMV)[1]
[13]  F←0
[14]  DFV←NCρ0
[15]  GV←N1ρ0
[16]  DGM←(N1,NC)ρ0
[17]  NCC←NTC←C←0
[18]  NRC←NR-1
[19]  MIN3S1
[20]  SCV←CV
[21]  L←F+LAMV+.×GV
[22]  DSV←DLV←DFV+(⍉DGM)+.×LAMV
[23]  DCV←-DLIM+.×DLV
[24]  DF←DCV+.×DLV
[25]  SDF←RATIO×|DF
[26]  D3←(-DF)*0.5
[27]  'ITERATION= ';NCC;'    TRIAL= ';NTC;'    F= ';F;'    L= ';L;'    ALP
      A= ';C;'    DL= ';DF;'    |DLV|= ';D3
[28]  'CV'
[29]  CV
[30]  →L21×ι0=N1
[31]  'GV'
[32]  GV
[33]  'LAMV'
[34]  LAMV
[35] L21:→L1×ιFMAX>F
[36]  'RESET CV'
[37]  →0
[38] L1:→L2×ιDFMIN<D3
[39]  →L25
[40] L2:DC←2×(FMIN-L)÷DF
[41] L3:DC←ALMIN⌈DC⌊ALMAX
[42] L5:NTC←NTC+1
[43]  A←C
[44]  FA←L
[45]  DFA←DF
[46]  C←A+DC
[47]  CV←SCV+C×DCV
[48]  MIN3S1
[49]  L←F+LAMV+.×GV
[50]  DLV←DFV+(⍉DGM)+.×LAMV
[51]  DF←DCV+.×DLV
```

```
[52]  D1←DLV+.×DLV
[53]  D2←D1*0.5
[54]  →L18×⍳SDF≥|DF
[55]  'ITERATION= ';NCC;'    TRIAL= ';NTC;'    F= ';F;'    L= ';L;'    ALPH
      A= ';C;'    DL= ';DF;'    |DLV|= ';D2
[56]  'CV'
[57]  CV
[58]  →L22×⍳0=N1
[59]  'GV'
[60]  GV
[61] L22:→L8×⍳FMAX>F
[62] L6:→L7×⍳NT>NTC
[63]  →L12
[64] L7:DC←RMIN×DC
[65]  C←A
[66]  L←FA
[67]  DF←DFA
[68]  →L5
[69] L8:→L9×⍳DFMIN<D2
[70] L25:NRC←NCC
[71]  →L18×⍳0<+/(0<(NEC↑|GV)-GMAX),0<(NEC↓GV)-GMAX
[72]  'DESIGN COMPLETE'
[73]  →L20
[74] L9:→L13×⍳0<DF
[75] L10:→L12×⍳NT≤NTC
[76]  →L11×⍳((1-1÷RMAX)×DFA)≤DF
[77]  DC←DC×RMAX
[78]  →L5
[79] L11:DC←DC×DFA÷DFA-DF
[80]  →L5
[81] L12:'TRIAL LIMIT'
[82]  →L20
[83] L13:B←C
[84]  FB←L
[85]  DFB←DF
[86] L14:D1←DFA+DFB+3×(FA-FB)÷DC
[87]  D2←2×DFA+D1
[88]  D3←DFA+DFB+2×D1
[89]  →L15×⍳(1E¯5×|D2)≤|D3
[90]  C←B-(DFA+2×D1)×DC÷D2
[91]  →L16
[92] L15:D2←((D1*2)-DFA×DFB)*0.5
[93]  C←B-(DFB+D1-D2)×DC÷D3
[94] L16:CV←SCV+C×DCV
[95]  MIN3S1
[96]  L←F+LAMV+.×GV
[97]  DLV←DFV+(⍉DGM)+.×LAMV
[98]  DF←DCV+.×DLV
[99]  D1←DLV+.×DLV
[100] D2←D1*0.5
[101] →L18×⍳SDF≥|DF
[102] NTC←NTC+1
[103] 'ITERATION= ';NCC;'    TRIAL= ';NTC;'    F= ';F;'    L= ';L;'    ALPH
      A= ';C;'    DL= ';DF;'    |DLV|= ';D2
```

Table A-2 (continued)

```
[104] 'CV'
[105] CV
[106] →L23×ι0=N1
[107] 'GV'
[108] GV
[109]L23:→L6×ιFMAX≤F
[110] →L12×ιNT≤NTC
[111] →L17×ι0>DF
[112] DC←C-A
[113] →L13
[114]L17:A←C
[115] FA←L
[116] DFA←DF
[117] DC←B-C
[118] →L14
[119]L18:SCV←DLV-DSV
[120] DSV←DLIM+.×SCV
[121] DLIM←DLIM-DSV∘.×DSV÷DSV+.×SCV
[122] DCV←C×DCV
[123] DLIM←DLIM+DCV∘.×DCV÷DCV+.×SCV
[124] 'DLIM'
[125] DLIM
[126] →L19×ιNRC>NCC
[127] NRC←NRC+NR
[128] DC←(NECρ1),(0<NEC↓GV)∨0<NEC↓LAMV
[129] A←0
[130]L26:N2←+/DC
[131] →L27×ιNC≥N2
[132] DC[N1-A]←0
[133] A←A+1
[134] →L26
[135]L27:→L19×ι0=N2
[136] N3←N2-NEC
[137] N4←2×N3
[138] N5←N3×N3
[139] N6←N4×NEC
[140] DFA←NEC↓DC/LAMV
[141] B←DC/[1] DGM
[142] FA←B+.×DLIM
[143] A←(DC/-GV)+FA+.×DLV
[144] B←FA SP⍉B
[145] FA←(N2,N4)ρ(N6ρ0),,⍉(N4,N3)ρ(N5ρ0),N5ρ¯1,N3ρ0
[146] N6←N2+N4
[147] D1←(INV B)+.×-A
[148] D1←(NEC↑D1),(-DFA)⌈NEC↓D1
[149] D2←(2×DFA+NEC↓D1)*0.5
[150] D3←NEC↓A+B+.×D1
[151] SDF←RATIO×(A+.×A)*0.5
[152] DF←N6ρ0
[153]L28:D1←D1+N2↑DF
[154] D2←D2+N3↑N2↓DF
[155] D3←D3+(-N3)↑DF
[156] DF←(A+(B+.×D1)-(NECρ0),D3),(D3×D2),(0.5×D2×D2)-DFA+NEC↓D1
```

```
157] DFB←(DF+.×DF)*0.5
158] 'ITERATION= ';NCC;'    TRIAL= ';NTC;'    |DLV|= ';DFB
159] 'DLAMV'
160] D1
161] →L29×ιDFMIN≥DFB
162] →L29×ιDFB≤SDF
163] →L12×ιNTC≥NT
164] NTC←NTC+1
165] FB←(N5ρ1,N3ρ0)\D2
166] FB←(,⍉(N4,N3)ρ((N5ρ1,N3ρ0)\D3),FB),,⍉(N4,N3)ρFB,N5ρ0
167] DF←(INV(N6,N6)ρ(,⍉(N6,N2)ρ(,B),,⍉FA),,⍉(N6,N4)ρ(,FA),FB)+.×-DF
168] →L28
169]L29:D1←DC\(NEC↑D1),(-DFA)⌈NEC↓D1
170] LAMV←LAMV+D1
171] DLV←DLV+(⍉DGM)+.×D1
172] L←L+D1+.×GV
173]L19:DCV←-DLIM+.×DLV
174] DSV←DLV
175] SCV←CV
176] DF←DCV+.×DLV
177] SDF←RATIO×|DF
178] DC←C
179] C←NTC←0
180] NCC←NCC+1
181] D2←(DLV+.×DLV)*0.5
182] 'ITERATION= ';NCC;'    TRIAL= ';NTC;'    F= ';F;'    L= ';L;'    ALPH
     A= ';C;'    DL= ';DF;'    |DLV|= ';D2
183] 'CV'
184] CV
185] →L24×ι0=N1
186] 'GV'
187] GV
188] 'LAMV'
189] LAMV
190]L24:→L3×ιNI>NCC
191] 'ITERATION LIMIT'
192]L20:CV←SCV
   ∇


   ∇ QUAD;L;Y;Z;N
1]    'NI=  ';NI;'    MODE= ';MODE;'    DXMIN= ';DXMIN
2]    'AM'
3]    AM
4]    'BA'
5]    BA
6]    'PM'
7]    PM
8]    'RM'
9]    RM
10]   'XM'
11]   XM
12]   ⎕
13]   →L3×ι0≠MODE
14]   N←(ρAM)[1]
```

Table A-2 (continued)

```
[15]   Y←1 DXM XM
[16]   Z←⍳0
[17]   L←0
[18]  L2:L←L+1
[19]   Z←Z,0⌈(+/|Y[L;])-2×Y[L;L]
[20]   →L2×⍳L<N
[21]   XM←XM+((AM-RM+.×XM) INC BA) LIN2(N,N)⍴((N×N)⍴1,N⍴0)\Z
[22]  L3:L←0
[23]  L1:L←L+1
[24]   Y←1 DXM XM
[25]   Z←⌈/⌈/|Y
[26]   'ITERATION= ';L;'    ||DXM||= ';Z
[27]   'XM'
[28]   XM
[29]   →0×⍳Z<DXMIN
[30]   XM←XM+((AM-RM+.×XM) INC BA) LIN2 Y
[31]   →L1×⍳L<NI
     ∇

     ∇ Z←AM SP BM;M;K;L
[1]    M←(⍴AM)[1]
[2]    Z←(M,M)⍴0
[3]    K←0
[4]   L1:K←K+1
[5]    L←K-1
[6]   L2:L←L+1
[7]    Z[L;K]←Z[K;L]←AM[K;]+.×BM[;L]
[8]    →L2×⍳L<M
[9]    →L1×⍳K<M
     ∇

     ∇ SYN;N1;N2;ML;MD;L;M;NU;NX;NY;ND
[1]    'FM'
[2]    FM
[3]    'GM'
[4]    GM
[5]    'HM'
[6]    HM
[7]    'PA'
[8]    PA
[9]    ⎕
[10]   NU←(⍴GM)[2]
[11]   NX←(⍴FM)[1]
[12]   NY←(⍴HM)[1]
[13]   N1←(⍴PA)[1]
[14]   ND←N1-1
[15]   N2←N1+1
[16]   MD←⍳0
[17]   ML←ND-1
[18]  L1:L←ND-ML
[19]  L2:DM←(NY,NX)⍴0
[20]   M←0
[21]  L3:M←M+1
```

```
[22]    DM←(DM+.×FM)+PA[N2-M;;]+.×HM
[23]    →L3×ɩM<L
[24]    →L5×MD
[25]    EM←DM+.×GM
[26]    'EM';ML
[27]    EM
[28]    →L4×ɩML≤0
[29]    ML←ML-1
[30]    →L1
[31] L4:MD←1
[32]    L←N1
[33]    BM←EM
[34]    →L2
[35] L5:'DM'
[36]    DM
[37]    'EIM'
[38]    ☐←BM←(INV(⌽BM) SP BM)+.×⌽BM
[39]    'KM'
[40]    ☐←AM←BM+.×DM
[41]    'AM'
[42]    ☐←AM←FM-GM+.×AM
[43]    'BM'
[44]    ☐←BM←GM+.×BM
     ∇
```

BIBLIOGRAPHY

1. APL/360 Primer, GH20-0689-1, International Business Machines Corporation, 1969.
2. APL/360 User's Manual, GH20-0683-1, International Business Machines Corporation, 1969.

Printed in Great Britain by William Clowes & Sons Limited
London, Colchester and Beccles

Index

Page references in boldface type indicate definitions